Selected Titles in This Series

Volume

10 Jerrold E. Marsden, George W. Patrick, and William F. Shadwick, Editors
Integration algorithms and classical mechanics
1996

9 W. H. Kliemann, W. F. Langford, and N. S. Namachchivaya, Editors
Nonlinear dynamics and stochastic mechanics
1996

8 Larry M. Bates and David L. Rod, Editors
Conservative systems and quantum chaos
1996

7 William F. Shadwick, Perinkulam Sambamurthy Krishnaprasad, and Tudor Stefan Ratiu, Editors
Mechanics day
1996

6 Anna T. Lawniczak and Raymond Kapral, Editors
Pattern formation and lattice gas automata
1996

5 John Chadam, Martin Golubitsky, William Langford, and Brian Wetton, Editors
Pattern formation: Symmetry methods and applications
1996

4 William F. Langford and Wayne Nagata, Editors
Normal forms and homoclinic chaos
1995

3 Anthony Bloch, Editor
Hamiltonian and gradient flows, algorithms and control
1994

2 K. A. Morris, Editor
Control of flexible structures
1993

1 Michael J. Enos, Editor
Dynamics and control of mechanical systems: The falling cat and related problems
1993

FIELDS INSTITUTE COMMUNICATIONS

THE FIELDS INSTITUTE FOR RESEARCH IN MATHEMATICAL SCIENCES

Integration Algorithms and Classical Mechanics

Jerrold E. Marsden
George W. Patrick
William F. Shadwick
Editors

American Mathematical Society
Providence, Rhode Island

The Fields Institute for Research in Mathematical Sciences

The Fields Institute is named in honour of the Canadian mathematician John Charles Fields (1863–1932). Fields was a remarkable man who received many honours for his scientific work, including election to the Royal Society of Canada in 1909 and to the Royal Society of London in 1913. Among other accomplishments in the service of the international mathematics community, Fields was responsible for establishing the world's most prestigious prize for mathematics research—the Fields Medal.

The Fields Institute for Research in Mathematical Sciences is supported by grants from the Ontario Ministry of Education and Training and the Natural Sciences and Engineering Research Council of Canada. The Institute is sponsored by McMaster University, the University of Toronto, and the University of Waterloo and has affiliated universities in Ontario and across Canada.

1991 *Mathematics Subject Classification*. Primary 70-06, 70-08;
Secondary 65-06, 65L05.

Library of Congress Cataloging-in-Publication Data

Integration algorithms and classical mechanics / Jerrold E. Marsden, George W. Patrick, William F. Shadwick, editors.
p. cm. —(Fields Institute communications, ISSN 1069-5265; v. 10)
Includes bibliographical references.
ISBN 0-8218-0259-3 (alk. paper)
1. Mechanics, Analytic. 2. Differentiable dynamical systems. 3. Initial value problems. I. Marsden, Jerrold E. II. Patrick, George W. (George William), 1958– . III. Shadwick, W. F., 1951– . IV. Series.
QA851.I57 1996
531′.0151535–dc20 96-21236
CIP

---

10 9 8 7 6 5 4 3 2 1 01 00 99 98 97 96

Contents

Preface

The present volume is an outgrowth of the conference *Integration Algorithms and Classical Mechanics* that was held at the Fields Institute in October, 1993. It was jointly sponsored by The Fields Institute and CITA, the Canadian Institute for Theoretical Astrophysics. As can be seen from the variety of articles in the volume, the conference succeeded scientifically and brought together a healthy mix of mathematicians, physicists and engineers.

Many articles in the volume concern symplectic integration algorithms, owing to the recent surge of interest in that area. For orientation to this area, see the article by Channell and Neri. An interesting list of open problems in the area are contained in the article McLachlan and Scovel. Further general information may be found in the contributions of Sanz-Serna, Earn and Lichtenberg, and Kaasalainen and Binney.

Applications to multibody problems are discussed in the article by Barth and Leimkuhler, as well as in the contribution of Reich. For applications to (solar) astrophysics, there is the contribution of of Wisdom, Holman, and Touma (see also the contribution of McLachlan), and that of Koseleff, although these contributions are of general interest as well. Applications to the design of particle accelerators can be found the the contribution of Dragt and Abell.

The preservation of the symplectic form is one possibility; the other possibility is that of energy-momentum conserving algorithms. Some conserving algorithms may be found in the contribution of Lewis and Simo. The contribution of Armero and Simo explains some methods for magnetohydrodynamics, and the contribution of Stuchi and Vieira-Martins applies their new visualization method to the classical example of Hénon and Heiles.

Sadly, Juan Simo passed away in the time between the conference and the publication of this volume. Jerry Marsden and Francisco Armero have written a summary of the part of Juan's work that was directly related to this volume. This volume is dedicated to Juan.

Juan Carlos Simo
1952 - 1994

The Work of Juan Carlos Simo on Integration Algorithms in Mechanics

Juan had a deep and long interest in geometric mechanics and its application to problems in nonlinear elasticity. In fact, his work was more than an interest; he made *major* contributions to this subject. His deep knowledge of geometric mechanics was crucial to his development of time-stepping algorithms for the numerical integration of a number of systems in structural, solid and fluid mechanics. This insight is typified by his work in Simo and Marsden [1984], Simo, Lewis and Marsden [1991], Simo, Marsden, and Krishnaprasad [1988], Simo, Posbergh and Marsden [1990], Simo, Lewis and Simo [1990], Posbergh and Marsden [1991], and Lewis, Ratiu, Simo and Marsden [1992].

Juan was actively involved in the development of integration algorithms for Hamiltonian systems. A common thread throughout his research was the formulation of numerical schemes that inherit the conservation properties of the underlying mechanical system. The conservation of energy and momentum (both linear and angular) are two important conserved quantities not only as observables of the motion, but also for the important consequences that their conservation (especially the conservation of the energy) imply for the stability and accuracy properties of the algorithm.

Juan was always interested in the fact, due to Ge and Marsden [1988], that algorithms cannot in general preserve the momentum map, the energy and the symplectic structure. This implies that choices must be made and he contributed to the often lively debate concerning which choice is better for which type of application. Motivated by problems in infinite dimensional systems such as elasticity and fluids and others arising in structural mechanics, where one has to deal with infinitely many modes, Juan came to favor energy-momentum type algorithms. On the other hand, he certainly recognized that for other problems, such as those in celestial mechanics, symplectic integrators indeed perform very well.

The formulation of time-stepping algorithms for nonlinear geometrically exact models in structural mechanics can be found in Simo and VuQuoc [1988], along with many early references. This paper concentrated in the the case of rods undergoing overall large motions. Although no particular conservation properties were obtained at the time, the main geometric characteristics of the integration algorithms for nonlinear configuration manifolds like the rotation group can be found fully developed in this reference. Juan's special care in the proper consideration of the different geometric aspects involved in mechanical problems permeate this and future works. The derivation of a number of algorithms that conserve energy and/or momentum for systems like rigid body systems, geometrically exact models for rods and shells, and others followed in a number of subsequent papers (Simo and Wong [1989], Simo, Tarnow and Wong [1992] and Simo, Rifai and Fox [1992]). These results led to a formulation of discrete energy-momentum algorithms in Simo and Tarnow [1992]. The conservation properties are built into the method and, although this last reference considers the particular case of 3D nonlinear elastodynamics, the generality and simplicity of the construction proposed therein made possible the extension of discrete energy-momentum algorithms to more complex systems. In this way, this approach can be found generalized to a number of other cases of interest, including the systems of nonlinear rods and shells in Simo, Tarnow and Doblare [1993], Simo and Tarnow [1994] and general Lie groups in Lewis et al. [1992]. An analysis of the stability properties of this method were presented in Gonzales and Simo [1993][1995] and Gonzales [1996], which show in a rigorous manner the sense and context in which these conserving schemes in the numerical integration of stiff Hamiltonian systems with symmetry perform well. The formulation of higher order methods with similar conservation properties can be found in Tarnow and Simo [1994].

Juan's work on the formulation of conserving schemes for the Hamiltonian case had in Juan's work a direct translation to the case of dissipative systems. Namely, the dissipative structure of these systems (characterized in viscous fluid flows by the existence of an absorbing set and a global attractor) has to be inherited by the discrete integration scheme. In this context, efficient time-stepping algorithms for the infinite dimensional systems of incompressible Navier-Stokes and magnetohydrodynamics (MHD) that exhibit the strong dissipative properties of possessing a discrete absorbing set and attractor can be found in (Simo and Armero [1994]) and (Armero and Simo [1994][1996]). Similarly, dynamic algorithms for finite strain plasticity can be found treated in (Simo [1992]) in the context of solid mechanics. The formulation of unconditionally stable fractional step methods for the temporal integration of coupled (thermomechanical problems in solid mechanics presented in Armero and Simo [1992][1993]). As in the previous cases, these schemes retain the characteristic dissipative structure possessed by the continuum evolution problem.

References

Armero, F. and Simo, J. C. [1992], *A New Unconditionally Stable Fractional Step Method for Non-Linear Coupled Thermomechanical Problems*, Int. J. Num. Meth. Eng., **35**, 737–766.

Armero, F. and Simo, J. C. [1993], *A-Priori Stability Estimates and Unconditionally Stable Product Formula Algorithms for Non-Linear Coupled Thermoplasticity*, Int. J of Plasticity, **9**, 149–182.

Armero, F. and Simo, J. C. [1994], *Long-Term Dissipativity of Time-Stepping Algorithms for an Abstract Evolution Equation with Applications to the Incompressible MHD and Navier-Stokes Equations*, Comp. Meth. Appl. Mech. Eng., to appear.

Armero, F. and Simo, J. C. [1996], *Formulation of a new class of fraction-step methods for the incompressible MHD equations that retains the long-term dissipativity of the continuum dynamical system*, this volume.

Ge, Z. and Marsden, J. E. [1988], *Lie-Poisson integrators and Lie-Poisson Hamilton-Jacobi theory*, Phys. Lett. A, **133**, 134–139.

Gonzalez, O. [1996], *Design and Analysis of Conserving Integrators for Nonlinear Hamiltonian Systems with Symmetry*, Thesis, Stanford University.

Gonzalez, O. and Simo, J. C. [1995], *On the Stability of Symplectic and Energy-Momentum Algorithms for Nonlinear Hamiltonian Systems with Symmetry*, Comp. Meth. Appl. Mech. Eng., submitted.

Lewis, D., Ratiu,, T. S., Simo, J. C., and Marsden, J. E. [1992], *The heavy top, a geometric treatment*, Nonlinearity, **5**, 1–48.

Lewis, D. and Simo, J. C. [1990], *Nonlinear stability of rotating pseudo-rigid bodies*, Proc. Roy. Soc. Lon. A, **427**, 281–319.

Lewis, D. and Simo, J. C. [1995], *Conserving algorithms for the dynamics of Hamiltonian systems on Lie groups*, J. Nonlinear Sci., **4**, 253–299.

Simo, J. C. [1992], *Algorithms for static and dynamic multiplicative plasticity that preserve the classical return mapping schemes of the infinitesimal theory*, Comp. Meth. Appl. Mech. Eng., **98**, 41–104.

Simo, J. C. and Armero, F. [1994], *Unconditional Stability and Long-Term Behavior of Transient Algorithms for the Incompressible Navier-Stokes and Euler Equations*, Comp. Meth. Appl. Mech. Eng., **111**, 111–154.

Simo, J. C.and Gonzalez O. [1993], *Assessment of Energy-Momentum and Symplectic Schemes for Stiff Dynamical Systems*, Proc. ASME Winter Annual Meeting, New Orleans, December 1993.

Simo, J. C. and Lewis, D. [1995], *Conserving algorithms for the dynamics of Hamiltonian systems on Lie groups*, J. Nonlinear Sci., **4**, 253–299.

Simo, J. C., Lewis, D. R. and Marsden, J. E. [1991], *Stability of relative equilibria I: The reduced energy momentum method*, Arch. Rat. Mech. Anal., **115**, 15-59.

Simo, J. C. and Marsden, J. E. [1984], *On the rotated stress tensor and a material version of the Doyle Ericksen formula*, Arch. Rat. Mech. Anal., **86**, 213–231.

Simo, J. C., Marsden, J. E., and Krishnaprasad, P. S. [1988], *The Hamiltonian structure of nonlinear elasticity: The material, spatial, and convective representations of solids, rods, and plates*, Arch. Rat. Mech. Anal., **104**, 125–183.

Simo, J. C., Posbergh, T. A., and Marsden, J. E. [1990], *Stability of coupled rigid body and geometrically exact rods: block diagonalization and the energy-momentum method*, Physics Reports, **193**, 280–360.

Simo, J. C., Posbergh, T. A., and Marsden, J. E. [1991], *Stability of relative equilibria II: Three dimensional elasticity*, Arch. Rat. Mech. Anal., **115**, 61–100.

Simo, J. C., Rifai, M. S., and Fox, D. D. [1992], *On a stress resultant geometrically exact shell models. Part VI: Conserving algorithms for nonlinear dynamics*, Comp. Meth. Appl. Mech. Eng., **34**, 117–164.

Simo, J. C. and Tarnow, N. [1992], *The discrete energy momentum method. Conserving algorithms for nonlinear elastodynamics*, ZAMP, **43**, 757–792.

Simo, J. C. and Tarnow, N. [1994], *A New Energy Momentum Method for the Dynamics of Nonlinear Shells*, Int. J. Num. Meth. Eng., **37**, 2527–2549.

Simo, J. C., Tarnow, N., and Doblare, M. [1993], *Nonlinear Dynamics of Three-Dimensional Rods: Exact Energy and Momentum Conserving Algorithms*, Comp. Meth. Appl. Mech. Eng., to appear.

Simo, J. C., Tarnow, N., and Wong, K. K. [1992], *Exact energy-momentum conserving algorithms and symplectic schemes for nonlinear dynamics*, Comp. Meth. Appl. Mech. Eng., **100** 63–116.

Simo, J. C. and VuQuoc, L. [1988], *On the dynamics in space of rods undergoing large overall motions—a geometrically exact approach*, Comp Meth. Appl. Mech. Eng., **66**, 125-161.

Simo, J. C. and Wong, K. K. [1989], *Unconditionally stable algorithms for the orthogonal group that exactly preserve energy and momentum*, Int. J. Num. Meth. Eng., **31**, 19–52.

Tarnow, N. and Simo, J. C. [1994], *How to Render Second Order Accurate Time-Stepping Algorithms Fourth Order Accurate while Retaining the Stability and Conservation Properties*, Comp. Meth. Appl. Mech. Eng., **115** 233–252.

Fields Institute Communications
Volume **10**, 1996

Formulation of a New Class of Fractional-Step Methods for the Incompressible MHD Equations that Retains the Long-Term Dissipativity of the Continuum Dynamical System

F. Armero
Division of Applied Mechanics
Department of Mechanical Engineering
Stanford University, Stanford, California 94305 USA
Currently at: Department of Civil Engineering
University of California, Berkeley, CA USA 94720

J. C. Simo*
Division of Applied Mechanics Stanford University
Stanford, CA 94305

Abstract. This paper introduces a new class of fractional-step methods for the incompressible MHD equations that exhibits strong properties of nonlinear stability and long-term dissipativity. The algorithms are based on a new operator split of the continuum problem of evolution that preserves the bounded dissipativity of the coupled system. A particular algorithm is shown to exhibit optimal long-term dissipative properties. Furthermore, this algorithm possesses these strong properties while being linear in the sense that it involves the solution of a *linear* problem at each time step. Representative numerical simulations are presented to assess the performance of these algorithms in combination with mixed finite element methods.

1 Introduction

The main goal of the research presented in this paper is the formulation of efficient time-stepping algorithms for the system of incompressible magnetohydrodynamics (MHD). This system can be characterized as a *constrained coupled dissipative dynamical system,* which involves the solution of highly nonlinear initial-boundary value problems. The numerical solution of such an infinite dimensional system of evolution requires very often the consideration of large-scale/long-term

---

1991 *Mathematics Subject Classification.* Primary: 65M12; Secondary: 65M60, 76W05.
*Deceased.

simulations. Efficiency of an implicit time integration scheme in these circumstances necessarily involves *unconditional stability;* that is, if algebraic equations are to be solved as part of the solution process, stability should be achieved with independence of the time step. The implicitness of the scheme is actually a must in constrained problems as it is the case for the incompressible flows considered herein. Furthermore, algorithms that optimally approximate the long-term dissipative structure characteristic of the physical problem should be identified, since this structure is often related to physical processes of interest. As in the case of Hamiltonian systems where one of the main goals is the design of numerical algorithms that preserve a number of conservation properties characteristic of this type of systems, as illustrated in detail in this volume, we view the long-term dissipative structure of the continuum problem as a fundamental property to be inherited by the intergration schemes.

Two general strategies are available for the numerical solution of coupled problems: *monolithic* and *fractional-step methods* (also known as *staggered techniques* in the engineering literature). The first approach consists in solving simultaneously for all the fields involved in the problem. This usually results in very large systems involving a difficult generalization of numerical techniques developed for each individual field and leading, altogether, to a very costly and unefficient solution of the problem. On the other hand, in the second approach the system is partitioned into several smaller and simpler sub-systems that take advantage of the physical nature of the problem. This leads to a much lower computational cost and modularity in the implementation that allows a very efficient combination of methods and software previously developed. However, the main disadvantage of the standard staggered techniques is their restricted stability properties.

We introduce in this paper a new class of fractional-step methods for the incompressible MHD system. The analysis of the long-term dissipative properties of monolithic schemes for this system can be found in (Armero and Simo [1994]), extending the results of (Simo and Armero [1993]) for the incompressible Navier-Stokes equations. The approach presented here follows the classical method of fractional steps. Full account of this technique can be found in (Yanenko [1971], Marchuck [1982]) and references therein. The fractional-step methods introduced in this paper are based on an operator split of the continuum problem that preserves the dissipativity of the problem of evolution. This approach led in (Armero and Simo [1992][1993]) to the formulation of very efficient, stable and accurate, staggered algorithms for nonlinear coupled thermomechanical problems in solid mechanics. In this fashion, the resulting numerical schemes exhibit unconditional nonlinear stability and optimally represent the dissipative structure of the continuum problem. Furthermore, these properties are accomplished with *linear* schemes, in the sense that their implementation leads to the solution of a linear boundary-value problem in each time step, avoiding the prohibitive computational cost of nonlinear iterations. To the authors' knowledge, these methods are the first of their kind in the literature. Despite these good results, the accuracy of the algorithms introduced here for incompressible MHD is affected by a Δt-dependence of the steady-state solutions, as shown in Section 3.2.2. Strategies to avoid this drawback are currently under investigation.

An outline of the rest of the paper is as follows. Section 2 summarizes the governing equations of the problem together with a brief description of their dissipative structure. Section 3 considers the formulation and analysis of different operator splits and the resulting fractional-step methods. Section 4 assesses the numerical performance of the schemes in a number of numerical simulations. Finally, Section 5 includes some concluding remarks.

2 Notation and Problem Definition

This section provides a summary of the evolution equations governing incompressible magnetohydrodynamics and describes briefly some known results regarding their long-term dissipative structure. This structure dictates the proper concepts of numerical stability and dissipativity of the time-stepping algorithms under investigation in this work, as defined in Section 2.3.

2.1 The local governing equations Consider a domain $\Omega \subset \mathbb{R}^{n_{\text{dim}}}$ ($n_{\text{dim}} = 2$ or $n_{\text{dim}} = 3$) with smooth boundary Γ filled with an electrically conducting incompressible Newtonian fluid. We will denote by $\mathbf{u}(\mathbf{x},t)$ and $\mathbf{b}(\mathbf{x},t)$ the spatial velocity and the magnetic induction of the fluid, respectively, at a point $\mathbf{x} \in \Omega$ and time $t \in [0,T]$, for some given interval of time T. A number of physical hypotheses, valid for the problems of interest, reduce the general system of Maxwell equations coupled with the Navier-Stokes equations to the so-called MHD equations. Briefly, these equations represent a non-relativistic motion of a non-magnetizable quasi-neutral conductor at low frequencies. We refer to the standard literature on the subject for an elaboration of these matters (see, e.g., Landau et al [1984]). In non-dimensional form, the MHD system of equations reads

$$\left.\begin{aligned} \partial_t \mathbf{u} \; &+ \mathbf{u}\nabla\mathbf{u} = -\nabla p + \tfrac{1}{\text{Re}}\Delta\mathbf{u} + S\,\text{curl}\,\mathbf{b}\times\mathbf{b} + \mathbf{f}, \\ \partial_t \mathbf{b} \; &- \text{curl}\big(\mathbf{u}\times\mathbf{b}\big) = -\tfrac{1}{\text{Re}_\text{m}}\,\text{curl}\big(\text{curl}\,\mathbf{b}\big), \\ \text{div}\,\mathbf{u} &= \; 0, \qquad \text{div}\,\mathbf{b} = \; 0, \end{aligned}\right\} \quad \text{in } \Omega, \tag{2.1}$$

where $p(\mathbf{x},t)$ denotes the pressure field and $\mathbf{f}(\mathbf{x},t)$ an external body force. The physical constraint of a divergence free magnetic induction field has been added explicitly.

The different regimes arising in the problem at hand are characterized by the three non-dimensional numbers Re = Reynolds number $:= UL/\nu$, Re_m = magnetic Reynolds number $:= \sigma\mu_0 UL$ and S = coupling number $:= B^2/\mu_0\rho U^2$, for characteristic values of the velocity U, length L and magnetic induction B, density ρ, kinematic viscosity ν, scalar electric conductivity σ of the fluid and (free-space) magnetic permeability μ_0. Without loss of generality these material parameters are assumed constant in Ω, i.e., homogeneous fluid. With this notation the (non-dimensional) electric current density $\mathbf{j}$ and electric field $\mathbf{e}$ are obtained as $\mathbf{j} = S\;\text{curl}\,\mathbf{b}$ and $\mathbf{e} = \text{curl}\,\mathbf{b}/\text{Re}_\text{m} - \mathbf{u}\times\mathbf{b}$.

The system of equations (2.1) has to be supplemented with the initial conditions

$$\mathbf{u}(\mathbf{x},0) = \mathbf{u}_0(\mathbf{x}), \qquad \mathbf{b}(\mathbf{x},0) = \mathbf{b}_0(\mathbf{x}) \qquad \text{in } \Omega, \tag{2.2}$$

along with proper boundary conditions on $\Gamma \times [0,T]$. For simplicity the analysis below is focused to the boundary conditions

$$\mathbf{u} = 0, \quad \mathbf{b}\cdot\mathbf{n} = 0 \quad \text{and} \quad \mathbf{n}\times\text{curl}\,\mathbf{b} = 0 \qquad \text{on } \Gamma, \tag{2.3}$$

where $\mathbf{n}$ is the unit normal to Γ. Conditions $(2.3)_{(2,3)}$ correspond to a perfectly conducting boundary. In this situation, the proper functional setting for the solutions of the system (2.1) can be described by the introduction of the space $H = H_u \times H_b$ where

$$H_u = H_b := \{\eta \in L_2(\Omega)^{n_{\rm dim}} : \operatorname{div} \eta = 0 \text{ in } \Omega,\ \eta \cdot \mathbf{n} = 0 \text{ on } \Gamma\}. \tag{2.4}$$

Remark 2.1 The convective term can be written in general as $\mathbf{C}(\mathbf{u})\mathbf{u}$, where

$$\mathbf{C}(\mathbf{u})\eta = \mathbf{u}\nabla\eta = \mathbf{u}\nabla\eta + \tfrac{1}{2} \operatorname{div} \mathbf{u}\ \eta, \tag{2.5}$$

since for the continuum problem we explicitly have $\operatorname{div} \mathbf{u} = 0$. An important fact exploited in crucial manner in the algorithmic analysis presented below is the following skew-adjoint property of the convective term:

$$\langle \mathbf{C}(\mathbf{u})\eta_1, \eta_2 \rangle = -\langle \mathbf{C}(\mathbf{u})\eta_2, \eta_1 \rangle, \quad \forall \eta_1, \eta_2 \in H^1(\Omega)^{n_{\rm dim}} \tag{2.6}$$

where $\langle \cdot, \cdot \rangle$ denotes the standard $L^2(\Omega)$ pairing. For the first form of the convective term in (2.5), this relation is satisfied if $\mathbf{u} \in H_u$. On the other hand, the second form in (2.5) (known as the skew-symmetric form) implies that the skew-symmetry property (2.6) also holds for $\mathbf{u}$ not necessarily divergence-free, but required only to satisfy the Dirichlet boundary condition $\mathbf{u} \cdot \mathbf{n} = 0$. This feature has important consequences in the numerical solution of the problem since at the spatially discrete level the divergence-free constraint is only satisfied weakly as an approximation to the space H_u.

2.2 A-priori stability estimate and long-term dissipative structure The numerical stability analysis presented below exploits two basic properties of the system (2.1). First, this system satisfies an a-priori stability estimate expressed in terms of a decay in time of an energy-like norm. The proper notion of dissipativity in the general context of nonlinear infinite dimensional dynamical systems involves the additional notion of existence of an absorbing set in the corresponding functional space. It is an important fact that the system of MHD equations (2.1) possesses for $n_{\rm dim} = 2$ such a dissipative structure. Details of these results can be found in Sermange and Temam [1983] and Temam [1988]. See also Duvaut and Lions [1972]. Below we provide a brief review of these results since similar properties will be shown to hold for the time-stepping algorithms introduced in this paper.

The total internal energy $K(\mathbf{u}, \mathbf{b})$ of an incompressible electrically conducting viscous fluid is the sum of the kinetic energy and magnetic energy, leading to the expression

$$K(\mathbf{u}, \mathbf{b}) = \tfrac{1}{2}\|\mathbf{u}\|^2 + \frac{S}{2}\|\mathbf{b}\|^2, \tag{2.7}$$

where $\|\eta\|^2 := \langle \eta, \eta \rangle$ is the square of the $L^2(\Omega)$ norm. Clearly, $K(\cdot, \cdot)$ defines a norm in H. The evolution of the total energy along the flow defined by equations (2.1) can be easily computed using integration by parts as follows

$$\begin{aligned} \tfrac{d}{dt}K(\mathbf{u}, \mathbf{b}) &= \langle \mathbf{u}, \partial_t \mathbf{u} \rangle + S\ \langle \mathbf{b}, \partial_t \mathbf{b} \rangle \\ &= -\tfrac{1}{\mathrm{Re}}\langle \nabla\mathbf{u}, \nabla\mathbf{u} \rangle - \tfrac{S}{\mathrm{Re_m}}\langle \operatorname{curl} \mathbf{b}, \operatorname{curl} \mathbf{b} \rangle + \langle \mathbf{f}, \mathbf{u} \rangle \\ &\quad + S\ \langle \operatorname{curl} \mathbf{b} \times \mathbf{b}, \mathbf{u} \rangle + S\ \langle \mathbf{u} \times \mathbf{b}, \operatorname{curl} \mathbf{b} \rangle \\ &\quad + \langle \mathbf{C}(\mathbf{u})\mathbf{u}, \mathbf{u} \rangle + \langle p, \operatorname{div} \mathbf{u} \rangle, \end{aligned} \tag{2.8}$$

where we have used the boundary conditions (2.3). The last two terms in (2.8) vanish because of the skew-adjoint property of the convective term defined by (2.6), and the incompressibility constraint. The cancellation of the coupling terms is to be noted, since it is crucial for the dissipative property of the MHD system to hold. As shown in Section 3, the design of a unconditionally stable staggered algorithm for this system has to allow this cancellation of terms to take place. The final expression governing the evolution of the total energy can then be written as

$$\frac{d}{dt}K(\mathbf{u},\mathbf{b}) = -\|\nabla\mathbf{u}\|^2 - \frac{S}{\mathrm{Re_m}}\|\operatorname{curl}\mathbf{b}\|^2 + \langle\mathbf{f},\mathbf{u}\rangle. \tag{2.9}$$

To arrive at estimates governing the decay of $K(\mathbf{u},\mathbf{b})$, we consider separately the case of $\mathbf{f}=0$ and the general case of arbitrary $\mathbf{f}$ but independent of time.

i. A-priori stability estimate. For the case $\mathbf{f}=0$, equation (2.9) shows that the total energy decreases along the flow. Use of the Poincaré-type inequalities

$$\|\nabla\eta\| \geq \sqrt{\lambda_u}\ \|\eta\|\ , \quad \|\operatorname{curl}\eta\| \geq \sqrt{\lambda_b}\ \|\eta\| \quad \forall\eta\in H\cap H^1(\Omega)^{n_{\mathrm{dim}}}, \tag{2.10}$$

for $\lambda_u,\lambda_b>0$, yields the following sharp estimate for the energy decay in H

$$\frac{d}{dt}K(\mathbf{u},\mathbf{b}) \leq -2\,\tilde{C}\,K(\mathbf{u},\mathbf{b}) \quad\Longrightarrow\quad K(\mathbf{u},\mathbf{b}) \leq \exp[-2\,\tilde{C}\,t]\ K(\mathbf{u}_o,\mathbf{b}_o), \tag{2.11}$$

where $\tilde{C}=\min\{\lambda_u/\mathrm{Re},\lambda_b/\mathrm{Re_m}\}$.

ii. Absorbing set in H. The long-term asymptotic behavior for the general case in which $\mathbf{f}\neq 0$ can be characterized as follows. The use of inequalities (2.10) together with Young's inequality for the body force term in (2.8) can be shown to imply the estimate

$$\frac{d}{dt}K(\mathbf{u},\mathbf{b}) \leq -C\ K(\mathbf{u},\mathbf{b}) + \frac{\|\mathbf{f}\|^2}{2C}, \tag{2.12}$$

where $C=\min\{\lambda_u/\mathrm{Re},2\lambda_b/\mathrm{Re_m}\}$. Assuming $\mathbf{f}$ to be independent of time, we obtain the asymptotic estimate

$$\lim_{t\to\infty}\ \sup\ K(\mathbf{u},\mathbf{b}) \leq \frac{\|\mathbf{f}\|^2}{2C^2}. \tag{2.13}$$

after making use of the classical Gronwall's lemma. This estimate implies the existence of an absorbing set in H for $n_{\mathrm{dim}}=2$ where the existence of solutions for any time interval T can be proved. The flow of an initial bounded set is then contained in this set for a sufficiently long time, thus characterizing the long-term behavior of the solutions. Systems exhibiting this property are referred to as *bounded dissipative* (Hale [1988]).

Remark 2.2 The existence of a global attractor in H follows from estimate (2.13) if the flow is uniformly compact in the long-term. For the system (2.1), this property follows from the existence of an absorbing set in $V=V_u\times V_b$, with $V_u=H\cap H_0^1(\Omega)^{n_{\mathrm{dim}}}$ and $V_b=H\cap H^1(\Omega)^{n_{\mathrm{dim}}}$ ($n_{\mathrm{dim}}=2$), by the compactness of the inclusion $V\subset H$. We refer to Temam [1988] for further details, and to Armero and Simo [1994] and references therein for the implications of the existence of this set in the algorithmic context.

2.3 Nonlinear stability and algorithmic dissipativity Let $\{\mathbf{u}_n, \mathbf{b}_n\}_{n=0}^{\infty} \subset H$ be a sequence computed with a generic time-marching scheme approximating the solution of the system (2.1) in time; i.e., formally $\mathbf{u}(t_n) \simeq \mathbf{u}_n$, $\mathbf{b}(t_n) \simeq \mathbf{b}_n$ for some partition of the time interval with a time-step $\Delta t := t_{n+1} - t_n > 0$. In contrast with the more classical approach where the convergence properties of the numerical scheme are studied as $\Delta t \to 0$, we shall view the time discretization as defining a *time discrete dynamical system* in H or, more generally, a *generalized system of evolution* by formally considering all possible solution sequences $\{\mathbf{u}_n, \mathbf{b}_n\}_{n=0}^{\infty}$ generated from a given initial condition $\{\mathbf{u}_0, \mathbf{b}_0\}$, as in (Humphries and Stuart [1993]). The interest goes then to the characterization of the behavior of the solution as $n \to \infty$ for a *fixed* and finite Δt. From a practical point of view, this is the case of interest since the numerical solution of the system (2.1) requires very often the consideration of long-term simulations. In this setting, the following definitions of *nonlinear stability* and *dissipativity* of numerical schemes are introduced.

Definition 2.1 *(H-stability)* The algorithm is said to be H-stable if in the absence of forcing ($\mathbf{f} = \mathbf{0}$) the *energy decay* estimate

$$K(\mathbf{u}_{n+1}, \mathbf{b}_{n+1}) \leq K(\mathbf{u}_n, \mathbf{b}_n) \tag{2.14}$$

holds for $\Delta t \leq \Delta t_{\text{crit}}$ and for every solution sequence $\{\mathbf{u}_n, \mathbf{b}_n\}_{n=0}^{\infty}$ obtained from any $\{\mathbf{u}_0, \mathbf{b}_0\} \in H$. If estimate (2.14) holds for any time step $\Delta t > 0$, the algorithm is said to be unconditionally H-stable.

This formal definition of stability is motivated by the a-priori stability estimate (2.11) that shows the energy decay associated to the continuum dynamical system under investigation. This notion of stability is weaker that the property of nonlinear B-stability appropriate for contractive problems; (see Butcher [1975]). This contractivity does not hold for the continuum equations under investigation and it is, therefore, too stringent to impose it in their numerical temporal approximations. The reader is referred to (Hairer and Wanner [1991]) for a review of different notions of stability.

Definition 2.2 *(Long-term dissipative behavior)* The algorithm is said to be H-dissipative if every solution sequence $\{\mathbf{u}_n, \mathbf{b}_n\}_{n=0}^{\infty}$ satisfies for some $R_H = \widehat{R_H}\big(\|\mathbf{f}\|_H, \Delta t\big)$, and $\Delta t \leq \Delta t_{crit}$ the estimate

$$\lim_{n\to\infty} \sup K(\mathbf{u}_n, \mathbf{b}_n) \leq \tfrac{1}{2}\, R_H^2, \tag{2.15}$$

for all $\{\mathbf{u}_0, \mathbf{b}_0\} \in H$ and time-independent forcing terms $\mathbf{f} \in H$. The term unconditionally H-dissipative will indicate that (2.15) holds $\forall \Delta t > 0$ with R_H independent of Δt.

Definition 2.2 of dissipative numerical schemes is motivated by the long-term estimate (2.13) satisfied by the exact solution. Similar notions of numerical dissipativity can be found in (Hale et al [1988],Foias et al [1991], Elliot and Stuart [1992], Simo and Armero [1993] and Humphries and Stuart [1993]).

3 Fractional-Step Methods for Incompressible MHD

The formulation of staggered algorithms for evolution equations of the general type

$$\left.\begin{aligned} &\tfrac{d}{dt}\chi(t) + \mathbf{A}\chi(t) = \mathbf{f} \\ &\chi(0) \;\; = \chi_o \end{aligned}\right\} \quad \text{in } H \; , \tag{3.1}$$

in the context of the method of fractional steps exploits additive splits of the differential operator $\mathbf{A}$ of the form $\mathbf{A} = \mathbf{A}_1 + \mathbf{A}_2$ or, more generally, as the sum of a finite number of operators $\mathbf{A}_i$. Then, a time-stepping algorithm for the global problem is obtained formally as the product of two algorithms $\mathbb{K}_{i,\Delta t}$ consistent with each $\mathbf{A}_i$ respectively, e.g.,

$$\mathbb{K}_{\Delta t} = \mathbb{K}_{2,\Delta t}\ \mathbb{K}_{1,\Delta t}\ . \tag{3.2}$$

The stability and convergence of (3.2) is the essence of the *Lie-Trotter-Kato product formula* for the approximation of linear semigroups; see (Chorin et al. [1978]) and references therein. The product formula algorithm (3.2) is first order accurate. Stable second order accurate algorithms can be obtained by a double pass technique of the form

$$\mathbb{K}_{\Delta t} = \mathbb{K}_{1,\frac{\Delta t}{2}}\ \mathbb{K}_{2,\Delta t}\ \mathbb{K}_{1,\frac{\Delta t}{2}}\ , \tag{3.3}$$

with individual second order stable schemes $\mathbb{K}_{i,\Delta t}$, as proposed originally in Strang [1968].

The goal of this work is to analyze fractional-step methods for the incompressible MHD equations arising from different splits of the continuum problem. Section 3.1 describes standard schemes arising from what is denoted by the *Eulerian split*, leading to conditionally stable methods. Section 3.2 describes a new operator split of the MHD system, the *induction split*, which yields unconditionally stable staggered algorithms.

3.1 The Eulerian split Given the coupled system of equations (2.1), the obvious strategy to decouple the computation of the velocity $\mathbf{u}$ (together with the pressure p) and the magnetic induction $\mathbf{b}$ is to solve the momentum equation $(2.1)_1$ with a fixed $\mathbf{b}$, solving equation $(2.1)_2$ with the computed velocity afterwards. This strategy has been followed, for instance, by (Richtmyer and Morton [1967]). Gauss-Seidel iterative methods can also be devised based on the split; (Gunzburger et al. [1991]) proposed some of these methods for the stationary MHD problem. These methods can be understood as arising from the following *additive operator split* of the fully coupled problem

Problem 1: *(Navier-Stokes)*

$$\left.\begin{aligned} \partial_t \mathbf{u} &= -\mathbf{C}(\mathbf{u})\nabla\mathbf{u} + \tfrac{1}{\mathrm{Re}}\Delta\mathbf{u} - \nabla p + S\ \operatorname{curl}\mathbf{b}\times\mathbf{b} + \mathbf{f}, \\ \partial_t \mathbf{b} &= 0, \\ \operatorname{div}\mathbf{u} &= 0, \end{aligned}\right\}$$
$$\mathbf{u} = 0 \text{ on } \Gamma. \tag{3.4}$$

Problem 2: *(Electric induction at fixed velocity)*

$$\left.\begin{aligned} \partial_t \mathbf{u} &= 0, \\ \partial_t \mathbf{b} &= \operatorname{curl}\big(\mathbf{u}\times\mathbf{b}\big) - \tfrac{1}{\mathrm{Re_m}}\operatorname{curl}\big(\operatorname{curl}\mathbf{b}\big), \\ \operatorname{div}\mathbf{b} &= 0, \end{aligned}\right\}$$
$$\mathbf{b}\cdot\mathbf{n} = 0,\ \ \mathbf{n}\times\operatorname{curl}\mathbf{b} = 0 \text{ on } \Gamma.$$

This split and the fractional-step methods based on it are characterized as follows.

i. *Structure of the split.* The name *Eulerian split* arises from the fact that in Problem 1 the magnetic induction is fixed in an Eulerian reference system. This split does not preserve the dissipative structure of the original problem. Indeed, the evaluation of the change of the total energy along the solution of Problem 2 leads to

$$\frac{d}{dt}K(\mathbf{u},\mathbf{b})\Big|_{\text{Problem 2}} = -\frac{S}{\mathrm{Re_m}}\|\operatorname{curl}\mathbf{b}\|^2 + S\big\langle \operatorname{curl}(\mathbf{u}\times\mathbf{b}),\mathbf{b}\big\rangle \not\leq 0\ , \tag{3.5}$$

in general, since the last term (coming from the induced electric field by $\mathbf{u}$) cannot be bounded appropriately. After the developments of (2.8), we can immediately recognize the cause of the lack of stability of algorithms based on this split; that is, the absence of the cancellation of the coupling terms crucial to the dissipativity of the coupled problem as pointed out in Section 2.2.

Algorithms based on the Eulerian split (3.4) can still exhibit conditional stability in the context of the fully discrete problem. The terms breaking the dissipative structure of the problem (see equation (3.5)) can be bounded in this context by an *inverse estimate* of the form $\|\nabla\eta^h\| \leq (M/h)\|\eta^h\|$ (see Ciarlet [1978]). The non-uniformity of this estimate in the discretization size h ($h\to 0$) renders the algorithm conditionally stable at best.

ii. *Model algorithm.* A representative time-stepping algorithm based on the Eulerian split is summarized in Table 1. The Navier-Stokes equations of Problem 1 have been solved with a linear semi-implicit scheme as in Simo and Armero [1993], and similarly for Problem 2. In this manner, the final algorithm is linear since each step requires the solution of a linear boundary value problem. Extensions to generalized mid-point approximations can be easily devised. As pointed out above, despite the implicit character of the solution algorithms for each sub-problem, the final method is only conditionally stable at best. Linearized stability conditions for the fully discrete equations arising from a spatial finite element discretization of the problem can be found in Armero [1993]. These conditions are obtained via the energy method by exploiting the inverse estimate indicated above.

Step 1 - Mechanical phase: Given $\mathbf{u}_n$ and $\mathbf{b}_n$, solve for $\mathbf{u}_{n+1}$ in Ω

$$\left.\begin{aligned} \frac{1}{\Delta t}\left[\mathbf{u}_{n+1}-\mathbf{u}_n\right] &= -\mathbf{C}(\mathbf{u}_n)\nabla\mathbf{u}_{n+1}+\frac{1}{\mathrm{Re}}\Delta\mathbf{u}_{n+1}-\nabla\bar{p} \\ &\quad +S\ \operatorname{curl}\mathbf{b}_n\times\mathbf{b}_n+\mathbf{f}_{n+1}, \\ \operatorname{div}\mathbf{u}_{n+1} &= 0, \end{aligned}\right\} \tag{3.6}$$

with $\mathbf{u}_{n+1}=0$ on Γ.

Step 2 - Magnetic phase: Given $\mathbf{u}_{n+1}$ and $\mathbf{b}_n$, solve for $\mathbf{b}_{n+1}$

$$\left.\begin{aligned} \frac{1}{\Delta t}\left[\mathbf{b}_{n+1}-\mathbf{b}_n\right] &= \operatorname{curl}\left(\mathbf{u}_{n+1}\times\mathbf{b}_{n+1}\right)-\frac{1}{\mathrm{Re_m}}\operatorname{curl}\left(\operatorname{curl}\mathbf{b}_{n+1}\right), \\ \operatorname{div}\mathbf{b}_{n+1} &= 0, \end{aligned}\right\} \tag{3.7}$$

with $\mathbf{b}_{n+1}\cdot\mathbf{n}=0$ and $\mathbf{n}\times\operatorname{curl}\mathbf{b}_{n+1}=0$ on Γ.

Table 1 Representative fractional-step method based on the Eulerian split of the MHD equations.

3.2 The induction split Consider the following alternative operator split for the coupled MHD system

Problem 1: *(Electric induction)*

$$\left.\begin{aligned} &\partial_t\mathbf{u}=S\ \operatorname{curl}\mathbf{b}\times\mathbf{b}, \\ &\partial_t\mathbf{b}=\operatorname{curl}\left(\mathbf{u}\times\mathbf{b}\right)-\frac{1}{\mathrm{Re_m}}\operatorname{curl}\left(\operatorname{curl}\mathbf{b}\right), \\ &\operatorname{div}\mathbf{b}=0, \end{aligned}\right\}$$
$$\mathbf{b}\cdot\mathbf{n}=0,\ \mathbf{n}\times\operatorname{curl}\mathbf{b}=0\quad\text{on}\quad\Gamma. \tag{3.8}$$

Problem 2: *(Navier-Stokes)*

$$\left.\begin{aligned} &\partial_t\mathbf{u}=-\mathbf{C}(\mathbf{u})\nabla\mathbf{u}+\frac{1}{\mathrm{Re}}\Delta\mathbf{u}-\nabla p+\mathbf{f}, \\ &\partial_t\mathbf{b}=0, \\ &\operatorname{div}\mathbf{u}=0, \end{aligned}\right\}$$
$$\mathbf{u}=0\text{ on }\Gamma.$$

This split and the fractional-step methods based on it are characterized as follows.

i. *Structure of the split.* Contrary to the Eulerian split, the operator split defined by equations (3.8) is characterized by the solution of the Navier-Stokes system in the second phase. Problem 1 corresponds to the the full electric induction equation; hence the name of *induction split.* The Lorentz force term ($S\ \operatorname{curl}\mathbf{b}\times\mathbf{b}$)

determines the evolution of the velocity along this first phase, which leads to an initial velocity for Problem 2 including these coupling effects. The crucial cancellation of coupling terms indicated in Section 2.2 takes place in Problem 1, leading to unconditional H-stable fractional-step methods as shown below. The induced electric field and the electric resistance contribution to the evolution of the magnetic field are not partitioned, making this split appropriate for the resistive limit of interest herein; see Remark 3.2 below in this respect.

ii. *Model algorithm.* A general class of integration algorithms is obtained from the induction split by considering, for instance, generalized mid-point approximations of the different fields in equations (3.8). A representative time-stepping algorithm is summarized in Table 3.2 for a typical time step $[t_n, t_{n+1}]$. Semi-implicit approximations have been assumed for each sub-problem. In particular, the convective term uses the velocity $\mathbf{u}_n$ and not the intermediate field $\tilde{\mathbf{u}}_{n+1}$. The resulting algorithm consists in the solution of two linear boundary value problems, with the problem in the magnetic phase being symmetric. For the sake of brevity, the forthcoming analysis concentrates in this representative algorithm.

Remark 3.1 *An efficient numerical implementation.* Although a-priori both unknown fields evolve in the magnetic phase of the algorithm in Table 3.2, an efficient staggered implementation of this scheme can be simply obtained by eliminating $\tilde{\mathbf{u}}_{n+1}$ with the introduction of $(3.9)_2$ in $(3.9)_1$. In the context of a finite element solution of the problem this is equivalent to the enforcement of the update $(3.9)_2$ at each quadrature point, whereas equation $(3.9)_1$ is solved globally from its weak form. Similarly, the solution of the mechanical phase (3.10) is accomplished independently of the intermediate velocity $\tilde{\mathbf{u}}_{n+1}$. The final implementation involves then minimal changes from the standard schemes based on the Eulerian split, hence taking full advantage of the partitioned structure of the algorithm. Noting that the cost of a direct solution of the finite element equations is proportional to the number of nodal variables to the cube (pressure eliminated at element level) and the symmetry of (3.9), we can estimate the ratio *Cost monolithic/ Cost staggered* $= 2 \cdot 4^3/(2 \cdot 2^3 + 2^3) = 5.33$.

Remark 3.2 Alternative splits preserving the cancellation of the coupling terms can be devised. For instance, the so-called *Lagrangian split* is analyzed in Armero [1993]. This split coincides with (3.4) except that the induced electric field term $\operatorname{curl}(u \times \mathbf{b})$ appears in the induction equation of Problem 1 instead. This corresponds to fixing the magnetic induction in a Lagrangian frame (i.e., this field is frozen in the fluid motion), as physically happens in the non-resistive limit $\mathrm{Re}_{\mathrm{m}} = \infty$. The Lagrangian split is, therefore, more appropriate for this limit. Note that (inviscid) ideal MHD is a Hamiltonian system (Holm et al [1985]) for which the dissipative structure of interest in this work does not apply.

3.2.1 **Nonlinear stability and dissipative long-term behavior** The main result can be stated as follows.

Theorem 3.1 *The fractional-step method described in Table 2 is unconditionally H-stable and unconditionally H-dissipative.*

Proof The first step consists in obtaining the evolution of the total energy, similar to (2.9), in this time discrete setting. To this end, multiply equation $(3.9)_1$

Step 1 - Magnetic phase: Given $\mathbf{u}_n$ and $\mathbf{b}_n$, solve for $\mathbf{b}_{n+1}$ and $\tilde{\mathbf{u}}_{n+1}$ in Ω

$$\left.\begin{aligned} \tfrac{1}{\Delta t}\big[\mathbf{b}_{n+1}-\mathbf{b}_n\big] &= \operatorname{curl}\big(\tilde{\mathbf{u}}_{n+1}\times\mathbf{b}_n\big) - \tfrac{1}{\mathrm{Re_m}}\operatorname{curl}\big(\operatorname{curl}\mathbf{b}_{n+1}\big), \\ \tilde{\mathbf{u}}_{n+1} &= \mathbf{u}_n + \Delta t\ S\ \operatorname{curl}\mathbf{b}_{n+1}\times\mathbf{b}_n, \\ \operatorname{div}\mathbf{b}_{n+1} &= 0, \end{aligned}\right\} \quad (3.9)$$

$$\text{with } \mathbf{b}_{n+1}\cdot\mathbf{n} = 0 \text{ and } \mathbf{n}\times\operatorname{curl}\mathbf{b}_{n+1} = 0 \text{ on } \Gamma.$$

Step 2 - Mechanical phase: Given $\mathbf{b}_{n+1}$ and $\tilde{\mathbf{u}}_{n+1}$, solve for $\mathbf{u}_{n+1}$ in Ω

$$\left.\begin{aligned} \tfrac{1}{\Delta t}\big[\mathbf{u}_{n+1}-\tilde{\mathbf{u}}_{n+1}\big] &= -\mathbf{C}(\mathbf{u}_n)\nabla\mathbf{u}_{n+1} + \tfrac{1}{\mathrm{Re}}\Delta\mathbf{u}_{n+1} - \nabla\bar{p} + \mathbf{f}_{n+1}, \\ \operatorname{div}\mathbf{u}_{n+1} &= 0, \end{aligned}\right\}$$

$$\text{with } \mathbf{u}_{n+1} = 0 \text{ on } \Gamma. \quad (3.10)$$

Table 2 Representative fractional-step method based on the induction split of the MHD equations.

by $S\ \mathbf{b}_{n+1}$ to obtain after integration by parts the estimate

$$\begin{aligned} \tfrac{S}{2\Delta t}\big[\|\mathbf{b}_{n+1}\|^2 - \|\mathbf{b}_n\|^2\big] &+ \tfrac{S}{2\Delta t}\|\mathbf{b}_{n+1}-\mathbf{b}_n\|^2 \\ &= -\tfrac{1}{\mathrm{Re_m}}\|\operatorname{curl}\mathbf{b}_{n+1}\|^2 - \big\langle \underbrace{S\ \big(\operatorname{curl}\mathbf{b}_{n+1}\ \times\mathbf{b}_n\big)}_{=\frac{1}{\Delta t}\left(\tilde{\mathbf{u}}_{n+1}-\mathbf{u}_n\right)}, \tilde{\mathbf{u}}_{n+1}\big\rangle \\ &\le -\tfrac{1}{\mathrm{Re_m}}\|\operatorname{curl}\mathbf{b}_{n+1}\|^2 - \tfrac{1}{2\Delta t}\big[\|\tilde{\mathbf{u}}_{n+1}\|^2 - \|\mathbf{u}_n\|^2\big], \end{aligned} \quad (3.11)$$

after making use of the inequality $\big(a-b\big)\ a \ge \frac{1}{2}\big(a^2-b^2\big)$. Therefore, we can write

$$\begin{aligned} \tfrac{1}{\Delta t}\big[K\big(\mathbf{u}_{n+1},\mathbf{b}_{n+1}\big)\ &-K\big(\mathbf{u}_n,\mathbf{b}_n\big)\big] \\ &\le -\tfrac{1}{\mathrm{Re_m}}\|\operatorname{curl}\mathbf{b}_{n+1}\|^2 + \tfrac{1}{2\Delta t}\big[\|\mathbf{u}_{n+1}\|^2 - \|\tilde{\mathbf{u}}_{n+1}\|^2\big]. \end{aligned} \quad (3.12)$$

Similarly, for the mechanical phase we obtain after multiplying equation $(3.10)_1$ by $\mathbf{u}_{n+1}$

$$\frac{1}{2\Delta t}\big[\|\mathbf{u}_{n+1}\|^2 - \|\tilde{\mathbf{u}}_{n+1}\|^2\big] \le -\frac{1}{\mathrm{Re}}\|\nabla\mathbf{u}_{n+1}\|^2 + \big\langle\mathbf{f}_{n+1},\mathbf{u}_{n+1}\big\rangle, \quad (3.13)$$

where the contribution of the convective term vanishes by skew-symmetry as it does the pressure term contribution due to the incompressibility constraint $(3.10)_2$.

Combining the inequalities (3.12) and (3.13) yields the final estimate

$$\begin{aligned} \tfrac{1}{\Delta t}\quad &\big[K\left(\mathbf{u}_{n+1},\mathbf{b}_{n+1}\right) - K\left(\mathbf{u}_n,\mathbf{b}_n\right)\big] \\ &\le -\tfrac{1}{\mathrm{Re_m}}\|\operatorname{curl}\mathbf{b}_{n+1}\|^2 - \tfrac{1}{\mathrm{Re}}\|\nabla\mathbf{u}_{n+1}\|^2 + \big\langle\mathbf{f}_{n+1},\mathbf{u}_{n+1}\big\rangle, \end{aligned} \quad (3.14)$$

from which unconditional H-stability follows by noting that the left-hand side of (3.14) is non-positive for $\mathbf{f}_{n+1} = 0$.

The proof of H-dissipativity proceeds from (3.14) as follows. The use of the Poincaré's inequalities (2.10) and Young's inequality for the forcing term, reduces (3.14) to

$$\begin{aligned} \tfrac{1}{\Delta t}\big[K(\mathbf{u}_{n+1},\mathbf{b}_{n+1}) \; &-K(\mathbf{u}_n,\mathbf{b}_n)\big] \\ &\leq -\tfrac{\lambda_u}{2Re}\|\mathbf{u}_{n+1}\|^2 - \tfrac{S\lambda_b}{\mathrm{Re_m}}\|\mathbf{b}_{n+1}\|^2 + \tfrac{\|\mathbf{f}\|^2}{2(\lambda_u/Re)} \\ &\leq C\; K(\mathbf{u}_{n+1},\mathbf{b}_{n+1}) + \tfrac{\|\mathbf{f}\|^2}{2C}, \end{aligned} \tag{3.15}$$

where C is defined after (2.12) and $\mathbf{f} \equiv \mathbf{f}_{n+1}$ time-independent. Repeated use of this inequality leads to

$$K(\mathbf{u}_n,\mathbf{b}_n) \leq A^n\; K(\mathbf{u}_o,\mathbf{b}_o) + \frac{\Delta t}{(1+C\Delta t)}\frac{\|\mathbf{f}\|^2}{2C}\sum_{k=1}^{n-1} A^k, \tag{3.16}$$

where $A := \big[1 + C\;\Delta t\big]^{-1}$. Since $|A| < 1$, the first term in (3.16) is such that the limit $A^n\; K(\mathbf{u}_o,\mathbf{b}_o) \to 0$ as $n \to \infty$ while the last term forms a geometric series that can be summed to obtain

$$\lim_{n\to\infty}\,\sup\, K(\mathbf{u}_n,\mathbf{b}_n) \leq \frac{\Delta t\|\mathbf{f}\|^2}{2C\;(1+C\Delta t)}\;\frac{1}{1-A} = \frac{\|\mathbf{f}\|^2}{2C^2}\,, \tag{3.17}$$

where we have used the definition of the amplification factor A. Therefore, the fractional step method summarized in Table 3.2 is unconditionally H-dissipative. Estimate (3.17) coincides with its continuum counterpart in this semidiscrete context, showing the *optimal dissipativity* of this time-stepping scheme. □

The proof of Theorem 3.1 shows that the stability and dissipative properties of this algorithm do not require the intermediate field $\tilde{\mathbf{u}}_{n+1}$ to satisfy neither the divergence-free constraint nor any particular set of boundary conditions. Note that the final solution $\mathbf{u}_{n+1}$ and $\mathbf{b}_{n+1}$ satisfy the divergence-free conditions and respective correct boundary conditions. The following remarks should be added.

Remark 3.3 Estimate (3.17) is independent of Δt. This situation is to be contrasted with the results obtained by Foias et al [1991] for the Kuramoto–Sivashinsky equation, and (Elliot and Stuart [1992]) for reaction–diffusion systems, involving the same type of multistep approximations.

Remark 3.4 The argument can be extended to generalized mid-point approximations that preserve the crucial cancellation of coupling terms. See (Armero and Simo [1994]) for similar developments in the analysis of monolithic schemes. In this context, it should be noted that the simple argument behind Theorem 3.1 showing the dissipativity of the scheme in Table 2 does not apply, for instance, to the classical mid-point rule (a second order scheme). See the aforementioned reference for details.

Remark 3.5 The counterpart of Theorem 3.1 in the fully discrete context arising from a mixed finite element discretization of the problem can be found in Armero [1993]. The crucial skew-symmetry of the convective term is accomplished by employing $(2.5)_2$, whereas the use of a modified weak-form of the problem allows the above developments involving discrete fields not necessarily pointwise divergence-free.

3.2.2 **A modified differential equation: Δt-dependence of steady-state solutions** A useful technique to study the stability and accuracy of numerical schemes is the so-called *modified differential equation.* This method is based on the construction of a differential equation that the fully discrete numerical scheme approximates to a higher order than the original equation. In this section, we will take a similar approach and construct a modified equation for the algorithm summarized in Table 2. We shall proceed formally with the semidiscrete problem.

By using the familiar vector identity $(\mathbf{a} \times \mathbf{b}) \times \mathbf{c} = [\mathbf{b} \otimes \mathbf{c} - \mathbf{c} \cdot \mathbf{b}\ \mathbf{1}]\mathbf{a}$ and combining equations (3.9) and (3.10) to eliminate the intermediate field $\tilde{\mathbf{u}}_{n+1}$, we obtain

$$\left.\begin{aligned} &\tfrac{1}{\Delta t}[b_{n+1} - b_n] - \operatorname{curl}\left(\mathbf{u}_n \times \mathbf{b}_{n+1}\right) = -\operatorname{curl}\left(\eta_{\Delta t}(\mathbf{b}_n)\operatorname{curl}\mathbf{b}_{n+1}\right) \\ &\tfrac{1}{\Delta t}[u_{n+1} - u_n] + \mathbf{C}(\mathbf{u}_n)\nabla\mathbf{u}_{n+1} = \tfrac{1}{\mathrm{Re}}\Delta\mathbf{u}_{n+1} - \nabla\bar{p} + S\operatorname{curl}\mathbf{b}_{n+1} \times \mathbf{b}_n + \mathbf{f}_{n+1} \\ &\operatorname{div}\mathbf{u}_{n+1} = 0, \qquad \operatorname{div}\mathbf{b}_{n+1} = 0, \end{aligned}\right\} \tag{3.18}$$

where the *modified magnetic diffusivity* $\eta_{\Delta t}(\cdot) \in \mathbb{R}^{n_{\mathrm{dim}} \times n_{\mathrm{dim}}}$ is given by

$$\eta_{\Delta t}(\mathbf{b}) := \frac{1}{\mathrm{Re_m}}\ \left[\mathbf{1} + \Delta t\ S\ \mathrm{Re_m}\ \left(\mathbf{b} \cdot \mathbf{b}\ \mathbf{1} - \mathbf{b} \otimes \mathbf{b}\right)\right]. \tag{3.19}$$

It should be noted that $\eta_{\Delta t}(\cdot)$ is positive definite, since for all $\mathbf{a} \in \mathbb{R}^{n_{\mathrm{dim}}}$

$$\mathbf{a} \cdot \eta_{\delta t}(\mathbf{b})\mathbf{a} = \frac{1}{\mathrm{Re_m}}\ |\mathbf{a}|^2 + \Delta t\ S\ \left(|\mathbf{a}|^2|\mathbf{b}|^2 - (\mathbf{a} \cdot \mathbf{b})^2\right)\ \geq\ \frac{1}{\mathrm{Re_m}}\ |\mathbf{a}|^2; \tag{3.20}$$

hence adding positive dissipation in the algorithm to the original magnetic diffusivity $1/\mathrm{Re_m}$. The modified differential equation corresponding to the algorithm in Table 2 is then the original MHD system (2.1) with $1/\mathrm{Re_m}$ replaced by $\eta_{\Delta t}(\mathbf{b})$.

The linear equations (3.18) in the unknown fields $\mathbf{u}_{n+1}$ and $\mathbf{b}_{n+1}$ are decoupled. They correspond to a semi-implicit algorithm based on the Eulerian split described in Section 3.1 but with the order of the two phases reversed. Therefore, we conclude that *the induction split stabilizes the Eulerian split by introducing an algorithmic magnetic diffusivity $\eta_{\Delta t}(\mathbf{b}_n)$ given by (3.19).* However, since $\eta_{\Delta t}(\mathbf{b}_n)$ depends explicitly on Δt, the steady-state solutions computed with this algorithm will depend on the time step Δt. The negative consequence of this Δt-dependece is, as occurs with classical projection methods for the incompressible Navier-Stokes equations (see Simo and Armero [1993] and references therein), is that it deteriorates the numerical accuracy of the method when computing steady-state solutions with a large Δt. From the definition (3.19) of $\eta_{\Delta t}$ and noting that $\mathbf{b} = \mathfrak{O}(1)$ (non-dimensional), we conclude that accurate solutions are obtained for $\Delta t\ S\ \mathrm{Re_m} \leq \mathfrak{O}(1)$. Therefore, although the method is unconditionally stable, this restriction limits the size of the time-step Δt from a practical standpoint in these situations.

4 Representative Numerical Simulations

The goal of this section is to illustrate the performance of the time-stepping algorithms introduced above. Two representative simulations are considered involving liquid metal type fluids (mercury).

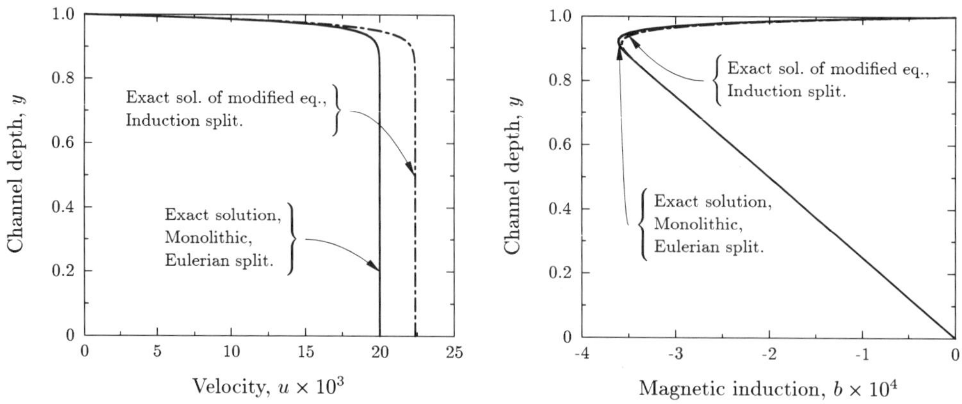

Figure 1 Hartmann flow. Velocity and magnetic induction profiles for different time-stepping schemes and $\Delta t = 0.01\ h$ (non-dimensional). The Eulerian split is stable for this time-step. ($Ha = 50$, $Pr_m = 1 \cdot 10^{-7}$)

4.1 Hartmann flow The goal of this example is to assess the stability and accuracy properties of the different time-stepping algorithms in a problem involving a final steady-state solution. The simple linear 1-D setting provided by the plane Hartmann channel flow is considered. This problem is the coupled counterpart of Poiseuille flow for the Navier-Stokes equations and, as in this case, it leads to a linear equation along the depth of the channel with a closed form solution for the steady-state. (See e.g. Landau et al. [1984]).

A plane channel of unit depth $a = 1$ is considered, with the non-slip boundary condition $u(\pm a) = 0$ along the walls. Let $\{x, y, z\}$ be a cartesian coordinate system defined such that the y-direction is normal to the channel walls, with x along the direction of the flow. A magnetic field B in the y-direction and a pressure gradient in the x-direction are imposed. Transversal (z-direction) open-circuit boundary conditions are assumed for the magnetic field. The fluid is assumed to be mercury, characterized by $Pr_m = 10^{-7}$. Numerical experiments are performed by discretizing the resulting 1-D coupled equations in the velocity u_x and magnetic induction b_x along the depth y of the channel. The spatial discretization consists of 100 C^0 linear elements for half the channel. The numerical performance of the staggered schemes based on the Eulerian split and the induction split is assessed. The monolithic backward-Euler scheme is also considered. Figure 1 depicts the velocity and magnetic induction profiles obtained by the different time-stepping algorithms for a small time-step $\Delta t = 0.01\ h$ (non-dimensional) and Hartmann number $Ha = 50$ ($Ha := \sqrt{S\mathrm{Re}\mathrm{Re}_\mathrm{m}} = B\ L\ \sqrt{\sigma/\rho\nu}$ is a measure of the strength of the applied magnetic field). As expected, the monolithic scheme exhibits a good numerical accuracy when compared with the exact solution. So does the Eulerian split for this small time-step. In contrast, the induction split yields a less accurate solution, especially in the velocity. As indicated in Section 3.2.2, this scheme exhibits steady-state solutions depending on the time-step size Δt through the introduction of a modified resistivity. In this linear 1-D case, a simple calculation shows that the exact solution of the modified equation can be obtained easily from the exact closed-form solution of the original system by considering the *modified Hartmann number*

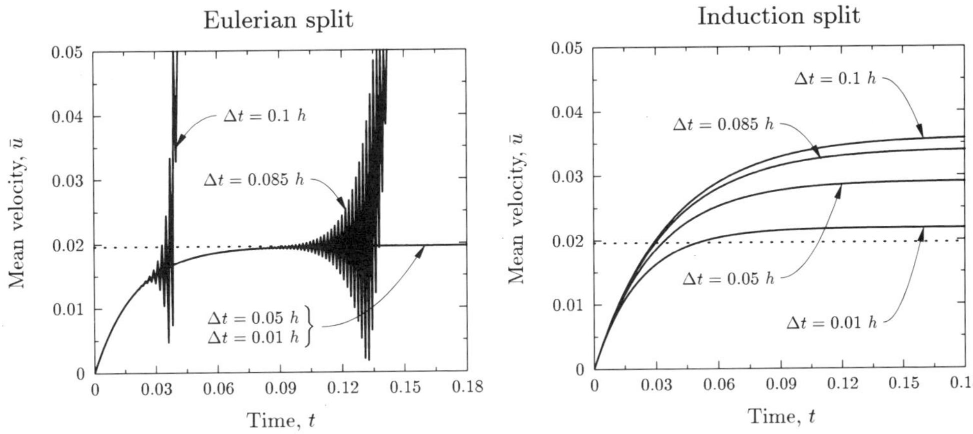

Figure 2 Hartmann flow. Evolution of the mean velocity versus time computed with the Eulerian and induction splits for different time-steps Δt (non-dimensional). - - - Exact steady solution. ($Ha = 50$, $Pr_m = 1 \cdot 10^{-7}$)

$Ha_{mod} = Ha/\sqrt{1 + \Delta t Ha^2}$ together with the magnetic induction field scaled by Ha^2_{mod}/Ha^2. The exact steady-state solution of the original and modified equations are depicted in Figure 1.

The Eulerian split and monolithic scheme converge to the same steady-state solution if small time-steps are used. However, as indicated in Section 3.1, the Eulerian split leads to unstable computations for larger time-steps. This is shown in Figure 2, where the evolution of the mean velocity ($\bar{u} = \int_0^a u_x(y)\ dy\ /a$) versus time is depicted for different time-steps Δt. The instability of the Eulerian split is apparent from these results, whereas the unconditional stability of the induction split is confirmed. As concluded from the analysis in Section 3.2.2, the stability gained by the induction split is achieved, however, at the expense of numerical accuracy in the steady state. The Δt dependence of the final solution leads to rather poor accurate steady-state solutions for large time-steps. As indicated in Section 3.2.2, this is the same situation as encountered in some traditional projection-type methods for the incompressible Navier-Stokes equations.

4.2 MHD flow past a circular cylinder The uncoupled flow of an incompressible fluid past a circular cylinder has become a standard benchmark problem for numerical solvers of the transient Navier-Stokes equations. At low Reynolds numbers, the flow develops two symmetric wakes past the cylinder. This solution becomes unstable for Reynolds numbers (based on the diameter) over 40, and periodic vortex shedding appears. The inclusion of the MHD coupling affects in a strong manner this flow pattern. The electromagnetic effects tend to eliminate the transient response and, altogether, the periodic solution of the uncoupled problem. The vorticity is eliminated from the mean flow in this resistive limit, with the appearance of the so-called Hartmann boundary layers. These effects are important for a number of technological applications (metal casting, thin film deposition, etc.) and make the study of this problem of a maximum interest. The goal of this example is to study these effects while assessing at the same time the performance of the Eulerian split in transient simulations.

The computational domain with imposed boundary conditions is depicted in Figure 3.a. To trigger the appearance of vortex shedding a geometric imperfection is introduced by lowering the bottom boundary. For the present simulations, the common value of Re $= 100$ is considered. The fluid is assumed at rest initially, with a sudden start. The case of an imposed in-plane magnetic field transversal to the flow is considered, with a characteristic value $\mathbf{b}_\infty = (0, 1)$ is assumed. This field is imposed on the whole boundary.

Figure 3.b depicts the finite element discretization considered for this problem. The spatial interpolation is based on the P2 $\oplus$ bubble/P1 triangle for the velocity and pressure field, respectively, with the magnetic induction field interpolated quadratically as the velocity. This element is depicted in Figure 3.a. The divergence-free constraints are enforced through an augmented Lagrangian formulation. This iterative process involves the introduction of a "magnetic pressure", which is interpolated linearly as well. We note that the pressure fields are discontinuous across elements, allowing the elimination of these fields at the element level. This fact results in a very cost-effective implementation. Only three augmentations are required in each time step to drive the discrete divergences to machine precision and are performed at no extra cost, since by the linearity of the schemes the system matrix is constant. Further details regarding the finite element solution of the problem can be found in (Armero and Simo [1994]).

The performance of the induction split algorithm of Table 2 is assessed. Solutions obtained in this case are compared with a monolithic (semi-implicit) linear scheme of this last reference. Figure 4 depicts the solution obtained for Re $= 100$, $Pr_m = 10^{-7}$ and different Hartmann numbers $Ha = 0$, 1, 2.5, 5, 10 with this staggered scheme. The streamlines and stationary streamlines at $t = 150$ have been included. The simulations are performed with a time increment of $\Delta t = 0.1$ time units to the final time $t \in [0, 150]$. Details of the solution near the cylinder have been included in Figure 5 and Figure 6.

The different phases of the solution are captured by the time evolution of the drag coefficient (C_D) and lift coefficient (C_L). This is depicted in Figure 7 for the two representative values of $Ha = 0$ and $Ha = 10$. In the first case (uncoupled problem), an initial phase characterized by a symmetric solution and a drop in the drag coefficient is observed. This solution becomes unstable at approximately $t \approx 50$, and vortex shedding appears. After a transient phase, the flow settles to a periodic solution. A period of $T = 6.0$ is observed for $Ha = 0$, in agreement with the experimental results of (Roshko [1954]). By increasing the Hartmann number, we observe a decrease of the amplitude of oscillations in both the drag and lift coefficients (not shown). The characteristic increase of the drag coefficient with the elimination of the vortex shedding can be observed for $Ha = 10$. It should be noted that in this case numerical experiments have shown that vortex shedding does not appear by introducing stronger imperfections or even with unsymmetric initial conditions. The lift coefficient vanishes identically, since the solution then obtained is symmetric. A shortening of the wakes past the cylinder with increasing Hartmann numbers can be observed in Figure 4 for this final symmetric solution, confirming the fact that the inclusion of the magnetic field in these conditions tends to eliminate all the disturbances from the state at infinity.

With respect to the numerical performance of the staggered scheme, a good agreement is observed with the values obtained by the monolithic scheme in this unsteady problem. As expected, the stronger the coupling the more discrepancy

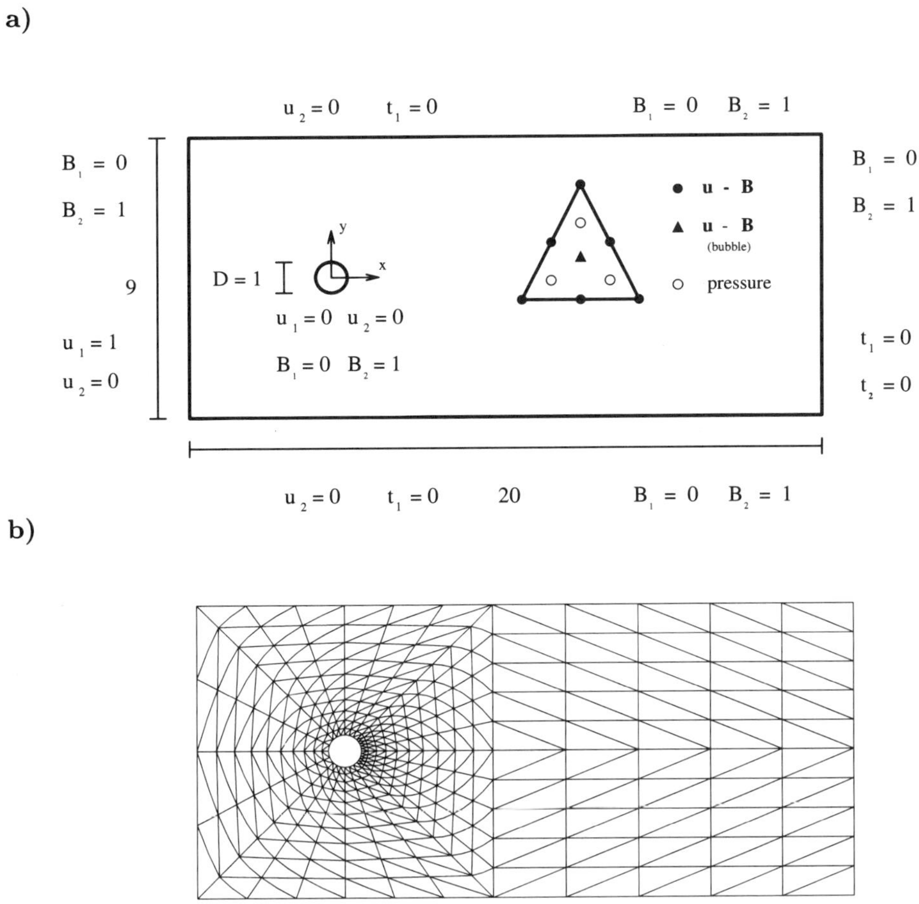

Figure 3 MHD flow past a circular cylinder. Problem description: a) Computational domain ($-4.5 \le x \le 15.5$, $-4.5 \le y \le 4.5$), assumed boundary conditions and definition of the finite elements. b) Finite element mesh (620 elements, 1302 nodes). The velocity $\mathbf{u}$ and magnetic induction $\mathbf{b}$ have quadratic interpolations with bubble (P2 $\oplus$ bubble). The pressures are linear and discontinuous across elements (P1).

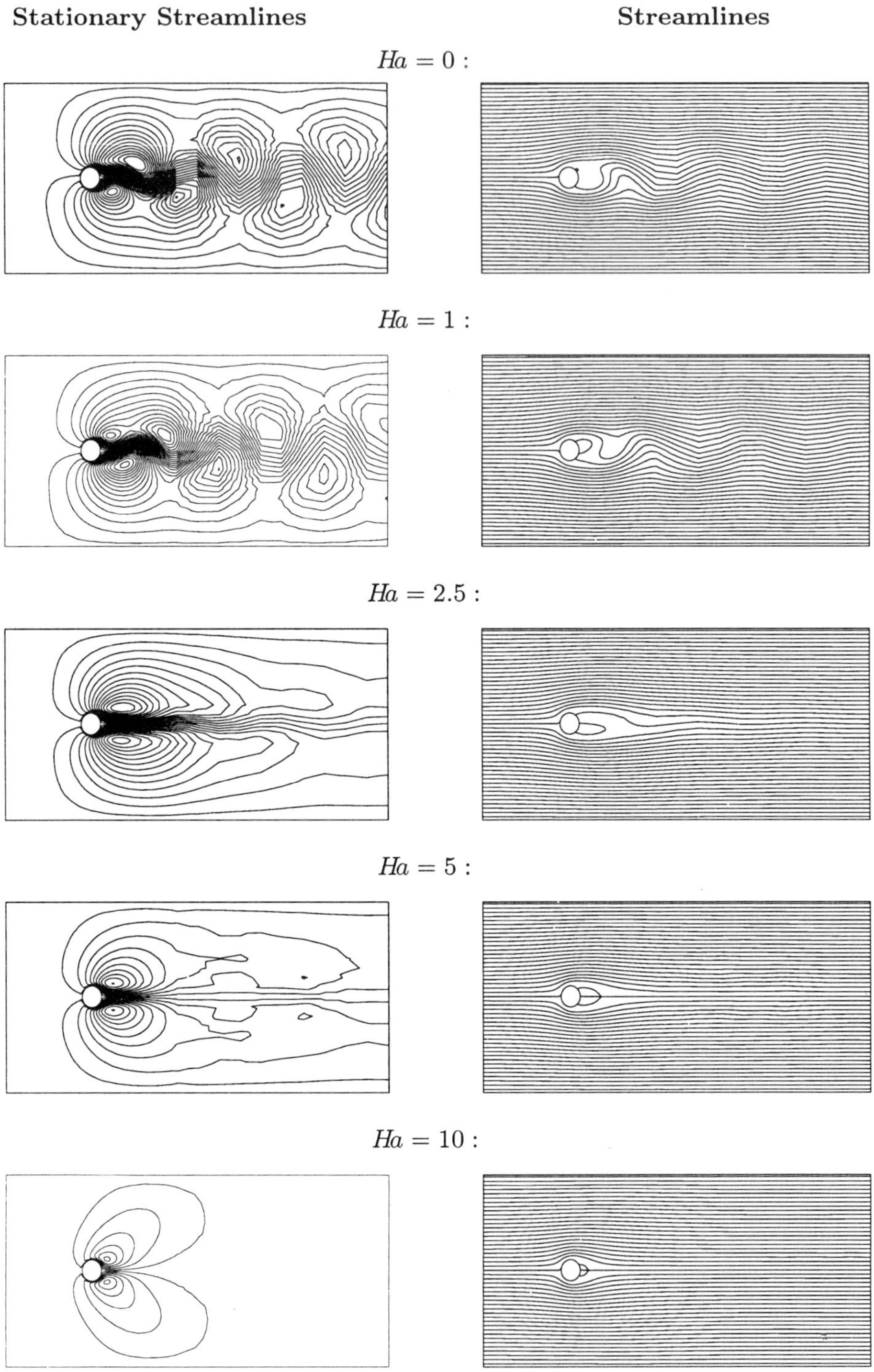

Figure 4 MHD flows past a circular cylinder. Stationary streamlines and streamlines at $t = 150$: $Re = 100, Pr_m = 10^{-7}$ and $Ha = 0, 1, 2.5, 5, 10$. The whole computational domain is shown.

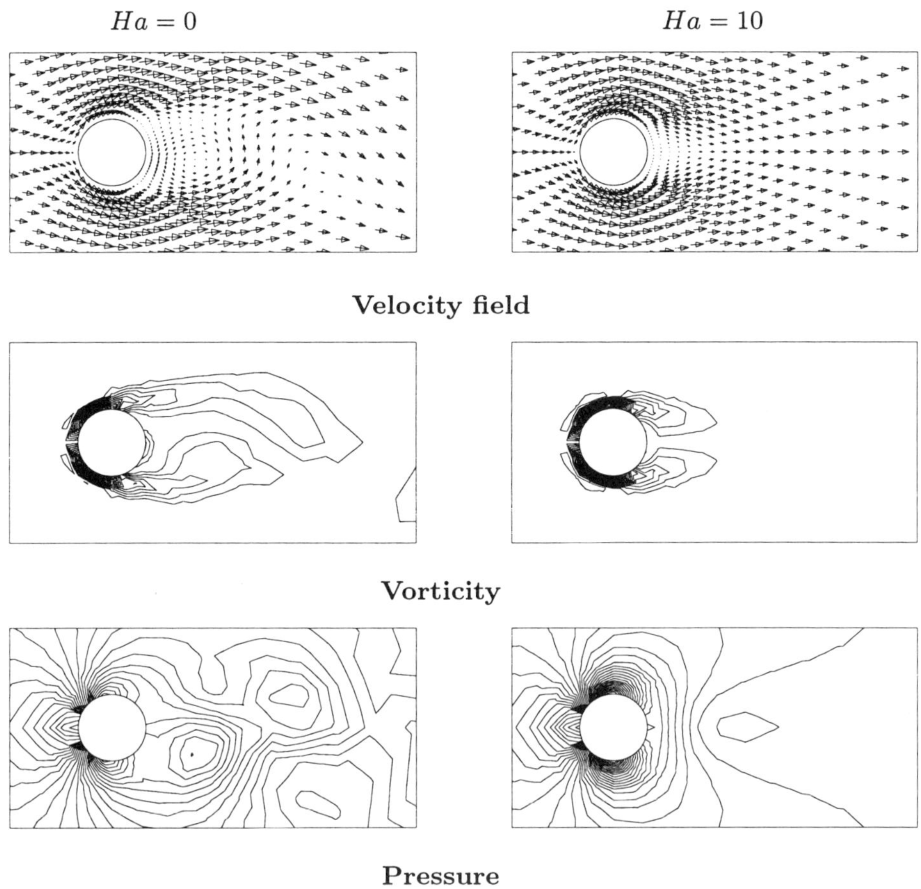

Figure 5 MHD flow past a circular cylinder. Vorticity, pressure and velocity fields at $t = 150$ for$Ha = 0$ and $Ha = 10$. Area shown: $-1.5 \leq x \leq 4.5$, $-1.5 \leq y \leq 1.5$.

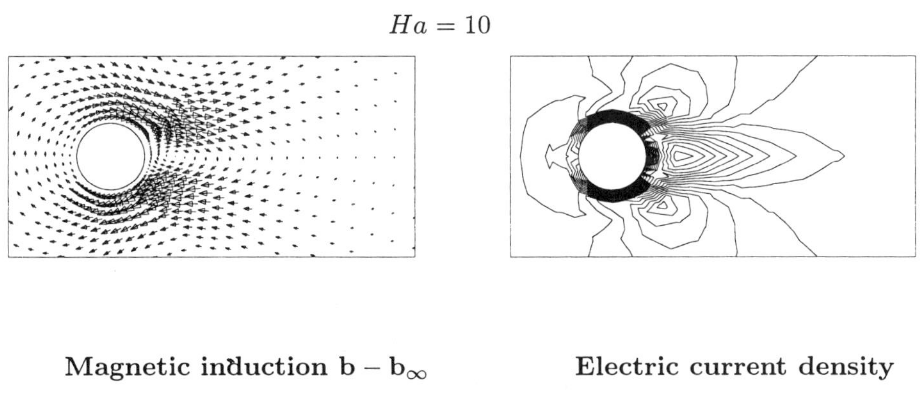

Figure 6 MHD flow past a circular cylinder. Details of the solution for $Ha = 10$ at $t = 150$. Area shown: $-1.5 \leq x \leq 4.5$, $-1.5 \leq y \leq 1.5$.

appears, but only a maximum relative difference of 2.0 % is found in the computed drag coefficients. The features of the solution obtained with the monolithic scheme are the same as the previously described; details are not included. The estimate of Remark 3.1 on the low computational cost of this split has been confirmed; that is the final solution is obtained with less than 1/5 the computational cost.

The different phases of the solution are captured by the time evolution of the drag coefficient (C_D) and lift coefficient (C_L). This is depicted in Figure 4.7 for the two representative values of $Ha = 0$ and $Ha = 10$. In the first case (uncoupled problem), an initial phase characterized by a symmetric solution and a drop in the drag coefficient is observed. This solution becomes unstable at approximately $t \approx 50$, and vortex shedding appears. After a transient phase, the flow settles to a periodic solution. A period of $T = 6.0$ is observed for $Ha = 0$, in agreement with the experimental results of Roshko [1954]. By increasing the Hartmann number, we observe a decrease of the amplitude of oscillations in both the drag and lift coefficients (not shown). The characteristic increase of the drag coefficient with the elimination of the vortex shedding can be observed for $Ha = 10$. It should be noted that in this case numerical experiments have shown that vortex shedding does not appear by introducing stronger imperfections or even with unsymmetric initial conditions. The lift coefficient vanishes identically, since the solution then obtained is symmetric. A shortening of the wakes past the cylinder with increasing Hartmann numbers can be observed in Figure 4.4 for this final symmetric solution, confirming the fact that the inclusion of the magnetic field in these conditions tends to eliminate all the disturbances from the state at infinity.

With respect to the numerical performance of the staggered scheme, a good agreement is observed with the values obtained by the monolithic scheme in this unsteady problem. As expected, the stronger the coupling the more discrepancy appears, but only a maximum relative difference of 2.0 % is found in the computed drag coefficients. The features of the solution obtained with the monolithic scheme are the same as the previously described; details are not included. The estimate of Remark 3.1on the low computational cost of this split has been confirmed; that is the final solution is obtained with less than 1/5 the computational cost.

5 Concluding Remarks

The stability and long-term dissipative properties of a new class of fractional-step methods for the incompressible MHD equations have been analyzed. A particular time-stepping algorithm involving the solution of two *linear* and *symmetric* boundary value problems at each time-step that exhibits unconditional stability and optimal long-term dissipativity has been identified. These results have been confirmed by representative numerical simulations based on mixed finite element methods.

6 Acknowledgements

Financial support for this research was provided by AFOSR under contract No. 2-DJA-826 with Stanford University. This support is gratefully acknowledged.

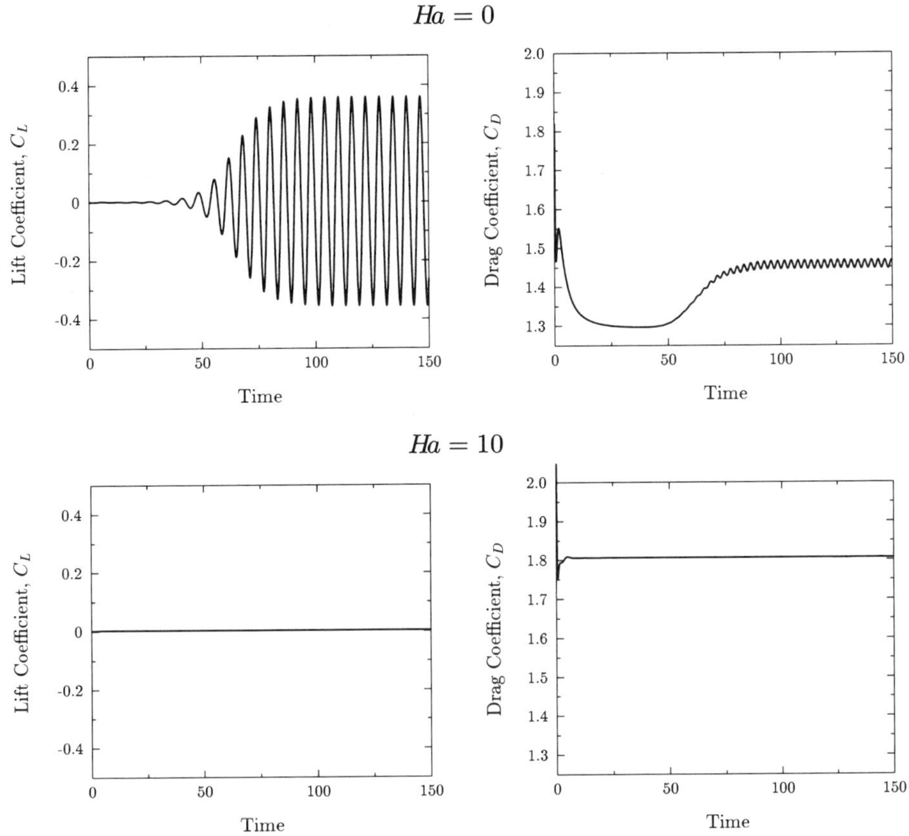

Figure 7 MHD flow past a circular cylinder. Evolution of the lift and drag coefficients: $Re = 100$, $Prm = 10^{-7}$ and $Ha = 0, 10$.

References

Armero, F. [1993], *Numerical Analysis of Dissipative Dynamical Systems in Solid and Fluid Mechanics, with a special emphasis on coupled problems*, Ph.D. Dissertation, Stanford University, Stanford CA. SUDAM Report 93-6

Armero, F. and Simo, J. C. [1992], *A new unconditionally stable fractional step method for nonlinear coupled thermomechanical problems* Int. Journal for Numerical Methods in Engineering, **35**, 737-766.

Armero, F. and Simo, J. C. [1993] *A–priori stability estimates and unconditionally stable product formula algorithms for nonlinear coupled thermoplasticity*, Int. Journal of Plasticity, **9**, 749-782.

Armero, F. and Simo, J. C. [1994], *Long-Term Dissipativity of Time-Stepping Algorithms for an Abstract Evolution Equation with Applications to the Incompressible MHD and Navier-Stokes Equations*, Computer Methods in Applied Mechanics and Engineering, in press.

Butcher, J. C. [1975], *A stability property of implicit Runge-Kutta methods*, BIT, **15**, 358-361.

Chorin, A. J., Hughes, T. J. R., McCracken, M. F. and Marsden, J. E. [1978], *Product formulas and numerical algorithms*, Comm. Pure Appl. Math., **31**, 205-256.

Ciarlet, P. G. [1978], *The Finite Element Method for Elliptic Problems*, North Holland, Amsterdam.

Duvaut, G. and Lions, J. L. [1972], *Inéquations en thermoélasticité et magnétohydrodyna- mique*, Archive for Rational Mechanics and Analysis, **46**, 241-279.

Elliot, C. M. and Stuart, A. M. [1992], *The global dynamics of discrete semilinear equations* IMA Preprint Series, **973**, University of Minnesota, Minneapolis, To appear in *SIAM Journal of Numerical Analysis* Dec. 1993.

Foias, C., Jolly, M. S., Kevrekidis, I. G., and Titi, E. S. [1991], *Dissipativity of numerical schemes* Nonlinearity, **4**, 591-613.

Gunzburger, M. D., Meir, A. J. and Peterson, J. [1991], *On the existence, uniqueness and finite element approximation of solutions of the equations of stationary, incompressible magnetohydrodynamics*, Mathematics of Computation, **56**, 523-563.

Hairer, E. and Wanner, G. [1991], *Solving Ordinary Differential Equations II, Stiff and Differential-Algebraic Problems*, Springer-Verlag, Berlin.

Hale, J. K. [1988], *Asymptotic Behavior of Dissipative Systems*, Mathematical Surveys and Monographs no. **25** American Mathematical Society, Providence, R.I.

Hale, J. K., Lin, X. B. and Raugel, G. [1988], *Upper semicontinuity of attractors for approximations of semigroups and partial differential equations*, Mathematics of Computation, **50**, 89-123.

Holm, D. D., Marsden, J. E., Ratiu, T. and Weinstein, A. [1985], *Nonlinear stability of fluid and plasma equilibria* Physics Reports, **123**.

Humphries, A. R. and Stuart, A. M. [1993], *Runge-Kutta methods for dissipative and gradient dynamical systems,* SIAM Journal of Numerical Analysis, to appear.

Landau, L. D., Lifshitz, E. M. and Pitaevskii, L. P. [1984], *Electrodynamics of Continuous Media,* Landau & Lifshitz Course of Theoretical Physics, Volume 8, Pergamon Press, Oxford.

Marchuk, G. I. [1982], Methods of Numerical Mathematics, 2nd edition, , Springer–Verlag, New York.

Richtmyer, R. D. and Morton, K. W. [1967], *Difference Methods for Initial–Value Problems,* 2nd edition, Interscience, New York.

Roshko, A. [1954], *On the Development of Turbulent Wakes from Vortex Streets,* NACA Report 1191.

Sermange, M. and Temam, R. [1983], *Some mathematical questions related to the MHD equations,* Comm. Pure Appl. Math., **36**, 635-664.

Simo, J. C. and Armero, F. [1993], *Unconditional stability and long-term behavior of transient algorithms for the incompressible Navier-Stokes and Euler equations,* Computer Methods in Applied Mechanics and Engineering, **111**, 111-154.

Strang, G. [1968] *On the construction and comparison of difference schemes,* SIAM Journal of Numerical Analysis, **5**, 506-517.

Temam, R. [1988], *Infinite-dimensional Dynamical Systems in Mechanics and Physics,* Applied Mathematical Sciences, **68**, Springer-Verlag, Berlin.

Yanenko, N. N. [1971], *Fractional–Step Methods,* Springer–Verlag, Berlin.

Fields Institute Communications
Volume **10**, 1996

Symplectic Methods for Conservative Multibody Systems

E. Barth
Courant Institute
New York University
New York, NY USA 10012

B. Leimkuhler
Department of Mathematics
University of Kansas
Lawrence, KS USA 66045

Abstract. Besides preserving the energy, the flow of a conservative multibody system possesses important geometric (*symplectic*) invariants. Symplectic discretization schemes that mimic the corresponding feature of the true flow have been shown to be effective alternatives to standard methods for many conservative problems. For systems of rigid bodies, the development of such schemes can be complicated or costly to implement, depending on the choice of problem formulation. In this article, we demonstrate that a special formulation of the multibody system (based on a particle representation) together with a symplectic discretization for constrained problems borrowed from molecular dynamics offers an efficient alternative to standard approaches. Numerical experiments illustrating this approach are described.

1 Introduction

In this paper, we consider efficient numerical integrators for systems of rigid bodies interconnected by various types of mechanical joints and subject to the forces of nature. Important applications include "dynamic models" [17] of undamped and mildly damped mechanisms and manipulators [38, 23].

Symplectic schemes have been found to be effective for diverse problems such as astronomical many-body systems [36] and molecular dynamics simulations [22]. We are particularly interested in two aspects of symplectic integrators: first, all other

1991 *Mathematics Subject Classification.* 65L05, 70-08, 70F20.

Key words and phrases. symplectic discretizations, constrained systems, differential-algebraic equations, multibody systems, molecular dynamics, modelling of mechanisms, modelling of manipulators.

The work of the first author was supported by a fellowship from the Kansas Institute for Theoretical and Computational Science. The work of the second author was supported by NSF grants DMS-9303223 and DMS-9205538 things being equal, they appear to provide realistic simulations on longer time intervals than nonsymplectic schemes, and second, the condition for being symplectic is connected to other important properties such as norm-preservation and angular momentum-preservation. All symplectic schemes are necessarily volume preserving. In some cases, as in our example of spinning top (Section 6), conservation of integrals proves critical to successful numerical simulation.

In standard treatments of classical mechanics [2], a rigid body is viewed as a collection of point masses with fixed interparticle distances. The natural description of the body as a system of cartesian coordinates for the locations of the point masses is typically replaced by a minimal coordinate representation, for example based on the introduction of quaternions. The idea of using cartesian ("natural") coordinates for the simulation of multibody systems was first discussed by [37, 6], with the aim of improving efficiency in real-time multibody simulation.

Our approach is based on a procedure we term *particulation* in which a given rigid body is replaced by a system of point particles while preserving its dynamical properties. Some tools for constructing such particle representations were first presented by Edward Routh in his classic treatise [32]. We describe a somewhat generalized version of his procedure, discuss the imposition of constraints, and show an equivalence under discretization of two particulations of the same body. Our technique is also related to the "cartesian approach" [10, 27] of modeling rigid polyatomic molecules. The principle contribution of this article is the recognition that the particle representation offers a simple, efficient means to symplectic discretization of rigid body dynamics.

We have also applied these techniques to treat systems of rigid bodies, showing explicitly how to handle forces of interaction and how various types of mechanical joints can be effectively modelled by quadratic length-type constraints. We touch on this briefly here, but more details can be found in [4].

To discretize the equations of motion, we use the RATTLE [1] variant of the SHAKE [34] discretization from constrained molecular dynamics, an efficient semi-explicit staggered method for the constrained equations that is both symplectic and angular momentum preserving. As an alternative to using the standard Gauss-Seidel iteration of [34] to solve the discretized equations, we have indicated in [5], in the context of molecular dynamics integration, how sparse matrix techniques can be applied to minimize the required computation.

For purposes of demonstration and evaluation, the techniques are implemented in an easy to use Matlab ToolBox **MDLab** which can be employed to study the particle formulation and various discretization schemes. We describe several experiments, including a spinning top and a seven bar chain.

An alternative to our elementary formulation in cartesian coordinates would be to develop algorithms for the motion of a rigid body formulated in $\mathbf{R}^3 \times SO(3)$ that are symplectic and/or integral conserving. This approach is discussed elsewhere in this volume (see, e.g. the article by Lewis and Simo), as well as in recent work of MacLachlan and Scovel [24]. Also in this volume, Reich discusses variations on the particulation approach and some connections between the two points of view.

2 Hamiltonian Systems, Symplectic Discretization, and Multibody Dynamics

A conservative system is one which posesses an energy functional H which is constant along trajectories. Hamilton's equations of motion are then

$$\begin{aligned}\dot{q} &= \frac{\partial H}{\partial p}\\ \dot{p} &= -\frac{\partial H}{\partial q}\end{aligned}$$

or,

$$\frac{d}{dt}\begin{pmatrix} q \\ p \end{pmatrix} = J\nabla H, \qquad J = \begin{bmatrix} 0 & I \\ -I & 0 \end{bmatrix}.$$

We can define the *flow map*[1] $\Phi_H^t : \mathbf{R}^{2n} \to \mathbf{R}^{2n}$ by

$$\frac{d}{dt}\Phi_H^t = J(\nabla H) \circ \Phi_H^t$$

The solution to the Hamiltonian system with initial values $(q(t_0), p(t_0)) = (q_0, p_0)$ is just

$$\begin{pmatrix} q(t) \\ p(t) \end{pmatrix} = \Phi_H^t \begin{pmatrix} q_0 \\ p_0 \end{pmatrix}.$$

The flow of a Hamiltonian system preserves the 2-form $dq \wedge dp = \sum dq^i \wedge dp^i$, the outer product of differentials. A mapping with this strong property is said to be *canonical* or *symplectic*. It can be shown that a necessary and sufficient condition for a mapping Φ to be symplectic is that the Jacobian Φ' satisfy

$$(\Phi')^T J \Phi' = J.$$

We will be interested in one-step discretizations i.e. methods which can be expressed in the form

$$\begin{pmatrix} q_{n+1} \\ p_{n+1} \end{pmatrix} = \Psi^h \begin{pmatrix} q_n \\ p_n \end{pmatrix}$$

for some mapping $\Psi^h : \mathbf{R}^{2n} \to \mathbf{R}^{2n}$. A symplectic discretization is one that preserves the same canonical 2-form: $dq_{n+1} \wedge dp_{n+1} = dq_n \wedge dp_n$. The "fundamental theorem" of symplectic discretizations is simply stated as follows: if the solution to a Hamiltonian system is approximated by a certain symplectic scheme with a single stepsize h, then there exists a formal series in powers of h for a perturbed Hamiltonian $\bar{H}(h)$ such that excluding rounding error, the flow map corresponding to $\bar{H}(h)$, $\Phi_{\bar{H}(h)}$, satisfies $\Phi^h_{\bar{H}(h)} = \Psi^h$ [39, 18]. The convergence of the formal series only occurs in a weak sense in the limit as $h \to 0$. Although little work has been undertaken on practical interpretations and limitations of this theoretical result, it does appear to imply a remarkable global stability of the energy for long term symplectic integrations, and this behavior is one reason for the recent surge in interest in symplectic schemes.

Another motivation for looking at symplectic schemes is that the strong conditions imposed by requiring symplecticity often also guarantee conservation of various integrals of the flow. A general discussion of symplectic discretization schemes can be found in [35].

[1]We are using a superscript lowercase "t" to indicate time here, whereas transpose will be signified by a superscript uppercase "T".

2.1 Constrained Multibody Systems In typical cases, the state of the system will be restricted to a submanifold of $\mathbf{R}^{2N}$ by algebraic constraints. In this paper, we will assume that our system only posesses holonomic position constraints of the form $g(q) = 0$, where $g : \mathbf{R}^N \to \mathbf{R}^m$ is a smooth map with a full rank Jacobian in the neighborhood of its zero-set $g^{-1}(0)$. One can then consider the modified Hamiltonian $\hat{H} = H + \lambda^T g$ augmented by multipliers λ determined by the constraints:

$$\begin{aligned} \dot{q} &= \frac{\partial \hat{H}}{\partial p} = \frac{\partial H}{\partial p} \\ \dot{p} &= -\frac{\partial \hat{H}}{\partial q} = -\frac{\partial H}{\partial q} - \frac{\partial g}{\partial q}^T \lambda \\ 0 &= g(q). \end{aligned}$$

The many-body Hamiltonian H has the form $H(p,q) = \frac{1}{2}p^T M^{-1} p + V(q)$ and arises when a system of point particles moves in potential V. M is a diagonal matrix containing the masses of the particles. If constraints are placed on positions of the particles only, then we arrive at Hamiltonian equations of the form

$$\dot{q} = M^{-1} p \tag{2.1}$$
$$\dot{p} = -\nabla_q V - g'(q)^T \lambda \tag{2.2}$$
$$0 = g(q). \tag{2.3}$$

In the traditional approach [26], the equations of motion of a system of rigid bodies are treated by introduction of some generalized coordinate system. Using the positions of the centers of mass and the three Euler angles describing the rotation of each body about its center of mass to parameterize the motion results in canonical equations with the augmented (constrained) Hamiltonian

$$H = \frac{1}{2} p^T \bar{M}^{-1} p + \tilde{V}(q) + \tilde{g}(q)^T \lambda \tag{2.4}$$

where p, q are the new coordinates involving positions of each body and orientation in space, $\bar{M} = \bar{M}(q) = diag(M, \tilde{I})$, with M a diagonal mass matrix and $\tilde{I}$ the inertial tensor which is a function of the Euler angles, and $\tilde{V}$ is V expressed in the new coordinate system. This formulation may have singular points (of $\tilde{I}$) which make it unsuitable for general numerical investigation [12]. The problem can be overcome with another choice of the generalized coordinates [12, 29], but a similar form of the equations of motion is obtained if other typical choices are employed [7]. Importantly, the mass matrix $\bar{M}$ is now position-dependent, unlike in the case of a system of particles in cartesian coordinates, and the Hamiltonian has become nonseparable.

2.2 Discretization of the Constrained Equations of Motion The equations (2.1)-(2.3) and the corresponding equations on the Hamiltonian (2.4) are systems of differential-algebraic equations (DAEs). Various techniques exist for handling these sorts of equations. For example, a traditional approach is to parameterize the constraint manifold to obtain ODE systems in local charts (e.g. using some automatic technique such as coordinate partitioning [33]). This technique can be performed while also preserving the canonical structure, and results in Hamiltonian systems of ODEs in the parameters which can then be discretized by symplectic schemes. However, as pointed out in [21], these techniques, although symplectic, do not result in a numerical solution that is the iteration of one and the same symplectic map, hence the fundamental theorem of symplectic discretization cannot be applied. Numerical experiments show a substantial drift in energy.

An alternative approach is to develop an extension of the vector field from the constraint manifold into the space $\mathbf{R}^{2N}$ which preserves the constraints as invariants of the flow and which reduces along the constraints to the original problem. This can be done via constraint differentiation, as discussed in [28, 11]. At first glance this technique appears to introduce some additional complexity over other formulations, while also typically producing a "drift" from the constraint manifold. However, it can be shown that certain combined choices of extension and discretization can eliminate the drift up to roundoff error, at least for quadratic constraints [20]. Moreover, the extensions can be viewed as partial analytical solution of the constrained problem, enabling the use of simplified techniques such as functional iteration for solving the discrete equations at each time step.

Another popular type of extension is commonly referred to as the penalty method [29, 6, 7]. In practical terms, rigid constraints are maintained by altering the vector field so that there is a strong resisting force to those motions that violate the constraints. In [29, 7], Hamiltonian formulations of the constrained equations of motion were integrated using this technique but without consideration of the symplectic and integral structures of the flow. A symplectic approach based on the penalty method could be constructed, but an important difficulty lies in the fact that the resulting differential equations will generally be made very stiff by these artificial forces, and the numerical solution of such problems will be unnecessarily complicated. In the Baumgarte approach, on the other hand, the extension is constructed in order that the constraint manifold is turned into an attracting set for the dynamics of the unconstrained problem. Here the chief problem lies in the proper selection of those artificial parameters that will determine the dynamical properties of the extension, but for underlying conservative dynamics, no such choice of a dissipative system can be appropriate.

Still another type of extension is based on the theory of constrained Hamiltonians of P. Dirac: an extended Hamiltonian vector field is constructed by considering the augmented Hamiltonian $\hat{H} = H + \lambda^T g + \mu^T g' H_p$ and choosing the multipliers so that $g = 0$ and $g' H_p = 0$ become (weak) invariants of the flow. This formulation was examined in [21], but is probably too costly to implement in its simplest form.

Instead of the above methods, we advocate the direct discretization of the constrained equations of motion. A simplified discrete problem results from treating the motion as that of a system of particles. For the constrained many-body Hamiltonian, we can use the following constrained variant of the leap-frog discretization:

$$\begin{aligned}
q_{n+1} &= q_n + hM^{-1}p_{n+1/2} && (2.5)\\
p_{n+1/2} &= p_n - (h/2)\nabla_q V(q_n) + g'(q_n)^T \lambda_n && (2.6)\\
g(q_{n+1}) &= 0 && (2.7)\\
p_{n+1} &= p_{n+1/2} - (h/2)\nabla_q V(q_{n+1}) + g'(q_{n+1})^T \mu_{n+1} && (2.8)\\
g'(q_{n+1})M^{-1}p_{n+1} &= 0. && (2.9)
\end{aligned}$$

This is the RATTLE [1] variant of the SHAKE discretization [34]; in this form, the solution of the discrete equations will automatically satisfy both the position and momentum constraints. It is straightforward to show under mild smoothness assumptions that these equations have a unique solution at each step provided the stepsize h is chosen sufficiently small. Starting from given data q_n, p_n (from the previous step), equations (2.5)-(2.7), which are a system of $2N + m$ equations, can be reduced to an m-dimensional nonlinear algebraic problem to be solved for λ_n. Next, (2.8)-(2.9) are seen to be a linear system for p_{n+1}, μ_{n+1}. The variables λ_n and μ_n are not propagated from step to step.

In molecular dynamics, SHAKE discretization is a common means for freezing chemical bonds; it was first shown to be convergent (2nd order accurate) and symplectic in [22], in both traditional and RATTLE formulations. Moreover, the scheme automatically preserves the angular momentum [40, 30]. In [24], the SHAKE discretization was applied in a somewhat different way to model the motion of the N-dimensional rigid body.

As a seemingly natural alternative, one might attempt to solve the reduced system (2.4) by a symplectic discretization. Jay [19] has shown that the constrained analogues of the Lobatto IIIa-b partitioned Runge-Kutta methods are symplectic for general Hamiltonians. Applied to $H = \frac{1}{2}p^T\bar{M}(q)^{-1}p + \tilde{V}(q) + \tilde{g}(q)^T\lambda$, we have

$$\begin{aligned}
q_{n+1} &= q_n + \frac{h}{2}(\bar{M}(q_n)^{-1} + \bar{M}(q_{n+1})^{-1})p_{n+1/2} \\
p_{n+1/2} &= p_n - \frac{h}{2}\nabla_q\tilde{V}(q_n) - \frac{h}{4}\,\nabla_q(p^T\bar{M}(q)^{-1}p)\big|_{(q_n,p_{n+1/2})} - \frac{h}{2}\tilde{g}'(q_n)^T\lambda_n \\
0 &= \tilde{g}(q_{n+1}) \\
p_{n+1} &= p_{n+1/2} - \frac{h}{2}\nabla_q\tilde{V}(q_{n+1}) - \frac{h}{4}\,\nabla_q(p^T\bar{M}(q)^{-1}p)\big|_{(q_{n+1},p_{n+1/2})} \\
&\quad - \frac{h}{2}\tilde{g}'(q_{n+1})^T\mu_{n+1} \\
0 &= \tilde{g}'(q_{n+1})\bar{M}^{-1}p_{n+1}.
\end{aligned}$$

These equations (which reduce to RATTLE for many-body Hamiltonians) are fully implicit, requiring the solution of a system of nonlinear equations of dimension $2N + m$ at each step (for q_{n+1}, $p_{n+1/2}$ and λ_n). Applied to the standard coordinates, direct symplectic discretization thus appears to be inordinately complex and computationally expensive.

Equations (2.5)-(2.9), on the other hand, are explicit in both q and p, and can be solved by efficient sparse-matrix techniques developed in [5] for molecular dynamics problems. We will discuss this numerical implementation briefly in §5, but we first describe how the many-body formalism provides a general, practical foundation for conservative multibody systems.

3 Particulation

The three dimensional rigid body has six degrees of freedom: three rotational and three translational. For any such body, an equivalent body representation can be developed using a collection of artificial massive particles whose relative positions are held fixed. We refer to each point along with its assigned mass as a *particle* and call the process of assigning particles to the bodies *particulation*. In effect, we are replacing arbitrary rigid bodies by rigid artificial "molecules." Our approach is analogous to—but slightly generalizes—a technique described in Routh [32].

Let B be a rigid body with mass M, center of mass q_{cm} and inertia tensor $I = (I_{ij})$. To model B we assign to the ith particle a mass m_i and a position $q_i = (x_i, y_i, z_i)^T$, $i = 1, 2, ..., k$ in order to preserve the total mass ($\sum m_i = M$), center of mass ($\frac{1}{M}\sum m_i q_i = q_{cm}$), the moments of inertia

$$\sum m_i(y_i^2 + z_i^2) = I_{1,1}, \qquad \sum m_i(x_i^2 + z_i^2) = I_{2,2}, \qquad \sum m_i(x_i^2 + y_i^2) = I_{3,3}$$

and the products of inertia

$$-\sum m_i x_i y_i = I_{1,2}, \qquad -\sum m_i x_i z_i = I_{1,3}, \qquad -\sum m_i y_i z_i = I_{2,3}$$

There are a total of 10 conditions to be met, with $4k$ free parameters.

A straightforward approach to placing the masses is to begin by diagonalizing the inertia tensor $I = (I_{ij})$, yielding the *axes of inertia* u_1,u_2,u_3. By placing three pairs of particles of equal mass symmetrically about the origin along the axes of inertia, we can reduce the number of unknowns to six: the masses and their distances from the origin. Moreover, this can be reduced even further by demanding either that all masses be of the same size or all be placed at the same distance from the origin. This leads to models with six mass particles, whereas four are sufficient to match any given inertial tensor. On the other hand, it is possible to reduce the equations for the six particles to equations for the motion of the center of mass and one particle from each pair.

A more elegant solution is the following particulation in four point particles (i.e. replacement by tetrahedral molecules). We assume that the body is oriented and located so that I is diagonal and the center of mass is the origin. Define four vectors, ξ_i, $i = 1, 2, 3, 4$ as follows: $\xi_1 = [x_1\sqrt{m_1}, x_2\sqrt{m_2}, x_3\sqrt{m_3}, x_4\sqrt{m_4}]^T$, $\xi_2 = [y_1\sqrt{m_1}, y_2\sqrt{m_2}, y_3\sqrt{m_3}, y_4\sqrt{m_4}]^T$, $\xi_3 = [z_1\sqrt{m_1}, z_2\sqrt{m_2}, z_3\sqrt{m_3}, z_4\sqrt{m_4}]^T$, and $\xi_4 = [\sqrt{m_1}, \sqrt{m_2}, \sqrt{m_3}, \sqrt{m_4}]$.

Next observe that

$$\xi_4^T \xi_1 = \sum_{i=1}^{4} m_i x_i$$

so orthogonality of ξ_4 and ξ_1 is equivalent to saying that the x-coordinate of the center of mass is zero. Similarly, $\xi_4^T\xi_2 = \xi_4^T\xi_3 = 0$ when the center of mass is the origin. Furthermore, we have relations

$$\xi_1^T \xi_2 = \sum_{i=1}^{4} m_i x_i y_i, \qquad \xi_1^T \xi_3 = \sum_{i=1}^{4} m_i x_i z_i, \qquad \xi_2^T \xi_3 = \sum_{i=1}^{4} m_i y_i z_i$$

implying ξ_1, ξ_2, and ξ_3 are mutually orthogonal precisely when the products of inertia vanish. Thus the particulation conditions reduce to orthogonality of the vectors ξ_i, $i = 1, 2, 3, 4$ together with the linear equations

$$\begin{aligned} \|\xi_2\|^2 + \|\xi_3\|^2 &= I_1 \\ \|\xi_1\|^2 + \|\xi_3\|^2 &= I_2 \\ \|\xi_1\|^2 + \|\xi_2\|^2 &= I_3 \\ \|\xi_4\|^2 &= M \end{aligned}$$

for the squared lengths of ξ_i. To find a solution, compute the squared lengths d_i of the ξ_i from:

$$\begin{bmatrix} d_1^2 \\ d_2^2 \\ d_3^2 \end{bmatrix} = \begin{bmatrix} 0 & 1 & 1 \\ 1 & 0 & 1 \\ 1 & 1 & 0 \end{bmatrix}^{-1} \begin{bmatrix} I_1 \\ I_2 \\ I_3 \end{bmatrix} = \frac{1}{2} \begin{bmatrix} I_2 + I_3 - I_1 \\ I_1 + I_3 - I_2 \\ I_1 + I_2 - I_3 \end{bmatrix}.$$

(By the "triangle inequality" [2], the solution is necessarily positive.)

Now select ξ_4 with squared length M (e.g. $m_i = M/4$, $i = 1,2,3,4$), and compute a Householder QR [16] decomposition of the 4×1 matrix ξ_4. The second through fourth columns of Q will be an orthonormal basis for the orthogonal complement of span$\{\xi_4\}$. Define ξ_1, ξ_2, ξ_3 by scaling the last three columns of Q by the computed lengths d_i. Finally, solve for the particle positions.

Although we have illustrated the procedure for a body with $n = 4$ particles, nothing changes for $n > 4$ except that the orthogonal complement of span$\{\xi_4\}$ has dimension $(n-1) > 3$, the result being additional freedom in the choice of the three vectors orthogonal to ξ_4.

In many cases, as when forces of interaction between bodies act on contact points of a body, we may wish to specify some k of the particle positions:

$$\begin{bmatrix} x_i \\ y_i \\ z_i \end{bmatrix} = \begin{bmatrix} \bar{x}_i \\ \bar{y}_i \\ \bar{z}_i \end{bmatrix}$$

for $i = 1 \ldots k$. Denoting the three chosen vectors from span$\{\xi_4\}^\perp$ as η_1,η_2 and η_3 we observe that the x-coordinates of the particles are given by

$$\begin{aligned} x &= \operatorname{diag}(\sqrt{m_1} \ldots \sqrt{m_n})^{-1} \xi_1 \\ &= \operatorname{diag}(\sqrt{m_1} \ldots \sqrt{m_n})^{-1} d_1 \eta_1. \end{aligned}$$

Similarly for the y and z-coordinates. Our task is to find vectors $\eta_j \in$ span$\{\xi_4\}^\perp$ for which the resulting coordinates satisfy the constraints. We can find $\eta_1 = \frac{\hat{\eta}_1}{\|\hat{\eta}_1\|}$ by solving the underdetermined linear system

$$\begin{bmatrix} A_1 \\ \xi_4^T \end{bmatrix} \hat{\eta}_1 = \begin{bmatrix} \bar{x} \\ 0 \end{bmatrix}$$

where $\bar{x}$ is the vector of the k specified x-coordinates and A_1 is the $k \times n$ matrix consisting of the first k rows of the matrix $d_1 \operatorname{diag}(\sqrt{m_1} \ldots \sqrt{m_n})^{-1}$. Setting $\xi_1 = d_1\eta_1$ we find $\xi_2 = d_2\eta_2$ and $\xi_3 = d_3\eta_3$ by successively finding the normalized solutions of the systems

$$\begin{bmatrix} A_2 \\ \xi_1^T \\ \xi_4^T \end{bmatrix} \hat{\eta}_2 = \begin{bmatrix} \bar{y} \\ 0 \\ 0 \end{bmatrix} \quad \text{and} \quad \begin{bmatrix} A_3 \\ \xi_1^T \\ \xi_2^T \\ \xi_4^T \end{bmatrix} \hat{\eta}_3 = \begin{bmatrix} \bar{z} \\ 0 \\ 0 \\ 0 \end{bmatrix}.$$

In some cases, one might simply elect to impose certain constraints on the particle positions or mass distribution, adding additional particles as needed in order for the system of equations to have an acceptable solution. The resulting system of (possibly underdetermined) linear and quadratic equations can be solved by a generalized Newton iteration [8] to obtain the particulated body. We use the

a) Conical body particulated to preserve symmetry

b) Conical body particulated using minimization scheme

Figure 1 Two particulations of the cone.

theory of convex sets to note the following general result about the relationship of m_i and q_i to q_{cm}.

Theorem 3.1 *A point r is in the interior of a finitely generated convex set C_m with (minimal) generating set $\{r_1, r_2, \ldots, r_m\}$ if and only if $\exists \alpha_i \in (0,1)$ such that $\sum \alpha_i r_i = r$ and $\sum \alpha_i = 1$.*

The proof is an adaptation of the well know barycentric coordinates theorem [31].

For our application, the theorem and the center of mass condition above imply that $m_i > 0, \forall i$ exactly when q_{cm} is inside the polyhedral convex set generated by the q_i. Further, if q_{cm} is on or outside the polyhedron, $m_i \leq 0$ for some i. While negative masses may be nonphysical, they pose no problem for the analysis here. (They might cause difficulties for the numerical solution of the discretized equations, however.)

As an example, consider a top as the solid generated by revolving the line $z = 2x, 0 \leq x \leq 1$ about the z axis. Assuming unit mass density, we can integrate over the conical volume to obtain $M = \frac{2\pi}{3}$, $r = [00\frac{3}{2}]^T$, and the inertial constants $I_z = \frac{\pi}{3}, I_x = I_y = \frac{17\pi}{10}$, and all products of inertia equal to zero. We model the cone with four particles. The sixteen parameters giving positions and masses are determined by imposing the 10 conditions above. We resolve all but one of the remaining degrees of freedom by requiring that one particle coincide with vertex of the cone, and that the three particles which are not fixed have equal mass. So constrained, the solution of the underdetermined system shares the symmetry properties of the original body (Figure 1a).

Alternatively, we could choose some objective function to minimize subject to the ten constraints. (For a survey of constrained minimization techniques, see [13].) To illustrate, for the conical body above we minimized the summed squared distances of four uniformly massive particles from the center of mass resulting in the particulation of Figure 1b.

3.1 Regularity A particulation consists of a set of point positions and point masses, along with a vector of constraints of the form $g_i(q) = 0$. We assume that the constraint set is solvable. A system of $n \geq 3$ particles in $\mathbf{R}^3$ has $3n$ degrees of freedom. The rigid body has six degrees of freedom. Each constraint removes one degree of freedom, thus if such a system is to be made rigid, $3n - 6$ rigid length constraints will be needed. On the other hand, for the system of equations to be well-posed and to enable the use of standard techniques to solve the discrete equations, we must have that the jacobian g' of the constraint equations $g = 0$ be full rank. When these two conditions are satisfied we say that the particulation is *regular*.

An example of an irregular particulation is the complete graph on five nodes, since the nine constraints shown in Figure 2 already make the structure rigid. Another example is the case of three particles connected by rods of lengths a,a and $2a$, hence defining a degenerate triangle.

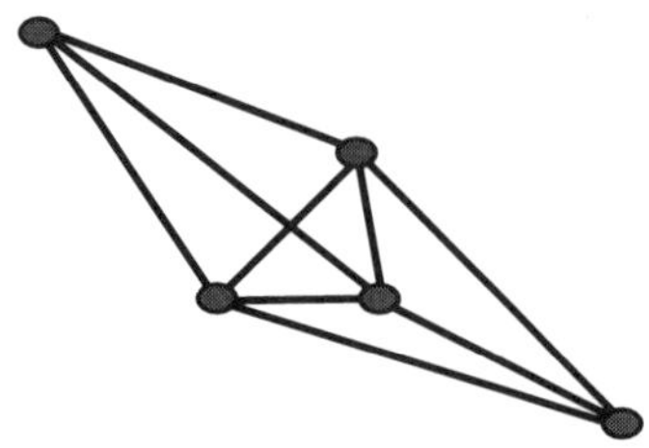

Figure 2 A five particle body.

3.2 Planar Bodies In certain types of mechanical applications, the rigid bodies may be planar, e.g. rigid plates. If the body lies in the xy plane, its moments will satisfy $I_1 + I_2 = I_3$ [2]. This is a necessary and sufficient condition, thus all particulations of a planar body must also be planar. In $\mathbf{R}^2$, a system of n particles is made rigid by the imposition of $2n - 3$ constraints. If we consider a particulation which is rigid in the plane, and then view it as an object in $\mathbf{R}^3$, it will no longer be rigid if $n > 3$, since $3n-6 \geq 2n-3$, $n > 3$. This means that additional constraints must be added if $n > 3$ in order to rigidify the body. However, such a particulation will then become a degenerate three dimensional body and thus will be irregular. Hence *the only particulations of planar bodies which are regular in* $\mathbf{R}^3$ *are those consisting of a single mass point, a pair of masses linked by a rigid rod, and a triangle.* We can of course specialize the derivation for the tetrahedral particulation given above to the planar case.

In molecular dynamics, many molecules including the important heme group in myoglobin [25] and benzene [10] are *nearly* rigid and planar. Such molecules might on occasion be replaced by artificial three particle models to simplify dynamics, although to represent an improvement over standard techniques [27], an efficient physical basis for distributing the many-body potentials over the artificial particles would need to be found.

In cases where nonplanar but nearly planar structure is important to the dynamics of the system, the price will be ill-conditioning of the nonlinear systems which must be solved at each timestep. An open question is to determine how one could obtain the optimal particulation of a given rigid body (i.e. one which effectively minimizes the ill-conditioning of the nonlinear systems).

3.3 Equivalent Particulations: Reductions Any two particulations of a rigid body yielding identical inertial tensors, centers of mass, and total mass will result in identical continuous-time dynamics. It is natural to ask if and when the corresponding property will be observed for the numerical dynamics. A priori, there is no reason to expect that two equivalent particulations will yield identical numerical solutions.

We first need a basis for comparison. Suppose that we are given a particulation consisting of four or more points $q = (q_1, q_2, \dots)$ not lying in a plane, and hence forming a basis for $\mathbf{R}^3$. The location of the center of mass at any time is defined by $\frac{1}{M}\sum m_i q_i = q_{cm}$ and each of the three axes of inertia u_i can be computed as linear combinations of the vectors $r_i = q_i - q_{cm}$.

Thus we can write in short, for some matrix A,

$$w(t) := \begin{pmatrix} q_{cm}(t) \\ u_1(t) \\ u_2(t) \\ u_3(t) \end{pmatrix} = Aq(t).$$

Now if $\hat{q}(t)$ is the position vector corresponding to any other particulation of the same rigid body, we can write, similarly,

$$w(t) = \hat{A}\hat{q}(t)$$

for some matrix $\hat{A}$. What this means is that $R(q, \hat{q}) := Aq - \hat{A}\hat{q}$ vanishes along the simultaneous motion of *both* particulations. Indeed, if p and $\hat{p}$ are the corresponding momenta of the two systems, we have

$$\frac{d}{dt}R(q, \hat{q}) = A\dot{q} - \hat{A}\dot{\hat{q}} = AM^{-1}p - \hat{A}\hat{M}^{-1}\hat{p}$$

defining this expression as $S(p, \hat{p})$, we see that $S = 0$ is a *linear integral invariant* of the flow of the combined system; such invariants are maintained by all of the common classes of discretization schemes, including, for example, Multistep, Runge-Kutta and Partitioned Runge-Kutta methods. It can also be seen that the linear secondary invariant $R = 0$ will also be maintained up to a weak instability.

We summarize with a theorem:

Theorem 3.2 *Assuming the discretized equations are solved exactly, equivalent particulations of the same nonplanar rigid body result in identical dynamics as long as the discretization scheme preserves linear first integrals.*

This discussion indicates an alternate set of coordinates: we could instead track the motion of the center of mass together with that of the three axes of inertia. Another way to view this is that we employ the particulation mentioned above consisting of three pairs of particles of equal mass placed symmetrically about the center of mass along the axes of inertia, but we then reduce the equations of motion to equations for the center of mass and three of the particles. Yet another choice would be to take two points on the body, and define, as in [37], a pair of arbitrary vectors fixed in the body and emanating from the two points, and then develop equations of motion in the resulting set of four basis elements.

4 Systems of rigid bodies

Arbitrarily complex systems of rigid bodies can, in principle, be treated with a particle-based approach. The basic questions are (i) how to incorporate various standard types of interactions between the bodies and (ii) how to distribute an applied force acting on the body to the system of particles. We treat the second issue here; in [4], it was shown that the at least the common holonomic joints of multibody dynamics can be represented in the particulated formulation in terms of algebraic constraints retaining the form of quadratic length constraints.

Given a force $\mathbf{F}$ acting on a body at the point given by a radial vector r from the center of mass, it is natural to choose a particulation which includes a particle at r. If such a particulation is not convenient, $\mathbf{F}$ must be distributed throughout the particulated body. To see how this is done, we decompose $\mathbf{F}$ into components parallel and perpendicular to r, say $\mathbf{F}_t$ and $\mathbf{F}_R$. Explicitly,

$$\mathbf{F}_T = \frac{rr^T}{\|r\|^2}\mathbf{F}$$

and

$$\mathbf{F}_R = \mathbf{F} - \mathbf{F}_T.$$

These are the translational and rotational components of $\mathbf{F}$. The torque about the center of mass is then given by $N = r \times \mathbf{F}_R$.

For a particulation with k particles, each with radial vector r_i, the distributed forces $\mathbf{F}_i$ must satisfy the following relations

$$\begin{aligned} N &= \sum_{i=1}^{k} r_i \times \mathbf{F}_i \\ \mathbf{F}_T &= \sum_{i=1}^{k} \frac{r_i r_i^T}{\|r_i\|^2}\mathbf{F}_i. \end{aligned}$$

5 Structure of the Nonlinear Equations at a Timestep

In this section, we consider the form of the discretized equations of motion when particulation is used to describe the system. We may assume that the constraint equations $g(q) = 0$ are just quadratic length constraints of the form $(r_i - r_j)^T(r_i - r_j) - L_{ij}^2 = 0$ which represent rods of length L_{ij} between particles i and j.

We briefly describe the techniques we propose for solving the nonlinear equations at each step of SHAKE discretization. In [5] these techniques are developed more fully in the context of molecular dynamics. In terms of q, the discrete equations are:

$$\begin{aligned} q_{n+1} &= q_n + hM^{-1}(p_n - (h/2)\nabla_q V(q_n) + g'(q_n)^T \lambda_n) \\ g(q_{n+1}) &= 0. \end{aligned}$$

For simplicity, we introduce the following notation

$$\begin{aligned} Q &= q_{n+1} \\ G &= g'(q_n) \\ \lambda &= h\lambda_n \\ \bar{Q} &= q_n + hM^{-1}(p_n - (h/2)\nabla_q V(q_n)). \end{aligned}$$

$\bar{Q}$ denotes the quantities on the right hand side which are known at a particular integration step. We may then write the equations as

$$\begin{aligned} Q &= \bar{Q} + M^{-1}G^T\lambda \\ g(Q) &= 0 \end{aligned}$$

or as a system in λ only

$$\tilde{g}(\lambda) = g(\bar{Q} + M^{-1}G^T\lambda) = 0.$$

Applying the Newton-Raphson iteration, we obtain

$$\lambda^{k+1} = \lambda^k - \tilde{g}'(\lambda^k)^{-1}\tilde{g}(\lambda^k)$$

where

$$\tilde{g}' = g'(\bar{Q} + M^{-1}G^T\lambda^k)M^{-1}G^T.$$

5.1 Symmetric Newton iteration Each iteration of the nonlinear solver requires the solution of a system of linear equations with the form given above. G and g' are the matrix of partial derivatives of the constraint equation evaluated at different stages of the iteration, hence they have identical nonzero structure but different entries. Since M^{-1} is diagonal, the jacobian of the system of nonlinear equations $g'M^{-1}G^T$ is thus *structurally symmetric* (the nonzero entries occur symmetrically). An alternative approach is to use the symmetric positive definite approximate Jacobian

$$GM^{-1}G^T.$$

In the presence of only rod constraints, the elements of the matrix are [5]

$$\begin{aligned}(GM^{-1}G^T)_{ll} &= \frac{L_{ij}^2}{m_i + m_j}\\ (GM^{-1}G^T)_{lm} &= \frac{\pm L_{ij}L_{jk}\cos(\theta_{lm})}{m_j}\end{aligned}$$

where rod l joins nodes i and j, rod m joins nodes j and k, and θ_{lm} denotes the angle between rods l and m.

Hence the diagonal entries are constant and the off diagonal entries (l, m) vary with the angle θ_{lm} between constraints l and m. For a particulated rigid body, these angles are constant. In systems of particulated bodies, the matrix consists of constant diagonal blocks corresponding to the bodies, coupled by nonconstant entries due to the changing angles between bodies. This suggests that for small motions between bodies the matrix may be held for several time steps, with updates only as needed. A key observation is that the structure of the matrix $GM^{-1}G^T$ depends only on the constraint equations. Further, nearly all of the entries depend solely on the (constant) distances between constrained particles and angles between constraints. Hence not only is the sparsity structure known in advance, but also the structure of entries which are constant throughout the integration. Any preprocessing required for the solution of the linear systems by sparse techniques, including special methods which take advantage of the constant entry structure, need be performed only once. Some examples of joints and the resulting sparsity patterns can be found in [4]. Further sparse matrix techniques for the symmetric linear systems are discussed in [5].

6 Numerical Experiments

A Matlab toolkit MDLab [3] has been developed to assist in the study of particulated models, symplectic discretizations, and efficient techniques for solving the discretized equations. The software is available by anonymous ftp (kitcs/MDLab.tar) from ftp.math.ukans.edu.

We have performed a wide range of experiments on small multibody systems using MDLab. Below, we study a pair of representative problems and compare the performance of two second order integrators: SHAKE and the two-step backward

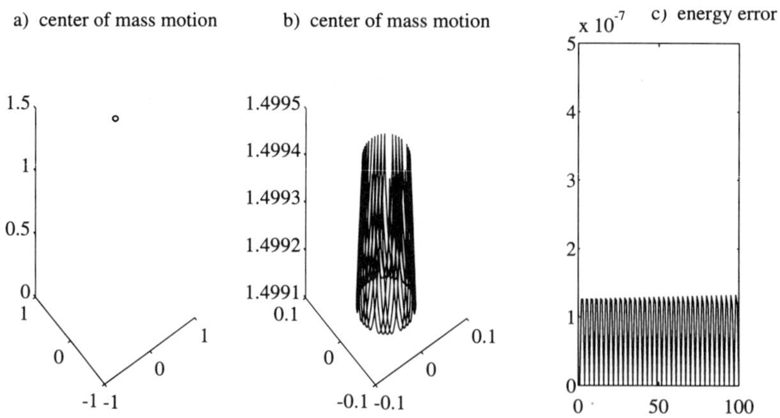

Figure 3 Leap-Frog Solution of Top on $[0, 100]$, $h = .05$.

difference formula BDF2, a popular nonsymplectic method for solving ODEs [14] and DAEs [9]. BDF2 was applied to the reduced-index formulation of equations (2.1)-(2.3) as in [15]. We verified correctness of our implementations and appropriateness of parameters such as stepsize and tolerance by monitoring the observed order of the integrators (two for both methods) and the constraint residuals (both position and velocity constraints were maintained to a small multiple of machine epsilon in all experiments).

6.1 Lagrange's Top We consider the conical body of Figure 1a as a top with the one vertex fixed, acting in a gravitational field. We particulate the top using four particles, one fixed at the vertex, and six rod constraints. An initial angular velocity was realized as consistent linear velocities of the three non-fixed points. No damping forces are considered, hence the system is conservative. The system was integrated with time steps of length $h = .05$ with SHAKE and BDF2. Figure 3 shows the relative energy error of the computed solution and the trajectory of the center of mass of the top for the SHAKE numerical solution. Over the interval [0,100], the motion can be seen to be stable. Precession and nutation, well known small scale motions from classical dynamics [26, 2], are well resolved for this value of h.

Figure 4 shows the relative energy error and the trajectory of the center of mass for the BDF2 solution with the time steps ($h = .05$) and ($h = .01$) over the interval $[0, 20]$. Energy loss is significant. Moreover, the plunge of the center of mass at $h = .05$ would correspond to contact of the side of the top with the ground, i.e. the top topples. In our experience, step sizes smaller by at least a factor of five are required to partially resolve the center of mass motion. Energy loss and slowed precession for $h = .01$ are still clearly evident.

We do not propose that the lack of symplecticity of BDF2 is, alone, the reason for the failure. Rather, we suspect that it is SHAKE's near-conservation of energy and the conservation of angular momentum that is important here. The presence of a noncentral gravitational force implies that angular momentum is not conserved. However, it is easy to see that in this example the third component of angular momentum *is* conserved since the gravitational force only acts in this direction. It seems that this conserved quantity is largely responsible for the regular motion of

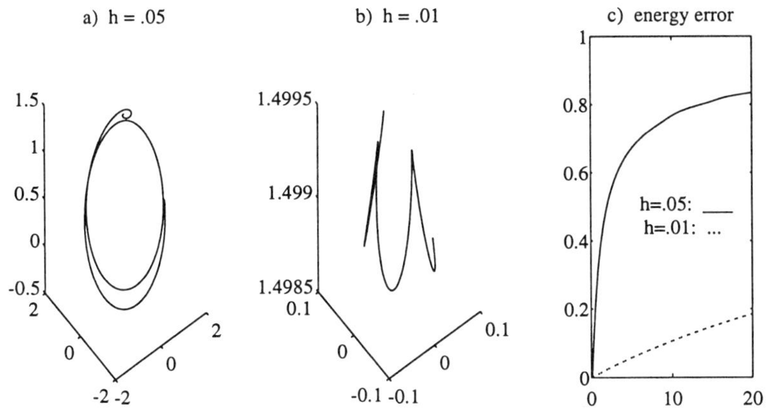

Figure 4 BDF2 Solution of Top on [0, 20].

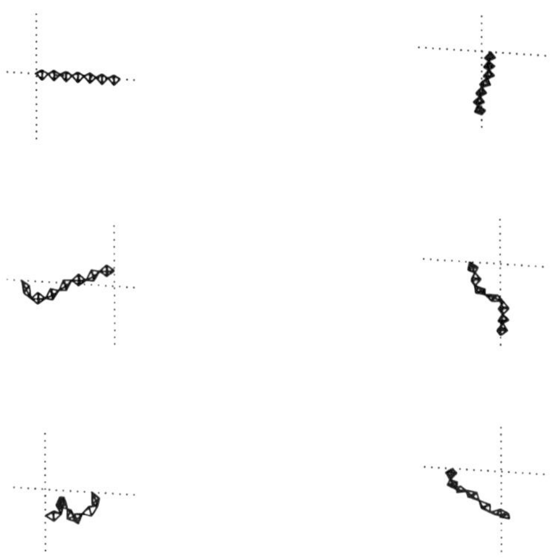

Figure 5 Motion of the 7-bar chain.

the center of mass, so it is critical that an angular momentum-conserving numerical scheme such as leapfrog/SHAKE be employed for numerical simulations.

6.2 Seven bar chain As a second example, we modelled a three dimensional seven bar chain with the ends of the chain constrained to slide along a pair of given fixed perpendicular axes in the xz plane (i.e. connected to the axes by prismatic joints). Our model used seven identical cylindrical bodies of uniform unit mass density moving in the presence of gravity. We particulate the cylinders with five particles, as in Figure 2.

From an initial linear configuration along the horizontal axis, the chain is allowed to swing in gravity. Using the SHAKE method, 2000 time steps of size $h = .05$ were computed to integrate over the interval [0,100]. Snapshots of the computed motion at various points in the interval [0, 50] are shown in Figure 5. Figure 6 shows the energy of the system throughout the integration. This numerical

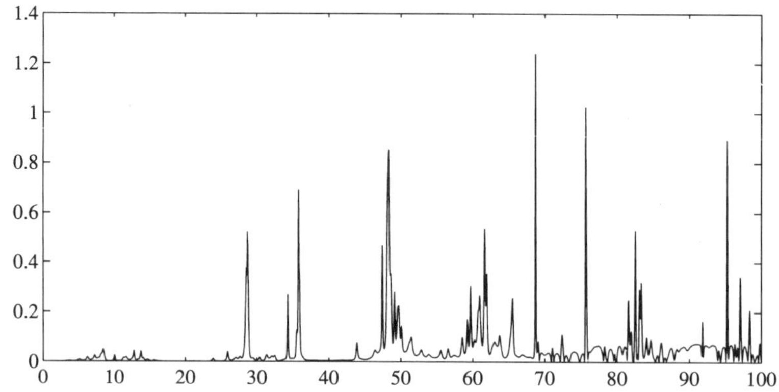

Figure 6 7-bar chain: relative energy error

experiment illustrates the property of symplectic methods suggested by the "fundamental theorem" stated in the introduction. Despite substantial variations from the exact energy, the approximate dynamics behave reasonably on long time intervals, i.e. as the solution of a some perturbed Hamiltonian system. Note particularly the sudden drastic changes in the energy at certain points during the integration. These correspond to shocks in the motion of the chain (i.e. like the "crack of a whip"). Although it seems surprising that the integrator partially recovers from these severe shocks, this example points up deficiencies in our understanding of the "nearby Hamiltonian" and also the need for variable stepsize symplectic integrators with good properties.

7 Conclusion

We have shown that the particle formulation provides an strong foundation for symplectic integration of the equations of motion for conservative multibody systems. Of course realistic comparisons with existing nonsymplectic techniques will have to await a careful implementation of the particulation technique in the considerable framework of a multibody simulator.

Acknowledgements

We would like to thank R. Byers, S. Reich and R. Skeel for helpful discussions during the preparation of this paper.

References

[1] Andersen, H. C. [1983], *Rattle: a 'velocity' version of the shake algorithm for molecular dynamics calculations*, J. Comp. Phys., **52**, 24-34.

[2] Arnold, V. I. [1975], *Mathematical Methods of Classical Mechanics.* Springer Graduate Texts in Mathematics No. 60, Springer-Verlag, NY.

[3] Barth, E. [1993], A user's guide to MBLab. *1993 MATLAB Conference proceedings*, electronic publication available from ftp.mathworks.com.

[4] Barth, E. [1994], *Numerical algorithms for constrained Hamiltonian systems.* Ph.D. Thesis, Department of Mathematics, University of Kansas.

[5] Barth, E., Kuczera, K., Leimkuhler, B., and Skeel, R. [1995], *Algorithms for constrained molecular dynamics*, J. Comp. Chem., **16**, 1192-1209.

[6] Bayo, E., Garcia de Jalon, J., Avello, A., and Cuadrado, J. [1991], *An efficient computational method for real-time multibody dynamics simulation in fully cartesian coordinates*, Computer Methods in Applied Mechanics and Engineering, **92**.

[7] Bayo, E., Jimenez, J. M. [1992], *On the use of canonical equations of motion for the dynamic analysis of constrained multibody systems*, In ASME Flexible Mechanisms, Dynamics, and Analysis, ASME DE- **47**, 405.

[8] Ben-Israel, A. [1965], *A modified Newton-Raphson method for the solution of systems of equations*, Israel. J. Math., **3**, 94–98.

[9] Brenan, K. E., Campbell, S. L. and Petzold, L. R. [1989], *Numerical Solution of Initial-Value Problems in Differential-Algebraic Equations.* North-Holland.

[10] Ciccotti, G., Ferrario, M., and Ryckaert, J.-P. [1982], *Molecular dynamics of rigid systems in cartesian coordinates: A general formulation*, Mol. Phys., **47**, 1253.

[11] Eich, E., Führer, C., Leimkuhler, B., and Reich, S. [1990], *Stabilization and projection methods for multibody dynamics*, Report A281, Helsinki U. of Tech., Helsinki.

[12] Evans, D. J. [1977], *On representation of orientation space*, Mol. Phys., **34**, 317.

[13] Fletcher, R. [1981], *Practical Methods of Optimization, Vol. 2: Constrained Optimization.* John Wiley and Sons.

[14] Gear, C. W. [1971], *Numerical Initial Value Problems in Ordinary Differential Equations.* Prentice-Hall.

[15] Gear, C. W., Gupta, G. K., and Leimkuhler, B. J. [1985], *Automatic integration of the Euler-Lagrange equations with constraints*, J. Comp. Appl. Math., **12 & 13**, 77–90.

[16] Golub, G. H., and Van Loan, C. [1989], *Matrix Computations*, 2nd. Ed. Johns Hopkins U. Press.

[17] Gorla, B. and Renaud, M. [1989], *Models of robot manipulators*, In Geometry and Robotics, Workshop (Goos, G. and Hartmanis, J. eds.), Toulouse, May, 1988, Lecture Notes in Computer Science No. 391, Springer.

[18] Hairer, E. [1994], *Backward error analysis of numerical integrators and symplectic methods*, Annals of Numerical Mathematics, **1**, 107-132.

[19] Jay, L. [1993], *Symplectic partitioned Runge-Kutta methods for constrained Hamiltonian systems*, Report, Department of Mathematics, University of Geneva.

[20] Leimkuhler, B. [1993], *Discretization and weak invariants*, preprint.

[21] Leimkuhler, B., and Reich, S. [1994], *Symplectic numerical integrators*, J. Comput. Phys., **112**, 117-125.

[22] Leimkuhler, B., and Skeel, R. D. [1994], *Symplectic numerical integrators for constrained Hamiltonian systems*, Accepted and to appear, Journal of Computational Physics.

[23] McClamroch, N. H. and Bloch, A. M. [1988], *Control of constrained Hamiltonian systems and applications to control of constrained robots*, In Dynamical Systems Approaches to Nonlinear Problems in Systems and Circuits (Salam, Fathi M. A., and Lev, Mark L. eds.), SIAM, 344-403.

[24] McLachlan, R., and Scovel, C. [1995], *Equivariant constrained symplectic integration*, J. Nonlin. Sc., **16**, 233–256.

[25] Mandelkern, L. [1983], *An Introduction to Macromolecules*, 2nd Ed. Heidelberg Science Library, Springer.

[26] Marion, J. [1965], *Classical Dynamics of Particles and Systems.* Academic Press.

[27] Meyer, M. [1990], *Simulation of plastic crystals.* In Computer Modelling of Fluids Polymers and Solids (Catlow, C. R. A., Parker, S. C. and Allen, M. P., eds.), NATO ASI Series C, **293**, Kluwer.

[28] Mrziglod, T. [1987], *Zur Theorie und numerischen Realisierung von Lösungmethoden bei Differentialgleichungen mit angekoppelten algebraischen Gleichungen*, Diplomarbeit, Math. Inst., Univ. zu Köln.

[29] Park, K. C., Chiou, J. C., and Downer, J. D. [1988], *Recent advances in multibody dynamics analysis procedures*, In Computational Methods in Bioengineering, BED-**9**, Springer.

[30] Reich, S., *Momentum conserving symplectic integrators.* Report, Department of Computer Science, University of British Columbia, 1993, submitted.

[31] Rockafellar, R. [1970], *Convex Analysis.* Princeton Univ. Press.

[32] Routh, E. [1905], *Dynamics of a System of Rigid Bodies*, 7th Ed. Macmillan and Co..

[33] Rheinboldt, W. [1986], *Numerical Analysis of Parameterized Nonlinear Equations.* University of Arkansas Lecture Notes in the Mathematical Sciences, **7**, John Wiley and Sons, N.Y.

[34] Ryckaert, J. P., Ciccotti, G., and Berendsen, H. J. C. [1977], *Numerical integration of the cartesian equations of motion of a system with constraints: molecular dynamics of n-alkanes*, J. Comp. Phys., **23**, 327-341.

[35] Sanz-Serna, J. M. [1992], *Symplectic integrators for Hamiltonian problems: An overview*, Acta Numerica, **1**, 243–286.

[36] Sussman, G. J. and Wisdom, J. [1992], *Chaotic Evolution of The Solar System*, Science, **257**, 56–62.

[37] Unda, J., Jimenez, J., Avello, A., and Garcia de Jalon, J. [1986], *Computational analysis of multibody systems using "natural" coordinates*, In Dynamics of Multibody Systems (Bianchi, G. and Schielen, W., eds.): IUATAM/IFToMM Symposium, Udine 1985. Springer.

[38] Wang, D. and Vidyasagar, M. [1987], *Algorithm for generating inertia matrices of a class of N-bar-linkage robots*, Modelling and Control of Robotic Manipulators, Proceedings of Winter Annual Meeting of ASME, Boston, 1987, ASME DSC-**6**, 111.

[39] Yoshida, H. [1993], *Recent progress in the theory and application of symplectic integrators*, Dyn. Astr., **56**, 27-43.

[40] Zhang, M.-Q. and Skeel, R. [1993], *Explicit canonical RKN methods preserve angular momentum.* Report, Department of Computer Science, University of Illinois.

Fields Institute Communications
Volume **10**, 1996

An Introduction to Symplectic Integrators

Paul J. Channell and Filippo R. Neri
Los Alamos National Laboratory
Los Alamos, NM 87545

1 Introduction

To begin let us recall some basic facts about Hamiltonian systems and symplectic maps; this material is just a synopsis of standard material in [14] and [1]. Given a $2n$ dimensional vector space (phase space) with coordinates (q,p) and a function $H : (q,p) \to R$, a dynamical system is Hamiltonian if the variables evolve according to

$$\frac{dq}{dt} = \frac{\partial H}{\partial p}, \tag{1.1}$$

$$\frac{dp}{dt} = -\frac{\partial H}{\partial q}, \tag{1.2}$$

Let us define a bilinear operation, $\{,\}$, the Poisson bracket, on pairs of functions on phase space, (f,g), by

$$\{f,g\} \equiv \sum_i \left(\frac{\partial f}{\partial q_i}\frac{\partial g}{\partial p_i} - \frac{\partial f}{\partial p_i}\frac{\partial g}{\partial q_i}\right) \tag{1.3}$$

Letting $z \equiv (q,p)$, the full collection of phase space variables, the Hamiltonian evolution equations can be written as

$$\frac{dz}{dt} = \{z,H\} \tag{1.4}$$

We can solve this equation iteratively, i.e. approximating the right hand side by a constant we can integrate to get

$$z_1 = z_0 + \{z_0,H\}t \tag{1.5}$$

Substituting this back into the right hand side of equation 1.4 and integrating again we get

$$z_2 = z_0 + \{z_0,H\}t + \{\{z_0,H\},H\}\frac{t^2}{2} \tag{1.6}$$

1991 *Mathematics Subject Classification.* Primary: 65L07; Secondary: 70H15.

Continuing in this way we get a formal solution

$$z = e^{t\{\cdot, H\}} z_0 \tag{1.7}$$

that will be useful in expressing general results later. In addition, Hamiltonian systems have certain important special properties; in particular:

1) Energy, i.e. the numerical value of the Hamiltonian function, is conserved in an autonomous Hamiltonian system.
2) If ϕ_t is the flow associated with a Hamiltonian system, then ϕ_t is a symplectic map (i.e. preserves a symplectic two-form ω, $\phi_t^* \omega = \omega$, see [1]) for all t.
3) Phase space volume is preserved by ϕ_t.
4) There is a list of invariants, Poincaré invariants, that are preserved by ϕ_t; these invariants are

$$\int\int \sum_i dq_i dp_i, \qquad \int\int\int\int \sum_{i \neq k} dq_i dp_i dq_k dp_k, \qquad \ldots .$$

Note that 3) is just the last in the chain of invariants in 4) and that 4) is equivalent to 2) so that in fact this list contains only two independent assertions, i.e. 1) there is a global invariant, the energy, and 2) there are differential invariants given by the symplectic form. For particular Hamiltonians there may be additional global invariants such as momentum or angular momentum, but in general there need not be. In numerical investigations one would like to use an algorithm that shares as many of these special properties of the system as possible; unfortunately, a theorem of Ge and Marsden [12] shows that one cannot have both properties, 1) and 2) above, without essentially integrating the trajectories exactly, i.e. exactly up to a reparameterization of time. A number of numerical schemes have been devised to exactly preserve the energy (see Simo's paper in these proceedings) and these schemes often do have performance superior to that of conventional integrators, and, in fact, sometimes superior to the performance of the symplectic integrators to be discussed here. The other alternative is to exactly preserve the differential invariants by using a symplectic integrator. Note that when using a symplectic integrator the extent to which energy is conserved can be used as a test of the performance of the integrator, since it is guaranteed not to be exactly preserved; if the energy deviates significantly from its initial value then the integration is in error. Likewise, preservation of the differential invariants can be used as a test of the quality of an energy-preserving algorithm, though the test is a bit harder to implement.

In this paper we will also consider time dependent Hamiltonian systems which, of course, do not preserve the energy but do preserve the differential invariants. We will treat these systems using the following simple trick. Given a time dependent Hamiltonian, $H(q,p,t)$, on a symplectic vector space and time, $R^{2n} \times R$, where q and p are canonically conjugate sets of variables, it is well known, [18], that such a system can be suspended as a time independent Hamiltonian system in a symplectic vector space R^{2n+2} simply by constructing the Hamiltonian

$$\tilde{H}(q,t,p,p_t) = p_t + H(q,p,t), \tag{1.8}$$

on the symplectic space with symplectic coordinates (q,t,p,p_t), where p_t is the coordinate canonically conjugate to t. Calling the new 'time' variable s, we see that $\tilde{H}$ is independent of s and that the resulting equation of evolution for t is

$$\frac{dt}{ds} = 1, \tag{1.9}$$

so that $t = s + s_0$ where s_0 is a constant. The evolution equations for (q,p) are just

$$\frac{dq}{ds} = \frac{\partial H}{\partial p}, \tag{1.10}$$

and

$$\frac{dp}{ds} = -\frac{\partial H}{\partial q}, \tag{1.11}$$

i.e. equations which, given the fact that $t = s + s_0$, are just the time dependent Hamiltonian equations in the original space. Though this trick of suspension may seem trivial, it makes the symplectic integration of time dependent Hamiltonian systems completely straightforward.

Symplectic integrators, i.e. symplectic maps that approximate the time δt flow of a Hamiltonian system, have been successfully applied to an assortment of Hamiltonian dynamical systems. Both implicit [8], [4], [9], [6], and explicit [25], [22],[11], [27], [24], [10], techniques of various types have been developed and exhibit very good stability for long time integration of these systems in most cases; in particular, the global geometrical properties of the systems seem to be fairly well preserved. The development of a symplectic integrator requires the ability to write down particular symplectic maps that can be used in various ways to approximate the flows of general Hamiltonian systems. Two ways of finding these symplectic maps are common: 1) a generating function is given; we treat integrators based on this technique in the next section; 2) explicitly integrable Hamiltonian systems are used; we consider integrators based on this technique in Section 3. Systems reducible to the previous ones and extensions to more general Hamiltonian systems are discussed briefly in Section 4. In the final section we summarize some of the arguments for the use of symplectic integrators as well as some of the caveats that apply.

2 Implicit integrators from generating functions

Implicit symplectic integrators derived from generating functions were first developed in an unpublished work of deVogelaere [8]. The general development which we briefly recapitulate here is given in [6].

Given two copies of a symplectic vector space with coordinates (q_0, p_0) and (q_1, p_1), consider a function $K(p_0, q_1)$. If a map, $\phi : (q_0, p_0) \to (q_1, p_1)$, is 'generated' by the solution of the equations

$$q_0 = -\frac{\partial K(p_0, q_1)}{\partial p_0}, \tag{2.1}$$

$$p_1 = -\frac{\partial K(p_0, q_1)}{\partial q_1}, \tag{2.2}$$

(i.e. (q_1, p_1) result from solving these equations), then ϕ is symplectic. Note that the first of these equations is implicit, but, given the solution of this first equation, the second equation is explicit. The proof of symplecticity of this map is difficult in a coordinate description, trivial in a geometric description, see [1]. Since any map given by a generating function is automatically symplectic, the principal problem in this case is to write down a generating function that generates a map that systematically approximates the given Hamiltonian flow of interest. This problem can be solved in general. Let us use a generating function, $K(p_0, q_1)$. The identity is generated by

$$K_{Id} = -p_0 \cdot q_1 \tag{2.3}$$

Letting

$$K = \sum_{m=0}^{\infty} \frac{t^m}{m!} K_m(p_0, q) \quad , \tag{2.4}$$

where

$$K_0 = K_{Id}, \tag{2.5}$$

then one finds [6]

$$K_1(p_0, q_1) = H(p_0, q_1), \tag{2.6}$$

$$K_2 = -\frac{\partial K_1}{\partial q} \cdot \frac{\partial H}{\partial p_0} \quad , \tag{2.7}$$

$$K_3 = -\frac{\partial K_2}{\partial q} \cdot \frac{\partial H}{\partial p_0} + \sum_{i,j=1}^{n} \frac{\partial K_1}{\partial q_i} \frac{\partial K_1}{\partial q_j} \frac{\partial^2 H}{\partial p_{0i} \partial p_{0j}} \quad , \tag{2.8}$$

etc. Note that at each order, higher derivatives of the Hamiltonian occur, i.e. at nth order, one has $(n-1)$th order derivatives of the Hamiltonian. Thus, in the transformations generated, an nth order transformation requires one to evaluate nth order derivatives of the Hamiltonian. We are thus assuming that these higher order derivatives are available; in general this will be true only if the Hamiltonian is given explicitly in terms of a symbolic formula. Even if the higher order derivatives of the Hamiltonian are available, in many cases it can be very difficult or time consuming to evaluate them; thus these implicit techniques have only a limited range of practical applicability. Though the generating function formulae seem complicated, they are completely straightforward and can be implemented easily in a Maple code that generates the appropriate FORTRAN code in principle; in practice the generation of the FORTRAN code can require large amounts of CPU time and memory, even for moderate order integrators. An example of the energy stability that typically results from symplectic integrators is shown in Figure 1 where the fractional energy error as a function of time is plotted for a particular initial condition $(x = 1.0, y = 1.0, p_x = 0.0, p_y = 1.0)$ with the Hamiltonian

$$H = p_x^2 + p_y^2 + \ln(1 + x^2 + y^4 + p_x^4 + p_y^2 + x^2 y p_x p_y^2). \tag{2.9}$$

The energy deviation rapidly achieves a maximum and a minimum value and then fluctuates between them, seemingly forever.

Time dependent Hamiltonian systems can be treated using the suspension trick. The same formulae can be used to derive the generating functions for symplectic integrators (after letting $\delta t \to \delta s$); in fact, the symbolic code of [6] (now translated to both MACSYMA and Maple) can be used without change simply by enlarging

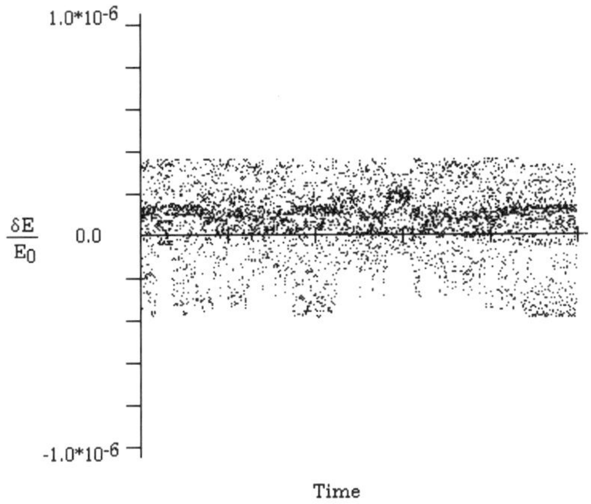

Figure 1 Fractional energy deviation, $\delta E/E_0$, using a second order, implicit, symplectic integrator generated by Maple. The Hamiltonian was $H = p_x^2 + p_y^2 + \ln(1 + x^2 + y^4 + p_x^4 + p_y^2 + x^2 y p_x p_y^2)$. The run parameters were final time $= 20000.0$, time step $= 0.001$. The initial conditions were:$x = 1.0$, $y = 1.0$, $p_x = 0.0$, $p_y = 1.0$.

the phase space dimension by 2 and using the extended Hamiltonian of Equation 1.8. The results of reference [4] for time dependent systems can be trivially rederived this way.

The principal advantages of implicit symplectic integrators are that they can be developed to any order and that they work in principle for any Hamiltonian that can be written down symbolically. Of course, the practical implementation of the higher order integrators is so difficult in most cases that only low order integrators can be used. In addition, implicit symplectic integrators tend to be slow. There are, however, strategies for speeding them up. Either one can use a Newton solver for the implicit equations rather than simple iteration, though usually this does not gain a great deal and involves evaluating higher derivatives and matrix inverses, or one can use predictor-corrector techniques using a conventional predictor and symplectic corrector for improved initialization of the iteration to gain a factor of $2 - 3$ in speed. Thus the disadvantages can often be dealt with in particular cases, but they remain very compelling reasons to consider explicit symplectic integrators where possible.

3 Explicit Symplectic Integrators

Many programs over the years have employed explicit symplectic integrators, in particular the second order leap-frog algorithm, though the developers were often unaware of the symplectic nature of the algorithms; the recent development of explicit symplectic algorithms is due to Wisdom [25], and to Ruth [22]. Other references to higher order explicit algorithms include [10] [11] [23] [24] [26] [27]. The basic idea of explicit symplectic integrators, also called composition methods, is to decompose the time δt flow of a Hamiltonian system into a sequence of maps that are the flows of Hamiltonian systems that can be evaluated explicitly and are thus each exactly symplectic. The fact that symplectic diffeomorphisms form a group ensures that the resulting composition is symplectic.

Let us begin by considering Hamiltonians of the form

$$H(q,p) = H_1(q) + H_2(p) \tag{3.1}$$

The solution can be written as

$$z = e^{t\{\cdot,H_1(q)\}+t\{\cdot,H_2(p)\}} z_0 \tag{3.2}$$

To first order in t this can be approximated by

$$z = e^{t\{\cdot,H_1(q)\}} e^{t\{\cdot,H_2(p)\}} z_0 + O(t^2) \tag{3.3}$$

In other words, to first order, the dynamics of the full Hamiltonian for a time t is the same as following the dynamics due to $H_2(p)$ alone for a time t, then following the dynamics due to $H_1(q)$ alone for a time t. The important point is that $H_1(q)$ and $H_2(p)$ are, individually, exactly integrable so that the exponentials can be evaluated exactly just by writing down the exact solutions to the Hamiltonian equations.

The equations of motion due to H_2 are

$$\frac{dq}{dt} = \frac{\partial H_2}{\partial p} \quad \frac{dp}{dt} = 0, \tag{3.4}$$

with the obvious solutions

$$p = p_0 \quad q = q_0 + t\frac{\partial H_2}{\partial p} \tag{3.5}$$

The equations of motion due to H_1 are

$$\frac{dp}{dt} = -\frac{\partial H_1}{\partial q} \quad \frac{dq}{dt} = 0, \tag{3.6}$$

with the obvious solutions

$$q = q_0 \quad p = p_0 - t\frac{\partial H_1}{\partial q} \tag{3.7}$$

We can combine these two solutions to give the single symplectic transformation

$$q_1 = q_0 + t\frac{\partial H_2(p_0)}{\partial p} \quad p_1 = p_0 - t\frac{\partial H_1(q_1)}{\partial q} \tag{3.8}$$

This is called the symplectic Euler, or first order Ruth, algorithm. The transformation due to $H_2(p)$ is often called a 'drift', especially if $H_2(p) = p^2/2$, and the transformation due to $H_1(q)$ is usually called a 'kick'. The first order Ruth algorithm is then expressed by saying that, to first order, the dynamics is approximated by a drift followed by a kick. A second order symplectic integrator can be based on:

Theorem $e^{t\{\cdot,H_1(q)\}+t\{\cdot,H_2(p)\}} = e^{\frac{t}{2}\{\cdot,H_2(p)\}} e^{t\{\cdot,H_1(q)\}} e^{\frac{t}{2}\{\cdot,H_2(p)\}} + O(t^3)$

Note that the error term is one higher order in t than in Equation 3.3. The algorithm that results from this theorem can be paraphrased as a 'half drift' then a 'full kick' then a 'half drift' . This algorithm is called the second order Ruth algorithm, or the leapfrog method, or the midpoint rule (because the force is evaluated at the midpoint). Note that this algorithm is second order, but has the same number of force evaluations (often the most computationally expensive part of a simulation) as the first order Ruth algorithm. Note also that the proof (see [10]) did not rely on the forms of H_1 and H_2, only on the fact that the triple product had the appropriate addition of commutators to approximate the left hand side to second order. In fact, the proof did not rely on the structure of the bracket or on the fact that we are dealing with Hamiltonian systems.

One can now see trivially that if the Hamiltonian has the form

$$H = H_1 + H_2 + \cdots + H_n, \tag{3.9}$$

then we have the generalized midpoint rule

$$e^{t\{\cdot,H_1\}+t\{\cdot,H_2\}+\cdots+t\{\cdot,H_n\}} = e^{\frac{t}{2}\{\cdot,H_n\}}e^{\frac{t}{2}\{\cdot,H_{n-1}\}}\cdots e^{\frac{t}{2}\{\cdot,H_2\}}e^{t\{\cdot,H_1\}}e^{\frac{t}{2}\{\cdot,H_2\}}\cdots e^{\frac{t}{2}\{\cdot,H_{n-1}\}}e^{\frac{t}{2}\{\cdot,H_n\}} + O(t^3)$$

In other words, the midpoint rule works for Hamiltonians with any number of terms of any form. The proof is by induction, working from the center triple outwards, i.e. the inner three terms approximate the action of H_1+H_2, the next two enclosing terms give an approximation to $H_1 + H_2 + H_3$, etc. The generalized midpoint rule usually forms the basis of the composition methods, socalled because an integrator is composed of integrators for pieces of the Hamiltonian.

Note that for time dependent Hamiltonian systems the generalized midpoint rule is quite simple. Letting the time advance (in the 'time' variable s) due to the term p_t in Equation 1.8 be evaluated as the central term, the generalized midpoint rule can be paraphrased as: evaluate the advance due to each term for a half time step, then advance the time by a full step, then evaluate the advance due to the terms in reverse order for a half time step.

Let us skip the third order Ruth method (developed in [22]) and go to the fourth order algorithm using the Yoshida trick, [27]. Suppose we are able to find a map, $\phi(t)$, say, for example, the map given by the midpoint rule, that approximates the time t evolution of H to order $2n$; i.e.

$$\phi(t) = e^{t\{\cdot,H\}} + O(t^{2n+1}) \tag{3.10}$$

Suppose further that

$$\phi(t)\phi(-t) = Id \tag{3.11}$$

Defining

$$z_0 = \frac{1}{2-2^{1/(2n+1)}} \qquad z_1 = -\frac{2^{1/(2n+1)}}{2-2^{1/(2n+1)}} \tag{3.12}$$

then we have

Theorem $\phi(z_0t)\phi(z_1t)\phi(z_0t) = e^{t\{\cdot,H\}} + O(t^{2n+3})$

i.e. , this combination of time steps using the order $2n$ integrator gives an order $2n+2$ integrator. One has thus produced a higher order integrator by the composition of lower order integrators. The proof of this, see [27] or [10], is completely analogous to the proof of the midpoint rule; in particular, it does not rely on the detailed form of the Hamiltonian or the bracket.

Starting with the midpoint rule as a second order approximation ($n = 1$), the above result immediately gives a fourth order approximation ($n = 2$), from which we can get a sixth order approximation, etc. The generalized midpoint rule and Yoshida's trick thus give us a technique for producing higher order explicit symplectic integrators from explicit integrators for 'pieces' of the Hamiltonian. Thus, to produce higher order explicit symplectic integrators we need only look for integrable Hamiltonian systems from which other Hamiltonians can be built. The class of all polynomial Hamiltonians, a much larger class than the set of Hamiltonians for which the Ruth methods were originally developed, can be integrated explicitly with the help of the following [5], [13]:

Theorem *Any Hamiltonian of the form*

$$H = H(q_1^{m_1} p_1^{n_1}, \dots , q_N^{m_N} p_N^{n_N}) \tag{3.13}$$

where m_i and n_i are nonnegative integers, is integrable, with integrals, I_i, given by

$$I_i = q_i^{m_i} p_i^{n_i} \qquad i = 1 \dots N \tag{3.14}$$

This theorem is used to observe that any monomial Hamiltonian

$$H = q_1^{m_1} \dots q_N^{m_N} p_1^{n_1} \dots p_N^{n_N} \tag{3.15}$$

is integrable. In fact, it is explicitly integrable in terms of elementary functions, i.e. algebraic functions and exponentials. We can thus immediately write down an explicit symplectic integrator for any term of any polynomial Hamiltonian, and then use the generalized midpoint rule and Yoshida's trick to find a fourth order, explicit, symplectic integrator for any polynomial Hamiltonian, even if time dependent. An example of the stability of this algorithm for the time-dependent Hamiltonian

$$H = \frac{p^2}{2} + \frac{q^2}{2} + \alpha \sin(\omega t) \frac{q^2 p^2}{2} \tag{3.16}$$

is shown in Figure 2 for the initial conditions $q = 0.6, p = -0.9$ where the error in the augmented (i.e. higher dimensional) Hamiltonian of Equation 1.8 is plotted as a function of time. The smallness of the error for such a long run (100000 periods of the time dependent coefficient, $= 2\pi/\omega$) implies that the variation in the supplementary variable p_t is very nearly the negative of the variation of the numerical value of the Hamiltonian. A plot of p_t versus time is shown in Figure 3 where it can be seen that the variation of the numerical energy is very large. A surface of section plot for this run is shown in Figure 4 where a point is plotted after every 10 periods of the time dependent coefficient. Stability of the algorithm thus appears to be excellent, even in a strongly nonlinear regime.

The extension of the composition methods to produce explicit integrators for any polynomial Hamiltonian with arbitrary time dependence presented in this section so greatly extends the realm of applicability of the explicit symplectic methods that implicit methods will now be needed only rarely. Certainly, explicit algorithms

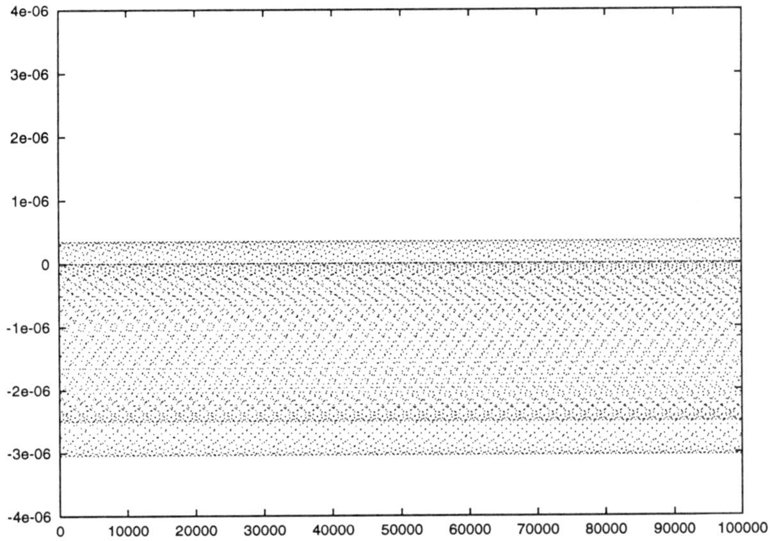

Figure 2 Fractional energy error for the extended Hamiltonian vs. time for a run with a general symplectic polynomial integrator. The Hamiltonian was $H = p^2/2 + q^2/2 + \alpha \sin(\omega t) q^2 p^2/2$. The run parameters were: final time $= 100000.0$, time step $= 0.05$, $\alpha = 1.6, \omega = 2\pi$ The initial conditions were: $q = 0.6$, $p = -0.9$. A point is plotted every ten periods of the time dependent coefficient.

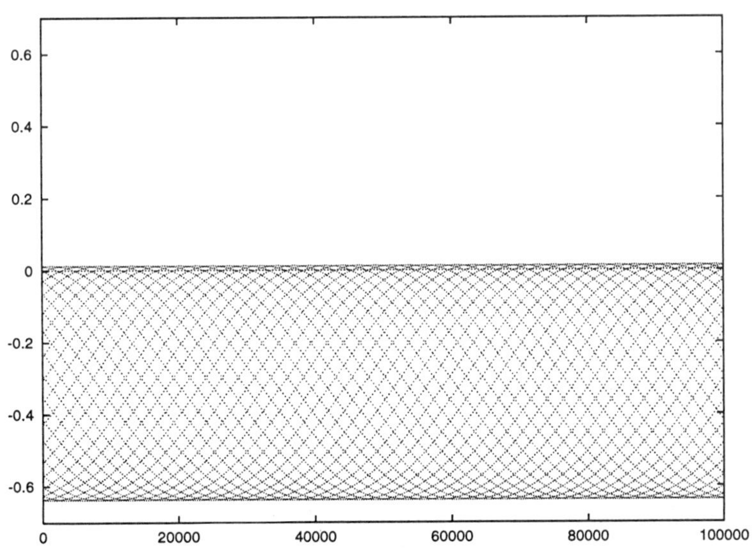

Figure 3 p_t versus time for the same run as Figure 2. The associated variation of the numerical energy (not the extended Hamiltonian) is thus appreciable.

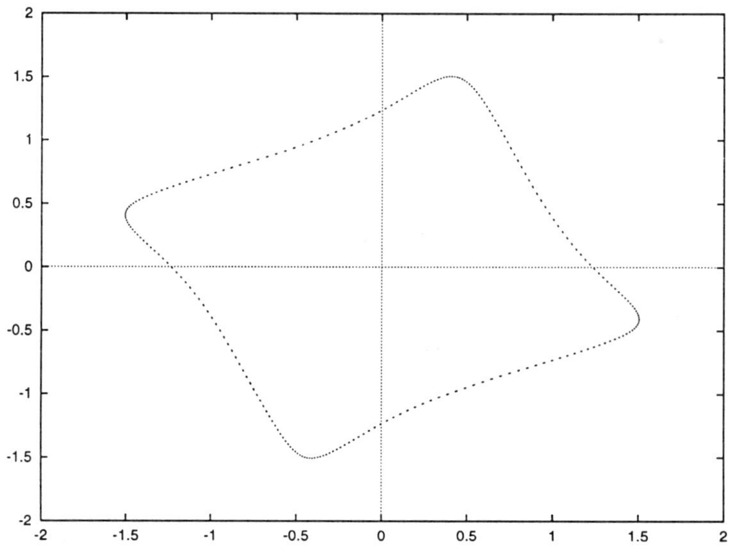

Figure 4 q versus p for the same run as Figure 2. The points do not fill a closed curve because the point is trapped in a high order resonance.

are much simpler to implement and usually are faster than equivalent implicit algorithms while retaining the desirable symplectic transformation properties and energy stability of the implicit algorithms.

4 Additional results

The motion of relativistic particles in a magnetic field obeys dynamics with the Hamiltonian

$$H = \sqrt{m^2c^4 + c^2(p - \frac{eA}{c})^2} \tag{4.1}$$

In order to find an explicit, symplectic integrator in this case one simply observes that energy conservation implies that this Hamiltonian and the Hamiltonian

$$\mathcal{H} = \frac{H^2}{2} \tag{4.2}$$

have the same geometric trajectories, the only difference being that time has to be rescaled by $t \rightarrow t/E$, where E is the value of the energy. If A can be expressed as a polynomial, then $\mathcal{H}$ is a polynomial Hamiltonian that can be integrated by the methods of the previous section.

Note that the composition methods can also be applied to derive unitary integrators for the time advance of quantum mechanical systems; recall the independence of the proof of the generalized midpoint rule and of Yoshida's theorem from the specific form of the bracket. Thus, the time ordered unitary time advance operator can be approximated as

$$Te^{\int_t^{t+h} H(t')dt'} = e^{hH\left(t+\frac{h}{2}\right)} + O\left(h^3\right) \tag{4.3}$$

where we have absorbed the usual factor of $-i$ into the definition of H and where T indicates the time ordering operator. The variable h is the size of the time advance. Using Yoshida's theorem this can be elevated to a higher order advance

$$Te^{\int_t^{t+h} H(t')dt'} = e^{hz_0 H\left(t+h-\frac{z_0}{2}h\right)} e^{hz_1 H\left(t+\frac{1}{2}h\right)} e^{hz_0 H\left(t+\frac{z_0}{2}h\right)} + O\left(h^5\right) \quad (4.4)$$

Thus, if one can find either a unitary approximation to the exponential, such as

$$e^{hH} = (1-\frac{h}{2}H)^{-1}(1+\frac{h}{2}H) + O\left(h^3\right), \quad (4.5)$$

or enough exactly integrable quantum mechanical systems to approximate the system of interest, composition methods can be used to produce higher order unitary time integration. For discrete quantum mechanical systems, such as spin systems, higher order unitary integrators can thus be immediately applied. For a general quantum mechanical system a preliminary, but very difficult and as yet unsolved problem, is to truncate the infinite dimensional description of the wave function to a manageable finite dimensional description in such a way as to preserve a unitary time evolution for the finite dimensional representation.

There are generalized Hamiltonian systems called Lie-Poisson and Poisson Hamiltonian systems; basically they result from a generalized bracket that is allowed to be degenerate, unlike the Poisson bracket which is nondegenerate. (The degeneracy leads to additional conserved quantities, called Casimirs, regardless of the form of the Hamiltonian.) Lie-Poisson and Poisson Hamiltonian systems are interesting because they include rigid body systems, the Vlasov-Poisson equations (collisionless Boltzman equations), the Vlasov-Maxwell equations, ideal fluid dynamics, ideal MHD, general relativity, ideal elastodynamics, and many others; see [19] for a review of the general theory and a treatment of some particular systems. It is natural to look for Lie-Poisson integrators for these systems that exactly preserve the bracket structure. Integrators based on the generating function techniqes of implicit symplectic integrators were found by Ge and Marsden [12], and extensions and improvements to their algorithm have been published in [7] and [3]. In some cases it is possible to find first order explicit integrators for these systems and then use the midpoint rule and Yoshida's trick to elevate them to explicit higher order Lie-Poisson integrators as was done for conventional Hamiltonian systems. These Lie-Poisson integrators exactly preserve all Casimirs of the system and seem to have the same excellent energy stability as the conventional symplectic integrators.

5 Summary

Symplectic integrators are numerical time advance algorithms that are elements of the same symplectic group of transformations as the flows of the original system. The fundamental motivation for using a symplectic integrator rather than a general algorithm was nicely stated by Hamming [16]:"...an algorithm which transforms properly with respect to a class of transformations is more basic than one that does not. In a sense the invariant algorithm attacks the problem and not the particular representation used..".

In addition to the general principle of invariance of Hamming, further support for the use of symplectic integrators can be found in a number of 'Lagrangian submanifold' stability theorems. In particular, if a Hamiltonian is nonlinear but nearly integrable, the Kolmogorov-Arnold-Moser (KAM) theorem [2], [20], asserts that any sufficiently small symplectic perturbation (such as that determined by a symplectic numerical algorithm) of the Hamiltonian dynamics preserves most of the incommensurate, nondegenerate, invariant tori of the original system, at least for small enough time step. In addition, unstable periodic orbits and lower dimensional tori [15] and their stable and unstable manifolds [17] are structurally stable and will be well represented by symplectic integrators. This global preservation of such structures is indeed observed with symplectic integrators [6].

Of course, the energy stability presented in our examples and observed by many others for most cases tried is yet more evidence that symplectic integrators can be expected to have superior performance in many instances. That energy stability should hold for very long (but not infinite) times is a result of Neishtadt [21], at least for analytic Hamiltonians, which states that the numerical map is exponentially close to the time δt map of a Hamiltonian system with a Hamiltonian function that is close (as a power of δt depending on the order of the integrator) to the original one.

Naturally, one should not simply assume that because one is using a symplectic integrator the results are accurate. Certainly, for large time steps symplectic integrators can give very poor results, and the definition of large time step depends on the Hamiltonian. If the Hamiltonian is resonant (i.e. degenerate), or singular or nearly singular, the time steps required to obtain reliable behavior may be so small as to be impractical. In these cases either conventional or energy-conserving algorithms may have significantly better performance than symplectic integrators. In addition, if the time span of the integration is relatively short, so that absolute accuracy rather than global fidelity is the primary concern, then there is no advantage to using a symplectic integrator and in these cases one will usually want to use a well developed and much better debugged conventional package. Nonetheless, if one is interested in the global behavior of trajectories resulting from very long time integrations of Hamiltonian systems, then symplectic integration algorithms may offer significant advantages over other techniques and should be seriously considered.

References

[1] Abraham, R. and Marsden, J. [1978], *Foundations of Mechanics, second edition.* Benjamin/Cummings Publishing Company, Reading, Mass.

[2] Arnold, V.I. and Avez, A. [1968], *Ergodic Problems of Classical Mechanics.* Benjamin, New York.

[3] Benzel, S., Ge, Z., and Scovel, C. [1993], *Elementary construction of higher order Lie-Poisson integrators.* Physics Letters A, **174**, 229–232.

[4] Channell, P. J. [1983], *Symplectic integration algorithms*, Los Alamos National Laboratory Internal Report AT-6, ATN-83-9.

[5] Channell, P. J. [1986], *Explicit Integration of Kick Hamiltonians in Three Degrees of Freedom*, Los Alamos National Laboratory Internal Report AT-6 : ATN-86-6.

[6] Channell, P. J. and Scovel, J.C. [1990], *Symplectic integration of Hamiltonian systems*, Nonlinearity, **3**, 231–259.

[7] Channell, P. J. and Scovel,J.C. [1991], *Integrators for Lie-Poisson dynamical systems*, Physica D, **50**, 80–88.

[8] De Vogelaere, R. [1956], *Methods of integration which preserve the contact transformation property of the Hamiltonian equations.* Department of Mathematics, University of Notre Dame report **4**.

[9] Feng, K. [1986], *Difference schemes for Hamiltonian formalism and symplectic geometry*, J. Comput. Math., **4**, 279–289.

[10] Forest, E., Bengtsson, J., and Reusch, M.F. [1991], *Application of the Yoshida-Ruth techniques to implicit integration and multi-map explicit integration*, Phys. Lett. A, **158**i, 99–101.

[11] Forest, E. and Ruth, R. [1990], *Fourth-order symplectic integration*, Physica D, **43** 105–117.

[12] Ge, Z. and Marsden, J. [1988], *Lie-Poisson Hamilton-Jacobi theory and Lie-Poisson integrators*, Phys. Lett. A, **133**, 135–139.

[13] Gjaja, I. [1994], *Monomial Factorization of Symplectic Maps*, Particle Accelerators, **43(3)**, 133-144.

[14] Goldstein, H. [1980], *Classical mechanics*, Second edition. Addison-Wesley, Reading.

[15] Graff, S.M. [1974], *On the Conservation of Hyperbolic Invariant Tori for Hamiltonian Systems*, J. Diff. Equations, **15** 1.

[16] Hamming, R.W. [1986], *Numerical Methods for Scientists and Engineers*, Second edition, p. 73, Dover, New York.

[17] Hirsch, M.W., Pugh, C., and Shub, M. [1977], *Invariant Manifolds*, Lec. Notes in Math., Vol. 583, Springer, New York.

[18] Lichtenberg, A. J. and Lieberman, M. A. [1982], *Regular and Stochastic Motion.* Springer-Verlag, New York.

[19] Marsden, J. E., Weinstein, A., Ratiu, T., Schmid, R. and Spencer, R.G., *Hamiltonian Systems With Symmetry, Coadjoint Orbits, and Plasma Physics*, Proc. IUTAM-ISIMM Symposium on Modern Developments in Analytical Mechanics, Atti Accad. Sci. Torino, Suppl., **117**, 289–340.

[20] Moser, J. [1973], *Stable and Random Motions in Dynamical Systems.* Ann. of Math. Studies, **77**, Princeton.

[21] Neishtadt, A.I. [1984], *The separation of motions in systems with rapidly rotating phase*, J. Appl. Math. Mech., **48**, 133–139,

[22] Ruth, R. [1983], *A canonical integration technique.* IEEE Transactions on Nuclear Science, **30**, 2669–2671.

[23] Saha, P. and Tremaine, S. [1992], *Symplectic integrators for solar system dynamics*, Astron. J., **104**, 1633–1640.

[24] Sanz-Serna, J. M. [1992], *Symplectic integrators for Hamiltonian problems: An overview*, Acta Numerica 1992, 243–286, Cambridge University Press, Cambridge.

[25] Wisdom, J. [1982], *The Origin of the Kirkwood Gaps: A Mapping for Asteroidal Motion Near the 3/1 Commensurability*, Astron. J., **87**, 577–593.

[26] Wisdom, J. and Holman, M. [1991], *Symplectic maps for the N-body problem*, Astro. J., **102** 1528–1538.

[27] Yoshida, H. [1990], *Construction of higher order symplectic integrators*, Phys. Lett. A, **150**, 262–268.

Fields Institute Communications
Volume **10**, 1996

Symplectic Maps and Computation of Orbits in Particle Accelerators

Alex J. Dragt and Dan T. Abell
Physics Department and Center for Theoretical Physics
University of Maryland
College Park, MD 20742
USA

Abstract. To good approximation, particularly for proton machines, the orbits of charged particles in accelerators are described by Hamiltonian flows. In large storage rings charged particles are expected to be held routinely for eight hours or more. During this time such particles make 10^8 or more revolutions, 10^{10} betatron oscillations, and 10^{12} passages through bending, focusing, and orbit correction magnets. To ensure the proper performance of a storage ring, it is necessary to determine particle orbit behavior and stability over this large number of revolutions. As a result of the large number of oscillations and magnet passages involved, the usual numerical integration methods do not work for this purpose. An alternate approach is to compute an approximate one-turn map, and then iterate this map. A description will be given of how such maps can be computed and the importance of representing and iterating them in such a manner that the symplectic condition is maintained.

1 Introduction

The European Large Hadron Collider (LHC) project has recently been approved for construction at CERN. It is designed to answer the next set of questions concerning the fundamental nature of matter. We will describe below the considerations having to do with orbit stability in the LHC. Similar considerations also hold for other large proton storage rings such as the Tevatron at Fermilab and any projected future storage rings designed to operate at energies beyond that of the LHC.

The LHC will be a proton-proton colliding beam machine consisting of two storage rings made of superconducting magnets. Each ring will be approximately 30 km in circumference, and will be required to hold protons for 8 hours or more. Since these protons travel at nearly the speed of light c, during this time they will

---

1991 *Mathematics Subject Classification.* Primary: 58F05; Secondary: 34C35, 58F08.

make N trips around the ring (turns) with N given by the relation

$$\begin{aligned} N &= \text{8hrs/(circulation time)} \simeq 8 \times 60 \times 60\text{sec}/(30 \text{ km/c}) \\ &\simeq 8 \times 60 \times 60 \times 3 \times 10^8/(30 \times 10^3) \simeq 3 \times 10^8. \end{aligned}$$

During each turn, a typical stored proton in an LHC ring will make approximately 70 betatron oscillations. Thus, during a storage time of 8 hours, there will be approximately 2×10^{10} betatron oscillations. Finally, each ring will contain approximately 5000 magnets for bending, focusing, and linear and nonlinear orbit corrections. Consequently, during a storage time of 8 hours, a proton in the LHC will make approximately 1.5×10^{12} magnet passages. As indicated above, similar figures hold for other storage rings as well.

To ensure the proper performance of a storage ring, it is necessary to determine particle orbit behavior and stability over these large numbers of betatron oscillations and magnet passages. We note that determining orbit stability over 10^{10} oscillations is roughly equivalent to studying the stability of the solar system since the Big Bang. Because of the excessive computation that would be required and the attendant problem of roundoff error, the usual numerical integration methods do not work for this purpose. An alternate approach is to compute an approximate one-turn map, and then to iterate this map.

By intent, an ideal storage ring has a closed orbit called the *design orbit.* Suppose canonical coordinates are introduced for single-particle phase space in such a way that the initial conditions for the design orbit correspond to all coordinates being zero. These coordinates are conveniently taken to be $q_x, q_y, \tau, p_x, p_y, p_\tau$. Here q_x, q_y denote transverse deviations from the design orbit, and p_x, p_y are the associated conjugate momenta. The quantity τ is the deviation in the time of flight. Finally p_τ, which can be shown to be canonically conjugate to τ, is the negative of the energy deviation from the design energy [17, 18]. For further discussion it is convenient to denote these six phase-space coordinates by a six-dimensional vector z,

$$(z_1, z_2, z_3, z_4, z_5, z_6) = (q_x, q_y, \tau, p_x, p_y, p_\tau).$$

Also, upon occasion, we will find it convenient to use coordinates $q_1, q_2, q_3, p_1, p_2, p_3$ defined by the relation

$$(q_1, q_2, q_3, p_1, p_2, p_3) = (q_x, q_y, \tau, p_x, p_y, p_\tau).$$

Suppose a proton is launched with *initial* conditions z^i. Then, after one turn or trip about the storage ring, it will have *final* conditions z^f. Since the initial conditions uniquely determine the final conditions, there exists a one-turn map $\mathcal{M}$ with the property

$$z^f = \mathcal{M} z^i. \tag{1.1}$$

This map will be called the *one-turn transfer* map [18, 25].

In addition, we assume that the transfer map $\mathcal{M}$ remains unchanged from turn to turn. (It can be shown that this is possible even if there are radio frequency bunching cavities in the ring. However, it is not true if there is AC ripple in the magnet power supplies.) Then the effect of many turns (say 10^8) can in principle be found by repeated application of $\mathcal{M}$.

In this paper we describe and/or explore various methods for approximating $\mathcal{M}$ in such a way that the approximation is accurate, symplectic, and easy to iterate.

2 Symplectic Condition

Suppose we now focus our attention on the orbit of a particle in a storage ring and consider how small variations in the initial conditions z^i produce small variations in the final conditions z^f. The connection between these variations is described by the Jacobian matrix

$$M_{ab}(z^i) = \partial z^f_a / \partial z^i_b. \tag{2.1}$$

Note that the Jacobian matrix generally depends on the orbit in question and thus, as indicated, is a function of the initial conditions z^i.

In addition, we assume that the orbit in question results from following a Hamiltonian flow. (This is a very good approximation for proton machines, for which synchrotron radiation effects can be neglected, and is a very useful first approximation for electron machines.) Then it can be shown that the Jacobian matrix $M(z^i)$ must satisfy the *symplectic condition*,

$$M(z^i)JM^T(z^i) = J, \ \forall z^i. \tag{2.2}$$

Here M^T denotes the transpose of M, and J is the matrix

$$J = \begin{pmatrix} 0 & I \\ -I & 0 \end{pmatrix}. \tag{2.3}$$

Each entry in J is a 3×3 block: 0 denotes a 3×3 zero matrix, and I denotes a 3×3 identity matrix [16, 18].

Observe that in principle the left-hand side of (2.2) should depend on z^i since M depends on z^i. However, the symplectic condition requires that in fact the combination MJM^T be independent of z^i. Thus, the symplectic condition places strong restrictions on M and, correspondingly, on the underlying transfer map $\mathcal{M}$. We will call any matrix satisfying (2.2) a *symplectic matrix*. [We remark for later reference that 6×6 symplectic matrices form a group called $Sp(6)$.] Similarly, we will call any map whose Jacobian matrix satisfies (2.2) a *symplectic map*. Note that the symplectic conditon (2.2) is quadratic and therefore nonlinear [16, 18].

There is another useful way of viewing the symplectic condition. Suppose f and g are any two functions of z^i. Then the Poisson bracket of f and g is defined by the relation

$$[f, g]_{z^i} = \sum_j (\partial f / \partial q^i_j)(\partial g / \partial p^i_j) - (\partial f / \partial p^i_j)(\partial g / \partial q^i_j).$$

Thanks to the properties of J, see (2.3), the Possion bracket can also be written in the form

$$[f, g]_{z^i} = \sum_{cd} (\partial f / \partial z^i_c) J_{cd} (\partial g / \partial z^i_d).$$

In particular, we find the so called *fundamental* Poisson brackets

$$[z^i_a, z^i_b]_{z^i} = J_{ab}.$$

Now compute the Poisson bracket $[z^f_a, z^f_b]$ where z^f_a and z^f_b are regarded as functions of z^i. Then, from the Jacobian matrix definition (2.1) and the symplectic condition

(2.2), we find the result

$$\begin{aligned}[z_a^f, z_b^f]_{z^i} &= \sum_{cd} (\partial z_a^f / \partial z_c^i) J_{cd} (\partial z_b^f / \partial z_d^i) \\ &= \sum_{cd} M_{ac} J_{cd} M_{bd} = (MJM^T)_{ab} = J_{ab} = [z_a^f, z_b^f]_{z^f}. \end{aligned} \quad (2.4)$$

We see that symplectic maps preserve the fundamental Poisson brackets. From this property it can be shown that symplectic maps preserve all Poisson brackets [16].

3 Taylor Maps and Symplectic Jets

Suppose we attempt to represent $\mathcal{M}$ explicitly by expanding the final conditions z^f as a Taylor series in the initial conditions z^i. Then we obtain a relation of the form

$$z_a^f = \mathcal{M} z_a^i = K_a + \sum_b R_{ab} z_b^i + \sum_{bc} T_{abc} z_b^i z_c^i + \sum_{bcd} U_{abcd} z_b^i z_c^i z_d^i + \cdots . \quad (3.1)$$

Thanks to work originally done by Poincaré, we know that under quite general conditions such an expansion exists and has a nonzero domain of convergence [2]. The constant terms K_a in (3.1) describe translations in phase space caused by imperfections in the ring arising from errors of misplacement, misalignment, and mispowering of various magnets. (Note that for an ideal machine $z^i = 0$ should lead to $z^f = 0$ since by construction these are the initial conditions for the *closed* design orbit.) The matrix R describes the linear properties of the ring. Finally, the higher order coefficients T, U, etc. describe nonlinear properties of the ring [16].

Suppose the Taylor representation (3.1) is used to compute the Jacobian matrix (2.1). Because this matrix must satisfy the symplectic condition, the various coefficients in the Taylor representation (3.1) are *not* independent, but rather are interrelated by a complicated collection of nonlinear conditions. For this reason, a Taylor series is not a particularly convenient way to describe or parameterize a symplectic map.

In actual practical calculation, it is possible to obtain the Taylor coefficients in (3.1) only up to some order. Imagine the map $\mathcal{M}$ is approximated by truncating the series (3.1) at some order, say the order up to which the coefficients are known, or perhaps some lower order. Such a truncated Taylor map is called a *jet.*[1] In particular, since the map $\mathcal{M}$ is assumed to be symplectic, its truncated Taylor approximation is called a *symplectic jet.* Now truncation is equivalent to setting all coefficients beyond some order exactly equal to zero. However, because of its nonlinearity, the symplectic condition interrelates coefficients of different orders. Thus, truncating a Taylor series map generally *violates* the symplectic condition.

Suppose we define a truncated Taylor map $\mathcal{T}_m$ by writing

$$\bar{z}_a^f = \mathcal{T}_m z_a^i = \sum_{\ell=0}^{m} t_a(\ell, z^i). \quad (3.2)$$

[1]More precisely, a jet is an equivalence class of maps, with the equivalence relation being that of having equal derivatives at some point (say the origin) through some order. See [30].

Here each function $t_a(\ell, z^i)$ is a homogeneous polynomial of degree ℓ in the variables z^i. Such a truncated Taylor map is called an *m-jet.* If we use such a map we find, in contrast to (2.4), the Poisson bracket result

$$[\mathcal{T}_m z_a^i, \mathcal{T}_m z_b^i]_{z^i} = [\tilde{z}_a^f, \tilde{z}_b^f]_{z^i} = J_{ab} + r_{ab}(m, z^i), \tag{3.3}$$

where $r_{ab}(m, z^i)$ denotes a remainder, generally nonzero, consisting of terms of degree m through degree $(2m-2)$. It can be shown that (3.3) is a necessary and sufficient condition for an m-jet to be a symplectic m-jet [16, 18, 21].

4 Lie Methods

Lie methods provide another way of representing symplectic maps. Let $f(z) = f(q,p)$ be some specified function of the phase-space variables z. Associated with each f is a *Lie operator*, denoted by the symbol $:f:$, which is a differential operator defined by the rule

$$:f: \stackrel{def}{=} \sum_j (\partial f/\partial q_j)(\partial/\partial p_j) - (\partial f/\partial p_j)(\partial/\partial q_j).$$

In particular, if $:f:$ acts on any phase-space function g, one finds the result

$$:f: g = \sum_j (\partial f/\partial q_j)(\partial g/\partial p_j) - (\partial f/\partial p_j)(\partial g/\partial q_j) = [f, g].$$

Powers of $:f:$ are defined by taking repeated Poisson brackets. For example $:f:^2$ is defined by the relation

$$:f:^2 g = [f, [f, g]].$$

Finally, $:f:$ to the power zero is defined to be the identity operator,

$$:f:^0 g = g.$$

Since powers of $:f:$ have been defined, it is also possible to define power series in $:f:$. Of special importance is the power series $\exp(:f:)$. This object is called the *Lie transformation* associated with $:f:$. The Lie transformation is also an operator and is formally defined by the exponential series

$$\exp(:f:) = \sum_{\ell=0}^{\infty} :f:^\ell / \ell!. \tag{4.1}$$

In detail, the action of $\exp(:f:)$ on any function g is given by the rule

$$\exp(:f:)g = g + [f, g] + [f, [f, g]]/2! + \cdots . \tag{4.2}$$

Suppose $f(z^i)$ is any function of the initial phase-space variables z^i. Consider a set of final phase-space variables z^f defined by the rule

$$z_a^f = \exp(:f:)z_a^i. \tag{4.3}$$

It can be shown that the relation between z^i and z^f given by (4.3) is a symplectic map. Upon comparing (1.1) and (4.3), we can write the symplectic map $\mathcal{M}$ in the operator form

$$\mathcal{M} = \exp(:f:).$$

What we have learned is that Lie transformations produce an endless supply of symplectic maps [16, 18, 21].

Conversely, suppose $\mathcal{M}$ is any symplectic map having a Taylor series expansion of the form (3.1). Then, $\mathcal{M}$ can be written uniquely in the factored product form [16, 18, 21]

$$\mathcal{M} = \exp(:f_1:)\exp(:f_2:)\exp(:f_3:)\exp(:f_4:)\cdots . \tag{4.4}$$

Here each function f_m is a homogeneous polynomial of degree m in the variables z^i.

By definition, a knowledge of $\mathcal{M}$ is equivalent to a knowledge of the relation between the initial conditions z^i and the final conditions z^f. And, according to (4.4), a knowledge of $\mathcal{M}$ amounts to determining certain homogeneous polynomials f_1, f_2, f_3, f_4, etc. Moreover, in contrast to the terms K_a, R_{ab}, T_{abc}, etc., in the Taylor series (3.1), the polynomials f_m are all independent. In other words, any set of polynomials when employed in (4.4) produces a symplectic map. Conversely any symplectic map can be written uniquely in the form (4.4). Finally, the product (4.4) can be truncated at any stage to provide an approximation to $\mathcal{M}$, and the result will still be a symplectic map [16, 18, 21].

Each factor in (4.4) has a direct physical interpretation. It can be shown that the factor $\exp(:f_1:)$ reproduces the constant terms K_a in the Taylor expansion (3.1), and thus produces a simple translation in phase space. Next, the factor $\exp(:f_2:)$ produces a linear transformation corresponding to the matrix R. Finally, the factor $\exp(:f_3:)$ produces second (and possibly higher) order nonlinear terms corresponding to the coefficients T_{abc}, U_{abcd}, etc.; and the factor $\exp(:f_4:)$ produces third (and possibly higher) order nonlinear terms corresponding to the coefficients U_{abcd}, etc.; etc. In general, given any symplectic m-jet (3.2), the homogeneous Lie generating polynomials $f_1(z^i), f_2(z^i), \cdots f_{m+1}(z^i)$ are uniquely specified by the polynomials $t_a(0, z^i), t_a(1, z^i), \cdots t_a(m, z^i)$, and vice versa [16, 18, 21].

5 Symplectic Completion of Symplectic Jets

Let $\mathcal{M}$ be a one-turn transfer map. Use the factorization (4.4) to write $\mathcal{M}$ in the form

$$\mathcal{M} = \exp(:f_1:)\exp(:f_2:)\mathcal{N}, \tag{5.1}$$

where $\mathcal{N}$ is the nonlinear map

$$\mathcal{N} = \exp(:f_3:)\exp(:f_4:)\cdots . \tag{5.2}$$

Our goal is to iterate $\mathcal{M}$ to compute turn-by-turn behavior. In other words, we wish to compute the quantities z^ℓ defined by the relation

$$z^{\ell+1} = \mathcal{M}z^\ell, \text{ where } z^0 = z^i, \text{ with } \ell = 0, 1, 2, \cdots .$$

There is no difficulty in computing the effects of maps of the form $\exp(:f_1:)$ and $\exp(:f_2:)$ since they simply amount to translations in and linear transformations of phase space, respectively. Consequently, these operations can easily be carried out exactly (to machine precision) on a computer without any violation of the symplectic condition. The same is not true of $\mathcal{N}$, the nonlinear part of the map, and we shall therefore devote our attention to $\mathcal{N}$.

In general $\mathcal{N}$ is available only in the form of a symplectic m-jet approximation. For example, with some effort and the aid of Lie algebraic, differential algebraic, and special numerical integration methods, it is possible to obtain an m-jet in all six variables with $m \simeq 11$ for a storage ring such as the LHC. See [5, 7, 8, 10, 22, 40, 41, 51].

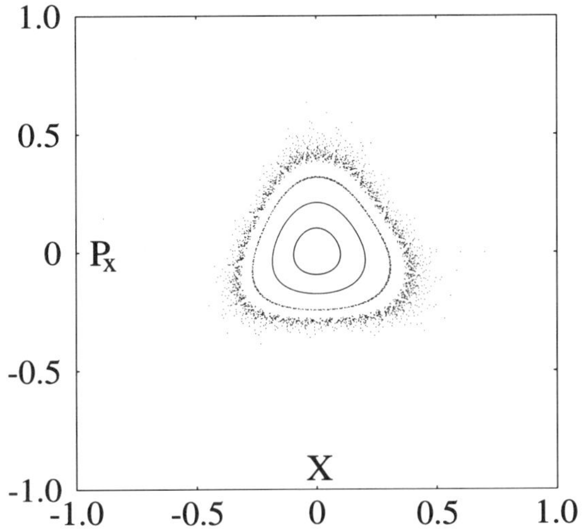

Figure 1 Truncated Taylor Approximation

As described in Section 3, an m-jet approximation is generally not symplectic. Figure 1 shows the result of iterating a simple two-dimensional symplectic map for which $\mathcal{N}$ has been approximated by a symplectic 8-jet. By contrast, Figure 2 shows the result of iterating the exact map.[2] In both cases, the map is applied repeatedly to seven different initial conditions. It can be shown that the 8-jet gives a good approximation to $\mathcal{N}$ when applied once. However, it is evident from a comparison of Figures 1 and 2 that (nonsymplectic) errors accumulate over several hundred iterations, and produce unsatisfactory results.[3] What we have learned from this and many other examples is that it is very important to respect the symplectic condition.

The stage has now been set for the presentation of a fundamental and very interesting problem: Given a symplectic m-jet (which in general is all we are able to compute), how can one find an exactly symplectic map that has this m-jet as a truncated Taylor approximation? We shall call this the problem of *symplectic completion* of symplectic jets.

---

[2]The exact map $\mathcal{M}$ is of the form (5.1) with $f_1 = 0, f_2 = -(66/180)(\pi/2)(p^2 + q^2)$, and $\mathcal{N} = \exp(: qp^2 :)$. The 8-jet map approximation consists of using the same f_1 and f_2, but $\mathcal{N}$ is replaced by the first 8 terms in its Taylor series expansion. See [24]. The points outside the (apparently) closed curves in Figure 2 iterate to infinity.

[3]The orbits under repeated iteration, for all seven initial conditions, move ever closer to the origin. That is, in some important vicinity of the origin, the truncated Taylor map is not area preserving (as a 2-dimensional symplectic map must be), but rather is area decreasing. For a further discussion of the need to symplectify Taylor maps, see [35].

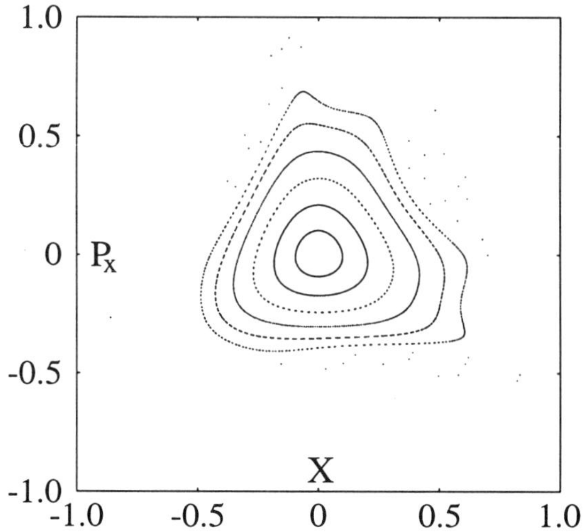

Figure 2 Exact Map

The methods of Section 4 provide one obvious solution to this problem. Suppose $\mathcal{N}$ is known to have the symplectic m-jet $\mathcal{T}_m$, with

$$\bar{z}_a^f = \mathcal{T}_m z_a^i = z_a^i + \sum_{\ell=2}^{m} t_a(\ell, z^i), \tag{5.3}$$

as a truncated Taylor approximation. As we outlined at the end of Section 4, we can construct the Lie generating polynomials $f_3(z^i), \cdots f_{m+1}(z^i)$ from the $t_a(\ell, z^i)$. Having done so, we can take as an *approximation* to $\mathcal{N}$ the symplectic map

$$\mathcal{N}_{\mathrm{a}} = \exp(:f_3:)\exp(:f_4:)\cdots\exp(:f_{m+1}:). \tag{5.4}$$

Evidently, $\mathcal{N}_{\mathrm{a}}$ as given by (5.4) provides a symplectic completion to the symplectic m-jet (5.3). However, the map (5.4) has the defect that it is very time consuming to evaluate when acting on z^i since, according to (4.1) or (4.2), the summation of an infinite series is required. What really interests us are symplectic completions of symplectic m-jets that can be evaluated relatively rapidly on a computer.

6 Topological Assumptions

According to Section 5, approximations to $\mathcal{M}$ or $\mathcal{N}$ that violate the symplectic condition, as m-jets do, are generally not satisfactory for the many iterations needed for long-term stability studies. (Of course, one might imagine taking m to be so large that the m-jet approximation would be symplectic to machine precision in the phase-space region of interest. In this case, however, the m-jet would be very time-consuming to evaluate because of the very large number of terms involved, and would therefore again be unsatisfactory.) Implicit in our discussion is the hope that symplectic approximations to $\mathcal{N}$ or $\mathcal{M}$ would be better.

There are two sanguine (perhaps almost to the point of naiveté) and yet remarkably tacit assumptions made in the design and construction of accelerators and storage rings. The first, made by machine builders, is that if two machines are nearly the same, their performance (including long-term orbit stability) should be nearly the same. Without this assumption, it would be impossible to proceed since construction errors are unavoidable at some level. The second, made by accelerator theorists, is that analytical/numerical models of machine behavior, despite their approximate nature, still have relevance for real machines. Without this assumption, it would be impossible to design machines. From a mathematical perspective, what is being assumed in either case is that if two symplectic maps are *close* (in some not yet precisely defined sense), then the behavior (including long-term behavior) of the systems described by these maps should be *close* (*i.e.*, nearly the same in again some not yet precisely defined sense). The purpose of this section is to formulate various topological assumptions (notions of closeness) in a more precise way.

Suppose $\mathcal{M}$ is written in the product form (5.1). Our attention will focus on the nonlinear part $\mathcal{N}$. We begin by defining the quantity

$$n(z) = \|\mathcal{N}z - z\|/\|z\|, \tag{6.1}$$

where $\| * \|$ denotes the usual Euclidean norm [20]. Since this quantity measures the relative size of the change in z produced by $\mathcal{N}$, we call it the *nonlinearity* of $\mathcal{N}$ at the point z. According to (5.3), $\mathcal{N}$ generally differs from the identity by terms of second order in the components of z, and hence $n(z)$ should scale as $\|z\|$. We shall be interested in maps $\mathcal{N}$ and domains of phase space $\mathcal{D}$ such that the nonlinearity $n(z)$ is not too large in $\mathcal{D}$. In other words, we might hope to achieve some success in treating nonlinear behavior if, say, $n < (1/4)$ in $\mathcal{D}$.

Let $\mathcal{T}_m$ be the symplectic m-jet of $\mathcal{N}$, and let $\mathcal{N}_a$ be an approximation to $\mathcal{N}$ obtained by *some* symplectic completion of $\mathcal{T}_m$. [Here we do not necessarily assume that $\mathcal{N}_{\mathrm{a}}$ is of the form (5.4).] Of course, of ultimate interest to us is how much $\mathcal{N}_a$ differs from $\mathcal{N}$. A measure of this difference is

$$\epsilon^a_m(z) = \|\mathcal{N}z - \mathcal{N}_{\mathrm{a}}z\|/\|z\|.$$

Since $\mathcal{N}_{\mathrm{a}}$ agrees with $\mathcal{T}_m$, and hence with $\mathcal{N}$, through terms of degree m, the quantity $\epsilon^{\mathrm{a}}_m(z)$ must scale as $\|z\|^m$. Unfortunately the quantity $\epsilon^a_m(z)$ is usually difficult to evaluate because $\mathcal{N}$ in general is not known exactly, or is very difficult to calculate exactly. It is sometimes possible, however, to obtain a useful bound on $\epsilon^a_m(z)$ by considering several other related quantities.

Consider the difference between $\mathcal{N}$ and $\mathcal{T}_m$ at a point z. It is conveniently measured by

$$\epsilon^T_m(z) = \|\mathcal{N}z - \mathcal{T}_m z\|/\|z\|.$$

Evidently $\epsilon^T_m(z)$ should scale as $\|z\|^m$. Again a knowledge of $\mathcal{N}$ is required, and therefore $\epsilon^T_m(z)$ is also difficult to evaluate. However, we notice that according to (3.3) the map $\mathcal{T}_m$ is generally not symplectic. Its departure from symplecticity at the point z is conveniently measured by

$$\delta^T_m(z) = \|[\mathcal{T}_m z_a, \mathcal{T}_m z_b] - J_{ab}\| = \|r_{ab}(m, z)\|.$$

Here $\| * \|$ denotes some kind of matrix norm: for example, the maximum column sum norm [13]. The quantity $\delta^T_m(z)$ also scales as $\|z\|^m$. In general, we expect the quantities $\epsilon^T_m(z)$ and $\delta^T_m(z)$ to be comparable,

$$\epsilon^T_m(z) \approx \delta^T_m(z). \tag{6.2}$$

It is useful to consider as well the difference between $\mathcal{N}_a$ and its underlying symplectic jet $\mathcal{T}_m$. To do so, we define

$$\epsilon_m^{Ta}(z) = \|\mathcal{T}_m z - \mathcal{N}_a z\|/\|z\|. \tag{6.3}$$

It is easily checked that $\epsilon_m^{Ta}(z)$ also scales as $\|z\|^m$.

We can now obtain the following inequalities:

$$\epsilon_m^{a}(z) \leq \epsilon_m^{T}(z) + \epsilon_m^{Ta}(z), \tag{6.4}$$
$$\epsilon_m^{Ta}(z) \leq \epsilon_m^{T}(z) + \epsilon_m^{a}(z), \tag{6.5}$$
$$\epsilon_m^{T}(z) \leq \epsilon_m^{a}(z) + \epsilon_m^{Ta}(z). \tag{6.6}$$

To prove (6.4), for example, we use the triangle inequality to write

$$\begin{aligned} \|\mathcal{N}z - \mathcal{N}_a z\| &= \|\mathcal{N}z - \mathcal{T}_m z + \mathcal{T}_m z - \mathcal{N}_a z\| \\ &\leq \|\mathcal{N}z - \mathcal{T}_m z\| + \|\mathcal{T}_m z - \mathcal{N}_a z\|. \end{aligned}$$

The inequality (6.4) is of particular interest. From (6.3) we see that $\epsilon_m^{Ta}(z)$ depends only on the jet $\mathcal{T}_m$ and its completion $\mathcal{N}_a$. Suppose we also believe (6.2), which claims that $\epsilon_m^{T}(z)$ can also be estimated from a knowledge solely of the jet $\mathcal{T}_m$. Then from (6.4a) we see that $\epsilon_m^{a}(z)$, which is the quantity of ultimate interest, can also be estimated simply from a knowledge of $\mathcal{T}_m$. In other words, to the extent that (6.2) is correct, one need not know $\mathcal{N}$ to estimate $\epsilon_m^{a}(z)$.

We are now ready to specify more precisely what would be desirable in an approximation $\mathcal{N}_a$ to the symplectic map $\mathcal{N}$ based on its symplectic jet $\mathcal{T}_m$. Select a phase-space domain $\mathcal{D}$ of interest, and define quantities $\epsilon_m^T, \delta_m^T, \epsilon_m^a$, and ϵ_m^{Ta} by the relations

$$\epsilon_m^T = \max_{z\in\mathcal{D}} \epsilon_m^T(z), \quad \delta_m^T = \max_{z\in\mathcal{D}} \delta_m^T(z), \quad \epsilon_m^a = \max_{z\in\mathcal{D}} \epsilon_m^a(z), \quad \epsilon_m^{Ta} = \max_{z\in\mathcal{D}} \epsilon_m^{Ta}(z).$$

What we then desire is that $\mathcal{N}_a$ satisfy the following *requirements*:

1. $\mathcal{N}_a$ should be exactly symplectic.
2. The quantity ϵ_m^{Ta} should be comparable to ϵ_m^T, which in turn is usually comparable to δ_m^T. That is, the symplectic completion process should make only minimal changes to $\mathcal{T}_m$. According to (6.4), this requirement also keeps ϵ_m^a comparable to ϵ_m^T.
3. The map $\mathcal{N}_a$ should be easy to evaluate exactly (to machine precision). That is, $\mathcal{N}_a z$ should be easily and exactly computable.

We observe that $\mathcal{N}_a$ as given by (5.4) satisfies requirement 1, seems in practice to satisfy requirement 2, and fails miserably to satisfy requirement 3. In the next sections we will describe other methods of approximation that come closer to satisfying all three requirements.

7 Use of Generating Function

Introduce the notation q, p for the components of z^i, and Q, P for the components of $\tilde{z}^f$. Let $F(Q,p)$ be any function of the variables Q, p. Then, as is well known, the solution to the implicit equations

$$q_j = \partial F/\partial p_j, \tag{7.1}$$

$$P_j = \partial F / \partial Q_j, \tag{7.2}$$

produces a symplectic map $\mathcal{N}_a$ with the property

$$\tilde{z}^f = \mathcal{N}_a z^i. \tag{7.3}$$

Suppose F, which will be referred to as a *generating function*, is of the form

$$F = \sum_j Q_j p_j + G^{m+1}(Q, p). \tag{7.4}$$

Here G^{m+1} is a polynomial in Q, p consisting of terms having degree 3 through degree $m + 1$. Then (7.1) and (7.2) take the form

$$Q_j = q_j - \partial G^{m+1} / \partial p_j = q_j - G^{m+1}_{p_j}(Q, p), \tag{7.5}$$

$$P_j = p_j + \partial G^{m+1} / \partial Q_j = p_j + G^{m+1}_{Q_j}(Q, p). \tag{7.6}$$

Again, the relations (7.5) and (7.6) are implicit. Imagine they are made explicit, to give $Q_j(q, p)$ and $P_j(q, p)$, and that they are then expanded in a Taylor series in the variables $z^i = (q, p)$. Now demand that this expansion have the symplectic jet $\mathcal{T}_m$ given by (5.3). In other words, if the Taylor series for $\mathcal{N}_a$ is truncated beyond terms of degree m, this truncated series should agree with (5.3). It can be shown that this demand can always be met, and that it specifies G^{m+1} uniquely. Consequently, the map $\mathcal{N}_a$ given by (7.1) through (7.4) provides a symplectic completion of the symplectic jet $\mathcal{T}_m$.

How well does this symplectic completion or *symplectification* work? Does it meet the three requirements set out in Section 6? By construction, it is exactly symplectic and therefore satisfies requirement 1. Experience indicates that in practice it also satisfies requirement 2, provided the nonlinearity $n(z)$ given by (6.1) is not too large. For example, Figure 3 shows the result of iterating the simple two-dimensional map of Figures 1 and 2 in the case that the 8-jet is symplectified using a G^9 generating function. Evidently Figures 2 and 3 are remarkably similar even when the nonlinearity is quite large.

Requirement 3 is somewhat more problematic. The calculation of G^{m+1} from $\mathcal{T}_m$ presents no particular problem, and in any case needs to be done only once. However, we observe yet again that (7.5) and (7.6) are implicit. What is done for each application of $\mathcal{N}_a$ is to solve (7.5) numerically to machine precision by use of Newton's method to give the $Q_j(q, p)$. Once the Q_j are known, the P_j are found (again numerically) by direct evaluation of (7.6).

Experience shows that in practice the solution of (7.5) by Newton's method is reasonably fast. For example, using a super computer, it is currently possible to perform about 5000 applications of $\mathcal{N}_a$ per second. Consequently, 10^8 applications of $\mathcal{N}_a$ requires about $(10^8/5000)$ seconds $\simeq 5.6$ hours. We see that by this method it is possible to follow the successive turns of a proton in a storage ring somewhat faster than real time.

We conclude that symplectic completion by the use of a generating function F of the form (7.4) is reasonably satisfactory.[4] Indeed, it is currently the only

[4]Generating functions were first used for the symplectification of accelerator maps in the code MaryLie. For further information, see [6, 9, 11, 14, 15, 19, 25, 37, 50]. Generating functions have been used in a somewhat similar context by Berg *et al.* [4]. They parametrize phase space using

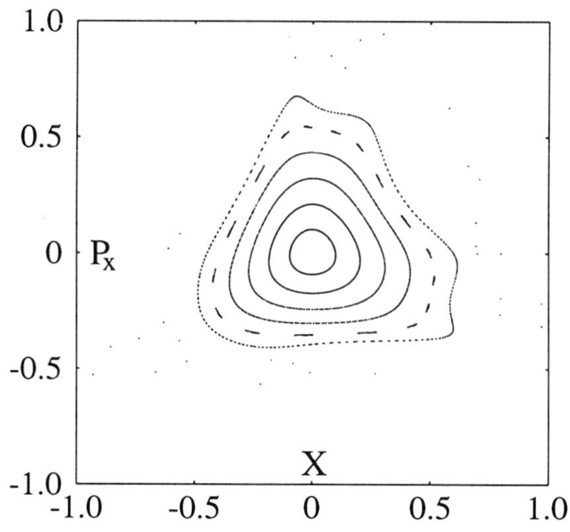

Figure 3 Generating Function Approximation

available procedure for symplectic completion that is even reasonably satisfactory. However, it is not entirely satisfactory. We wish to have a faster procedure. Also, the required solution of (7.5) by Newton's method is somewhat awkward since the Newton iteration process needs to be monitored for convergence. For these and other reasons it is worthwhile to search for other methods of symplectic completion.

8 Cremona Maps

We define a *Cremona* map to be a symplectic map whose Taylor series expansion terminates. In other words, a Cremona map is a polynomial symplectic map. Evidently the product of two Cremona maps is again a Cremona map. Also, it can be shown that the inverse of a Cremona map is again a Cremona map.[5] Consequently, Cremona maps form a group.

Cremona maps have the great virtue that they and their products and their inverses can be evaluated easily and exactly. The purpose of this section is to show that they can also be used to complete symplectic jets. Indeed, there are many such completions.[6] We will show that, given any symplectic jet $\mathcal{T}_m$, there is a Cremona map $\mathcal{C}$ whose truncated Taylor series agrees with $\mathcal{T}_m$. That is, when the Taylor series for $\mathcal{C}$ is truncated beyond terms of degree m, the result is $\mathcal{T}_m$.

Cremona completions satisfy requirement 1 of Section 6 by definition, and also satisfy requirement 3 since polynomials are easy to evaluate. In fact, we will see that there are Cremona completions whose evaluation requires relatively few operations.

---

action-angle variables and fit to tracking data a generating function having a Fourier representation in angles and spline fits in actions. Their aim is to study the long-term stability of orbits far from the origin of phase space.

[5]That the inverse of a Cremona map is again a Cremona map will be obvious for the Cremona maps to be described in this section. For some discussion of the more general case see [3, 31, 39, 49].

[6]The use of Cremona maps for accelerator physics was first proposed in [25] without any indication as to how to construct them. Subsequently and independently, a method for constructing what amounts to Cremona maps was proposed by J. Irwin [34]. His construction inspired our study of the relation (8.11).

What is not yet clear is whether Cremona completions can be found that also satisfy requirement 2.

We begin our discussion by generating a copious supply of Cremona maps. Consider the Lie relation

$$\begin{aligned} z_a^f &= \exp(:g:)z_a^i \\ &= (1+ :g: + :g:^2 /2! + \cdots)z_a^i. \end{aligned} \tag{8.1}$$

Suppose g has the property

$$:g:^2 \, z_a = 0, \text{ for all } a. \tag{8.2}$$

Then g will be called a *jolt*, and $\exp(:g:)$ will be called a *jolt map* [20]. If g is also a polynomial, then $\exp(:g:)$ is a Cremona map. This last assertion follows from three simple observations: First, according to Section 4, $\exp(:g:)$ is a symplectic map. Second, if g is a jolt, it follows from (8.2) that the exponential series (8.1) terminates to give the result

$$\begin{aligned} z_a^f &= (1+ :g:)z_a^i = z_a^i + :g: z_a^i \\ &= z_a^i + [g, z_a^i]. \end{aligned} \tag{8.3}$$

Finally, if g is a polynomial, then $[g, z_a^i]$ is a polynomial.

Suppose g is of the form

$$g = h(q) \tag{8.4}$$

where h is any function. Such a function g will be called a *kick*, and $\exp(:g:)$ will then be called a *kick map* [24]. It is easy to check that a kick is a special case of a jolt. First, we have the immediate result

$$:g: q_j = [h(q), q_j] = 0,$$

which implies

$$:g:^2 \, q_j = 0.$$

Second, we find the relations

$$:g: p_j = [h(q), p_j] = \partial h/\partial q_j = h_{q_j}(q),$$

$$:g:^2 \, p_j = :g: h_{q_j}(q) = [h(q), h_{q_j}(q)] = 0.$$

Consider Lie transformations $\mathcal{L}$ of the form

$$\mathcal{L} = \exp(:f_2:). \tag{8.5}$$

Then we have results

$$\mathcal{L} z_a = \sum_b R_{ab} z_b, \tag{8.6}$$

$$\mathcal{L}^{-1} z_a = \sum_b (R^{-1})_{ab} z_b, \tag{8.7}$$

where R is a symplectic matrix [16]. Since (8.6) is a linear relation, $\mathcal{L}$ will be referred to as a *linear* map. We observe that if g is a jolt, then $\tilde{g}$ defined by the equation

$$\tilde{g}(z) = \mathcal{L} g(z) = g(\mathcal{L} z) = g(Rz) \tag{8.8}$$

is also a jolt. Indeed, we have the relations

$$:\tilde{g}: = :\mathcal{L}g: = \mathcal{L} :g: \mathcal{L}^{-1}, \tag{8.9}$$

$$:\tilde{g}:^2 = \mathcal{L} :g:^2 \mathcal{L}^{-1}. \tag{8.10}$$

(In writing (8.8) and (8.9) we have made use of certain isomorophism properties of Lie operators with respect to addition and multiplication [16, 18, 21].) It follows from (8.2), (8.7), and (8.10) that we have the result

$$:\tilde{g}:^2 z_a = \mathcal{L} :g:^2 \mathcal{L}^{-1} z_a = 0.$$

Let us put together what we have learned so far. Suppose g is a polynomial kick, *i.e.*, any polynomial function of the form (8.4), and suppose we define $\tilde{g}$ by (8.8) for an arbitrary $\mathcal{L}$. From (8.8) we see that $\tilde{g}$ is also a polynomial. Moreover, we know $\tilde{g}$ is a jolt. It follows that $\exp(:\tilde{g}:)$ is a Cremona map.

We will now assemble tools to show that the supply of Cremona maps we have just found can be used to give symplectic completions of symplectic jets. Let $M(m,d)$ be the number of monomials of degree m in a d dimensional space. A simple combinatoric calculation gives the result

$$M(m,d) = M_{md} = (m+d-1)!/[m!(d-1)!].$$

Next we specialize to the case of $d = 3$. Let $Q_k^m(q)$ with $k = 1, 2, \cdots M(m,3)$ denote the monomials of degree m in the variables q_1, q_2, q_3 normalized by a multiplicative constant to be specified shortly. Consider this challenge: Suppose we are given a collection of homogeneous polynomials $h_3(z), h_4(z), \ldots h_n(z)$. Find a number $N(n)$, a set of linear maps $\mathcal{L}_\ell$ with $\ell = 1, 2, \ldots N(n)$, and a set of coefficients $a_{\ell k}^m$, such that the following relations hold,

$$h_m = \frac{1}{NM_{m3}} \sum_{\ell=1}^{N} \sum_{k=1}^{M_{m3}} a_{\ell k}^m \mathcal{L}_\ell Q_k^m; \quad m = 3, 4, \ldots n. \tag{8.11}$$

We will see that there is reason to believe that the relations (8.11) can be met, and we will give an estimate for $N(n)$ [42].

Discussion of the relations (8.11) is facilitated by the introduction of an inner product in the vector space of polynomials. Let $G_r^m(q,p)$ denote a general monomial in all the q's and p's defined by the rule

$$G_r^m(q,p) = [\mu_1!\mu_2!\mu_3!\nu_1!\nu_2!\nu_3!]^{-1/2} q_1^{\mu_1} q_2^{\mu_2} q_3^{\mu_3} p_1^{\nu_1} p_2^{\nu_2} p_3^{\nu_3}.$$

Here we have the homogeneity relation

$$\mu_1 + \mu_2 + \mu_3 + \nu_1 + \nu_2 + \nu_3 = m, \tag{8.12}$$

and the index r ranges from 1 to $M(m,6)$. Note that, by construction, there is a unique value of r for each set of exponents μ_i, ν_i satisfying (8.12). Similarly, Q_k^m will denote a special case of G_r^m where all p's have exponent zero:

$$Q_k^m = [\mu_1!\mu_2!\mu_3!]^{-1/2} q_1^{\mu_1} q_2^{\mu_2} q_3^{\mu_3}.$$

We will now introduce an inner product $\langle , \rangle$ in the vector space of polynomials by the rule

$$\langle G_{r'}^{m'}, G_r^m \rangle = \delta_{m'm} \delta_{r'r}. \tag{8.13}$$

It can be shown that $Sp(6)$ has a unitary subgroup $U(3)$, and that the inner product (8.13) is preserved by all transformations $\mathcal{L}$ of the form (8.5) providing the corresponding matrix R defined by (8.6) belongs to $U(3)$. See [16].

Now consider the homogeneous polynomial h_m. We have the expansion

$$h_m = \sum_{r=1}^{M_{m6}} c_r G_r^m, \tag{8.14}$$

where, with the aid of the inner product just defined, each expansion coefficient c_r is given by the relation

$$c_r = \langle G_r^m, h_m \rangle. \tag{8.15}$$

We also have the relation

$$\langle h_m, h_m \rangle = \sum_r c_r^2 = (c, c). \tag{8.16}$$

For a given m, insert the relation (8.11) into (8.15) to get the result

$$c_r = \frac{1}{NM_{m3}} \sum_{\ell=1}^{N} \sum_{k=1}^{M_{m3}} a_{\ell k}^m \langle G_r^m, \mathcal{L}_\ell Q_k^m \rangle. \tag{8.17}$$

Consider a vector space V^m having $N(n) \times M(m,3)$ dimensions. View the $a_{\ell k}^m$ for fixed m and varying ℓ and k as entries of a vector in V^m. Thus, any choice of the $a_{\ell k}^m$ (m fixed and ℓ, k varying) is equivalent to selecting some vector in V^m. We will call this vector a^m. Next, consider the quantities $\sigma_{\ell k}^r$ defined by the relations

$$\sigma_{\ell k}^r = \langle G_r^m, \mathcal{L}_\ell Q_k^m \rangle. \tag{8.18}$$

For each r (with ℓ, k varying), the quantities $\sigma_{\ell k}^r$ may also be viewed as entries of a vector σ^r in V^m. We will call the σ^r *sensitivity* vectors. With the aid of these definitions, we see that (8.17) is equivalent to the (more compact) relations

$$c_r = \frac{1}{NM_{m3}} \sum_\ell \sum_k \sigma_{\ell k}^r a_{\ell k}^m = \{\sigma^r, a^m\}; \;\; r = 1, 2, \ldots M(m,6). \tag{8.19}$$

Here the inner product in V^m, denoted by $\{,\}$, includes the factor $(NM_{m3})^{-1}$.

According to (8.19), the only part of a^m that contributes to the c_r is that which lies in the subspace of V^m spanned by the sensitivity vectors. (Any part of a^m that is perpendicular to all the σ^r makes no contribution to the c_r.) Consequently, we make the *ansatz*

$$a^m = \sum_{s=1}^{M_{m6}} \alpha_s \sigma^s, \tag{8.20}$$

where the α_s are coefficients yet to be determined. Upon combining (8.19) and (8.20) we find the relations

$$c_r = \sum_s \{\sigma^r, \sigma^s\} \alpha_s.$$

These relations can be written more compactly in the form

$$c = \Gamma \alpha, \tag{8.21}$$

where Γ, with entries Γ_{rs}, is the *Gram matrix* formed from the sensitivity vectors:

$$\Gamma_{rs} = \{\sigma^r, \sigma^s\}. \tag{8.22}$$

We see from (8.18) that for each m the matrix Γ depends only on the choice of the $\mathcal{L}_\ell$.

Assuming that Γ is nonsingular, we can solve (8.21) for α to give the relation

$$\alpha = \Gamma^{-1}c. \tag{8.23}$$

We will eventually be interested in bounds on a^m, and in arranging that a^m be as small as possible. [The ansatz (8.20) is already one step in the direction of minimizing the length of a^m, because adding to (8.20) vectors that are perpendicular to all the σ^r can only increase the length of a^m.] From (8.20) we have the result

$$\{a^m, a^m\} = \sum_{rs} \alpha_r \alpha_s \{\sigma^r, \sigma^s\} = (\Gamma\alpha, \alpha). \tag{8.24}$$

With the aid of (8.21) and (8.23), we can rewrite (8.24) in the form

$$\{a^m, a^m\} = (c, \Gamma^{-1}c). \tag{8.25}$$

According to (8.22) the Gram matrix Γ is real, symmetric, and positive. Therefore, it can be diagonalzed by a real orthogonal transformation, its eigenvalues are real and non-negative, and its eigenvectors can be normalized to form a real orthonormal set. Let γ_r with $r = 1, 2, \cdots M(m, 6)$ be the orthonormal eigenvectors of Γ, and let λ_r be the associated eigenvalues. Then the vector c has an expansion of the form

$$c = \sum_r \xi_r \gamma_r,$$

and the relation (8.25) can, with the aid of (8.16), be rewritten in the form

$$\begin{aligned}\{a^m, a^m\} &= (c, \Gamma^{-1}c) = \sum_r \xi_r^2/\lambda_r \\ &\leq (\lambda^{min})^{-1} \sum_r \xi_r^2 = (\lambda^{min})^{-1}(c, c) = (\lambda^{min})^{-1}\langle h_m, h_m\rangle. \end{aligned} \tag{8.26}$$

Here λ^{min} is the smallest eigenvalue of Γ. The relation (8.26) shows that if we wish to minimize the size of a^m, then we should add to the challenge (8.11) the further challenge of selecting the $\mathcal{L}_\ell$ in such a way that the smallest eigenvalue of Γ is as large as possible.

What can be said about the size of $N(n)$? All we have to offer at the moment is a simple dimension counting argument and the results of limited numerical experimentation. The most one could hope for is that the vector space spanned by the $\mathcal{L}_\ell Q_k^m$ would have the maximal dimension $M(m, 6)$, as required by (8.11), for $N(n)$ as small as possible. Now, the nominal dimension of the space spanned by the $\mathcal{L}_\ell Q_k^m$ is $N(n) \times M(m, 3)$. Therefore, we might hope to achieve the relation

$$N(n) = \max_m \{\text{Ceiling } [M(m, 6)/M(m, 3)]\}; \; m = 3, 4, \ldots n. \tag{8.27}$$

Here the *Ceiling* function is a function that rounds its argument up to the next integer. It is easily verified that the ratio $M(m,6)/M(m,3)$ increases monotonicaly with m. Consequently, the relation (8.27) is equivalent to the hope

$$N(n) = \text{Ceiling}\,[M(n,6)/M(n,3)]. \tag{8.28}$$

Table 1 lists the first few values of N given by (8.28). With regard to numerical

Table 1

n	$M(n,6)$	$M(n,3)$	$N(n)$
3	56	10	6
4	126	15	9
5	252	21	12
6	462	28	17
7	792	36	22
8	1287	45	29
9	2002	55	37
10	3003	66	46
11	4368	78	56
12	6188	91	68

experimentation, we have found that (8.28) holds at least for $n \leq 5$. That is, in these cases we have found choices of the $\mathcal{L}_\ell$ such that (8.11) holds for arbitrary $m \leq n$ and Γ is indeed invertible ($\lambda^{min} > 0$). Finally, we remark that there are group theoretic and quadrature formula considerations that suggest that (8.11) always has a solution, but these considerations alone do not seem sufficient to guarantee (8.28) [42]. We have already mentioned that $Sp(6)$ has a unitary subgroup $U(3)$, which in turn has a unitary unimodular subgroup $SU(3)$, which in turn has an orthogonal subgroup $SO(3)$. See [16]. In making group theoretic arguments, it seems helpful to choose the $\mathcal{L}_\ell$ in such a way that their corresponding R matrices [see (8.6)] belong to distinct elements of the coset space $SU(3)/SO(3)$.

The stage is set for the Cremona completion of a symplectic jet. To simplify notation, we will consider the completion of an arbitrary $(n-1)$-jet $\mathcal{T}_{n-1}$ of the form (5.3). In the spirit of (5.4), such a jet has the *Lie* completion $\mathcal{N}_L$ given by the relation

$$\mathcal{N}_L = \exp(:f_3:)\exp(:f_4:)\cdots\exp(:f_n:). \tag{8.29}$$

Our goal will be to construct from $f_3, f_4, \cdots f_n$ a Cremona map $\mathcal{C}$ having the jet $\mathcal{T}_{n-1}$. More precisely, and what amounts to the same thing, we will construct a Cremona map whose Lie factorization is of the form

$$\mathcal{C} = \mathcal{N}_L \exp(:g_{n+1}:)\exp(:g_{n+2}:)\cdots . \tag{8.30}$$

(Here, and in the discussion that follows, we will use the notation g_m to denote a *generic* and *unspecified* homogeneous polynomial of degree m. In other words, occurrences of the symbol g_m in different places *do not* necessarily refer to the same polynomial.) The key ingredient in this construction will be the relation (8.11) and its solution as given by the relations (8.14), (8.18), (8.20), (8.22), and (8.23).

To begin the construction of $\mathcal{C}$, set $h_3 = f_3$ in (8.11) and solve for the coefficients $a^3_{\ell k}$ using (8.20) and (8.23). Let $\mathcal{C}_3$ be the map defined by the relation

$$\mathcal{C}_3 = \prod_\ell \mathcal{L}_\ell \exp(:\frac{1}{NM_{33}} \sum_{k=1}^{M_{33}} a^3_{\ell k} Q^3_k :)\mathcal{L}_\ell^{-1}. \tag{8.31}$$

Evidently $\mathcal{C}_3$ is a Cremona map. Also, we have the Lie algebraic relation

$$\mathcal{L}_\ell \exp(:\frac{1}{NM_{33}} \sum_k a^3_{\ell k} Q^3_k :)\mathcal{L}_\ell^{-1} = \exp(:\frac{1}{NM_{33}} \sum_k a^3_{\ell k} \mathcal{L}_\ell Q^3_k :). \tag{8.32}$$

By employing (8.32) in (8.31) and using the Baker-Campbell-Hausdorff (BCH) formula for combining Lie exponents [16, 18, 21], we find that (8.31) can also be written in the form

$$\mathcal{C}_3 = \exp(:\frac{1}{NM_{33}} \sum_{\ell k} a^3_{\ell k} \mathcal{L}_\ell Q^3_k :) \exp(:g_4:) \exp(:g_5:) \cdots . \tag{8.33}$$

But, by construction, the first factor in (8.33) is $\exp(:f_3:)$. Consequently, we find that $\mathcal{C}_3$ has the Lie factorization

$$\mathcal{C}_3 = \exp(:f_3:) \exp(:g_4:) \exp(:g_5:) \cdots . \tag{8.34}$$

Let $\mathcal{R}_4$ be a *remainder* map defined by the relation

$$\mathcal{R}_4 = \mathcal{C}_3^{-1} \mathcal{N}_L. \tag{8.35}$$

Upon comparing (8.29) and (8.34) we see that (8.35) has a Lie factorization of the form

$$\mathcal{R}_4 = \exp(:h_4:) \exp(:g_5:) \cdots . \tag{8.36}$$

Next use in (8.11) the h_4 that appears in (8.36), and solve for the $a^4_{\ell k}$. Define a map $\mathcal{C}_{34}$ by the relation

$$\mathcal{C}_{34} = \prod_\ell \mathcal{L}_\ell \exp(:\frac{1}{NM_{33}} \sum_k a^3_{\ell k} Q^3_k + \frac{1}{NM_{43}} \sum_{k'} a^4_{\ell k'} Q^4_{\ell k'} :)\mathcal{L}_\ell^{-1}. \tag{8.37}$$

Again, $\mathcal{C}_{34}$ is a Cremona map. Some thought shows that it can be rewritten in the form

$$\mathcal{C}_{34} = \mathcal{C}_3 \exp(:h_4:) \exp(:g_5:) \cdots . \tag{8.38}$$

This result is found by manipulating the exponents in (8.37) with the aid of the BCH formula, by judicious insertion of factors of $\mathcal{L}_\ell^{-1}\mathcal{L}_\ell$, and by use of (8.11). Now rewrite (8.35) and (8.36) in the form

$$\exp(:h_4:) = \mathcal{C}_3^{-1} \mathcal{N}_L \exp(:g_5:) \cdots ,$$

and insert this result for $\exp(:h_4:)$ into (8.38) to see that

$$\mathcal{C}_{34} = \mathcal{N}_L \exp(:g_5:) \cdots .$$

At this point it should be obvious that the process we have begun can be repeated sufficiently often to give the desired result

$$\mathcal{C} = \mathcal{C}_{34\cdots n} = \mathcal{N}_L \exp(:g_{n+1}:) \cdots .$$

To satisfy the reader on this point, we will carry out one more step to show that the pattern we have outlined does indeed continue. Let $\mathcal{R}_5$ be the remainder map defined by the relation

$$\mathcal{R}_5 = (\mathcal{C}_{34})^{-1}\mathcal{N}_L.$$

It has a Lie factorization of the form

$$\mathcal{R}_5 = \exp(:h_5:)\exp(:g_6:)\cdots .$$

Use h_5 in (8.11) to obtain the coefficients $a^5_{\ell k}$. Define $\mathcal{C}_{345}$ by the relation

$$\begin{aligned}\mathcal{C}_{345} &= \prod_\ell \mathcal{L}_\ell \exp\left[\tfrac{1}{N}\left(:\tfrac{1}{M_{33}}\sum_k a^3_{\ell k}Q^3_k + \tfrac{1}{M_{43}}\sum_{k'} a^4_{\ell k'}Q^4_{k'}\right.\right.\\ &\left.\left.+\tfrac{1}{M_{53}}\sum_{k''} a^5_{\ell k''}Q^5_{k''}:\right)\right]\mathcal{L}_\ell^{-1}.\end{aligned} \tag{8.39}$$

Using the same procedure as before, $\mathcal{C}_{345}$ can be rewritten in the form

$$\mathcal{C}_{345} = \mathcal{C}_{34}\exp(:h_5:)\exp(:g_6:)\cdots .$$

Finally, we get the result

$$\mathcal{C}_{345} = \mathcal{N}_L\exp(:g_6:)\cdots .$$

We end this section by commenting on what has been accomplished. Consider again the requirements set forth in Section 6. By construction, requirement 1 is met. With regard to requirement 3, look at (8.39) from which the general form of $\mathcal{C}_{34\cdots n} = \mathcal{C}$ can be inferred. We see that to the extent that (8.27) is valid, remarkably few operations are required to evaluate $\mathcal{C}$. For example, according to Table 1, $\mathcal{C}$ for an 11-jet can be evaluated using 3×68 polynomial evaluations and 69 matrix multiplications of a six-vector.

Requirement 2 is more problematic. What we need to achieve is that the polynomials g_{n+1}, g_{n+2}, etc. occurring in (8.30) be small. These polynomials arise in the course of Lie factorizing $\mathcal{C}$. Examine the general form of $\mathcal{C}$, see (8.39), and think about the general form of the BCH formula. Evidently, the offensive g's involve products and powers of the coefficients $a^m_{\ell k}$, which are the entries in the vectors a^m. Now the motivation for attempting to keep the sizes of the a^m as small as possible can be explained: minimizing the a^m is a method, perhaps somewhat crude, of minimizing the offensive g's.

We are currently studying the problem of how to select the $\mathcal{L}_\ell$ so as to make λ^{min}, the smallest eigenvalue of Γ, as large as possible. [See (8.26).] Very recently we have been able to use group theory to compute the value of λ^{min} (and the complete spectrum of Γ) in the continuum limit $N \to \infty$. Because this limit corresponds to a sampling of all possible $\mathcal{L}_\ell$, the resulting λ^{min} is the best possible and represents a target for which to aim. In Table 2 we list these optimal values of λ^{min} in the cases where one wishes to factor polynomials of degree three, $\lambda^{min}(3)$, or degree four, $\lambda^{min}(4)$, in phase spaces of two, four, and six dimensions. Results have also been found for polynomials of higher degrees [1] .

Much of our initial work has focused on the simplest case of a two-dimensional phase space and polynomials of degree four. There we have found that the vast majority of sets of $\mathcal{L}_\ell$ (for example, sets selected randomly) lead to values of λ^{min} orders of magnitude smaller than the optimal value of 1/16. Simultaneously, the corresponding a^m are large, in accord with (8.26), and the offensive g's are also

Table 2

dim	$\lambda^{min}(3)$	$\lambda^{min}(4)$
2	1/8	1/16
4	1/32	1/80
6	1/80	1/240

large. Motivated by a study of quadrature formulas, we have also found a few sets of $\mathcal{L}_\ell$ having values of λ^{min} close or equal to 1/16. For these good sets of $\mathcal{L}_\ell$ the associated a^m are of moderate size, and the possibly offensive g's seem acceptably small—although not as small as we had hoped.[7] We have learned from the two-dimensional case that almost all sets of $\mathcal{L}_\ell$ are bad, good sets of $\mathcal{L}_\ell$ are rare, and very good sets are exceptional.

For the cases of particular interest to us - four-dimensional and especially six-dimensional phase spaces - we have verified that good sets of $\mathcal{L}_\ell$ are also exceedingly rare. However, motivated by group-theoretical arguments, we have found in the six-dimensional case one set of 108 $\mathcal{L}_\ell$ whose Γ spectra for polynomials of degrees three and four closely match those of the continuum limit. Based on this experience, we believe that the discovery of good sets of $\mathcal{L}_\ell$ in the six-dimensional case is related to the construction of quadrature formulas for the coset space SU(3)/SO(3). Furthermore, based on similar group-theoretical arguments, we believe it possible in the four-dimensional case to find good sets of $\mathcal{L}_\ell$ using quadrature formulas for the two-sphere, which is topologically equivalent to the coset space SU(2)/SO(2). Still, much work remains. In particular, we need to test the good set of $\mathcal{L}_\ell$ we have found for the six-dimensional case to see how well it works in practice (*e.g.*, are the offensive g's acceptably small). Moreover, since 108 $\mathcal{L}_\ell$ are many more than the required minimum of nine for polynomials of degree four (see Table 1), we intend to study how well this set works for polynomials of degree higher than four. More important, it seems that a proper understanding of how to choose good sets of $\mathcal{L}_\ell$ must be rooted in deep and unstudied properties of the symplectic group. We seek to understand those properties and thereby discover other (we hope smaller) good sets of $\mathcal{L}_\ell$.

9 Other Methods

Suppose the functions Q_k^m in (8.11) are replaced by the single function $(q_1)^m$. Then we might seek a number $N'(n)$, a set of linear maps $\mathcal{L}_\ell$, and a set of coefficients a_ℓ^m such that the following modified relations hold,

$$h_m(z) = \sum_{\ell=1}^{N'} a_\ell^m \mathcal{L}_\ell (q_1)^m; \quad m = 3, 4, \ldots n. \tag{9.1}$$

From (8.8) we have the relation

$$\mathcal{L}_\ell (q_1)^m = (\mathcal{L}_\ell q_1)^m = [t_\ell(1, z)]^m.$$

[7] We also chose some simple 2-dimensional symplectic maps and constructed their Cremona map approximations using a good set of $\mathcal{L}_\ell$. We found acceptable agreement between the behavior of the original maps and their Cremona approximations, including long-term behavior under repeated iteration. See [20, 23, 24].

Here $t_\ell(1,z)$ denotes some homogeneous polynomial (labelled by ℓ) of degree 1 in the variables z. It follows that demanding (9.1) is equivalent to finding a set of homogeneous first-degree polynomials $t_\ell(z,1)$ and coefficients a_ℓ^m such that we have the relations

$$h_m(z) = \sum_{\ell=1}^{N'} a_\ell^m [t_\ell(1,z)]^m; \quad m = 3,4,\ldots n. \tag{9.2}$$

In the case of a single value of m, the solution of (9.3) is a variant of the *Waring problem.* Since we require that (9.3) hold for several m values, we may view the solution of (9.3) as a *generalized* Waring problem.

A review of the steps in Section 8 that led from (8.11) to the construction of Cremona maps shows that Cremona maps can also be constructed starting from (9.1), or equivalently, (9.3). Thus, the construction of Cremona maps is equivalent to finding solutions to the generalized Waring problem. Unfortunately, although there is a literature on the Waring problem, not enough seems to be known about the subject at present to assist in the explicit construction of Cremona maps. See [27, 33, 36, 43, 44]. However, there do exist instructive counter-examples in this literature that illustrate that naive dimension-counting arguments, such as that which led to (8.28), can be misleading.

We close this section with a brief discussion of three other methods for the construction of symplectic maps that can be evaluated exactly and relatively easily. The first method is that of *monomial factorization* [32]. From the BCH formula it follows that the map $\mathcal{N}$ given by (5.2) can be rewritten in the form

$$\mathcal{N} = \left[\prod_{m=3}^{n} \prod_{r=1}^{M(m,6)} \exp(:b_r^m G_r^m:)\right] \exp(:g_{n+1}:) \cdots . \tag{9.3}$$

Here, as in Section 8, we use the notation g_{n+1} to denote an unspecified homogeneous polynomial of degree $n+1$. It is a remarkable fact that the monomially generated maps $\exp(:b_r^m G_r^m:)$ can be evaluated exactly in closed form [12, 32, 38, 48]. For example, in the two-dimensional phase space case, we have the result

$$\exp(:bq^r p^s:)q = q[1 + b(r-s)q^{r-1}p^{s-1}]^{[s/(s-r)]}, \tag{9.4}$$

$$\exp(:bq^r p^s:)p = p[1 + b(r-s)q^{r-1}p^{s-1}]^{[r/(r-s)]}, \tag{9.5}$$

when $r \neq s$, and

$$\exp(:bq^r p^s:)q = q\exp(-brq^{r-1}p^{r-1}), \tag{9.6}$$

$$\exp(:bq^r p^s:)p = p\exp(+brq^{r-1}p^{r-1}), \tag{9.7}$$

when $r = s$. Analogous results for the cases of higher dimensional phase spaces follow immediately from (9.4) through (9.7). Thus, given an $(n-1)$-jet $\mathcal{T}_{n-1}$ of the form (5.3), the approximate map $\mathcal{N}_\text{a}$ given by the relation

$$\mathcal{N}_\text{a} = \prod_{m=3}^{n} \prod_{r=1}^{M(m,6)} \exp(:b_r^m G_r^m:) \tag{9.8}$$

provides a symplectic completion that can be evaluated exactly and relatively easily.

How well does this symplectification meet the requirements of Section 6? Requirement 1 is obviously met. Although there are a large number of terms in the product (9.8), with careful programming it might also be possible to meet requirement 3 reasonably well. With regard to requirement 2, preliminary experience suggests that there may be difficulties: the g's appearing in (9.3) seem to be unacceptably large [32].

The last two symplectification methods to be discussed lead to an *integrable polynomial factorization.*[8] We have seen that all the monomially generated maps $\exp(:b_r^m G_r^m:)$ can be evaluated exactly in closed form. It can be shown that there are also simple linear combinations of monomials (all of the same degree) such that their associated Lie maps can also be evaluated exactly. That is, there are coefficients $d_r^{m,i}$, which define the polynomials

$$g_m^i = \sum_r d_r^{m,i} G_r^m,$$

such that the maps $\exp(:g_m^i:)$ can be evaluated exactly in closed form. Such polynomials are called *integrable.* Moreover, it can be shown that any homogeneous polynomial h_m can be written in terms of the g_m^i. More explicitly, for any h_m, there is a choice of the $d_r^{m,i}$ such that h_m has the representation

$$h_m = \sum_{i=1}^{N''(m)} g_m^i. \tag{9.9}$$

Values of $N''(m)$ for the first few values of m are listed in Table 3 below [47].

Table 3

m	$N''(m)$
3	8
4	20
5	42
6	79

Suppose we wish to make an integrable polynomial factorization of the map $\mathcal{N}_L$ given by (8.29). There are two possible procedures. The first is analogous to the procedure for constructing Cremona maps. For this procedure, set $h_3 = f_3$ in (9.9). Let $\mathcal{N}_3$ be the map defined by the relation

$$\mathcal{N}_3 = \prod_{i_3=1}^{N''(3)} \exp(:g_3^{i_3}:).$$

Then from the BCH formula we have the result

$$\mathcal{N}_3 = \exp(:f_3:)\exp(:g_4:)\exp(:g_5:)\cdots.$$

[8] The idea behind this method was first described under the designation "solvable map"; see [26]. It was subsequently discovered independently by Shi and Yan, and developed in much fuller detail; see [46, 47].

In analogy to the case of Cremona maps [see (8.29) and (8.35)], define a remainder map by the relation

$$\mathcal{R}_4 = \mathcal{N}_3^{-1}\mathcal{N}_L.$$

It has the factorization

$$\mathcal{R}_4 = \exp(:h_4:)\exp(:g_5:)\cdots . \tag{9.10}$$

Now use the h_4 of (9.10) in (9.9) to define a map $\mathcal{N}_{34}$ by the rule

$$\mathcal{N}_{34} = \prod_{i_3=1}^{N''(3)} \exp(:g_3^{i_3}:) \prod_{i_4=1}^{N''(4)} \exp(:g_4^{i_4}:).$$

It follows from the BCH formula that $\mathcal{N}_{34}$ satisfies the relation

$$\mathcal{N}_{34} = \exp(:f_3:)\exp(:f_4:)\exp(:g_5:)\cdots .$$

Evidently the process just illustrated can be repeated at will to give the approximate map

$$\mathcal{N}_{\text{a}} = \mathcal{N}_{34\cdots n}, \tag{9.11}$$

which satisfies the relation

$$\mathcal{N}_{\text{a}} = \mathcal{N}_L \exp(:g_{n+1}:)\exp(:g_{n+2}:)\cdots . \tag{9.12}$$

The second procedure for making an integrable polynomial factorization uses symplectic integration [47]. For this procedure, set $h_m = f_m$ for $m = 3, 4, \cdots n$, use (9.9), and construct $\mathcal{N}_{\text{a}}$ by writing

$$\mathcal{N}_{\text{a}} = \left[\exp(:\sum_{i_3=1}^{N''(3)} g_3^{i_3}:)\right]_{SI} \left[\exp(:\sum_{i_4=1}^{N''(4)} g_4^{i_4}:)\right]_{SI} \cdots \left[\exp(:\sum_{i_n=1}^{N''(n)} g_n^{i_n}:)\right]_{SI} . \tag{9.13}$$

Here the notation $[\ ]_{SI}$ means that the quantity in question is to be evaluated using a symplectic integrator of sufficiently high order to give the relation

$$\left[\exp(:\sum_{i_m=1}^{N''(m)} g_m^{i_m}:)\right]_{SI} = \exp(:f_m:)\exp(:g_{n+1}:)\exp(:g_{n+2}:)\cdots . \tag{9.14}$$

As a result of (9.14), $\mathcal{N}_{\text{a}}$ as given by (9.13) again satifies (9.12).

By construction $\mathcal{N}_{\text{a}}$, as given by either (9.11) or (9.13), is symplectic and therefore meets requirement 1 of Section 6. With careful programming, it might also be possible to meet requirement 3 as well. However, we note that the evaluation of $\mathcal{N}_{\text{a}}z$ for both monomial factorized and integrable polynomial factorized maps requires substantially more work than the evaluation of $\mathcal{C}z$ for the corresponding Cremona symplectification. For example, in the case of $n = 6$, the monomial factorization given by (9.8) contains 896 maps that must be evaluated, and inspection of (9.4) through (9.7) shows that some of these evaluations require the relatively slow process of computing roots or exponential functions. In the corresponding case of integrable polynomial factorization as given by (9.11), inspection of Table 3 shows that $\mathcal{N}_{\text{a}}$ contains 149 (= 8 + 20 + 42 + 79) maps. If symplectic integration is used with integrable polynomial factorization as in (9.13), 503 maps must be evaluated [47]. In both procedures, the maps that must be evaluated are at least as computationally intensive as those occurring in the case of monomial factorization. By contrast, reference to Table 1 for the case $n = 6$ shows that the corresponding Cremona symplectification, *if it can actually be achieved*, requires the evaluation of only 17 maps. Moreover, the evaluation of each map in Cremona symplectification requires only simple polynomial and matrix computations.[9]

With regard to meeting requirement 2 of Section 6, relatively little is yet known for the case of integrable polynomial factorization.[10] One might hope that as a reward for the extra work involved, particularly when symplectic integration is employed as in (9.13), the offensive terms $g_{n+1}, g_{n+2} \ldots$ would be negligible. However, this is not yet known to be true. Moreover, the case of monomial factorization seems to serve as a cautionary tale to the effect that additional work is not always rewarded [32].

As a final comment, we remark that many of the methods for symplectification that we have described, including the generating function and Cremona map methods, may possibly be used for the construction of *symplectic lattice maps*. Symplectic lattice maps are maps that can be regarded as the restriction of a symplectic map to a lattice in such a way that the map leaves the lattice invariant.[11] See [45]. Such lattice maps are of interest because they can be iterated indefinitely often without the introduction of roundoff error [29, 28]. To date no attempt has been made to construct and apply symplectic lattice maps in the field of accelerator physics. At least a preliminary exploration of such an approach would seem to be well worthwhile.

10 Conclusions

Symplectic maps are remarkably rich in their structure and properties. Although considerable progress has been made in the construction and manipulation of symplectic maps, there remains much to learn about their properties — including both their symplectic approximation and their behavior under repeated iteration.

Acknowledgements

This work was supported in part by DOE grant DEFG0196ER40949.

[9]At this point we take the opportunity to remark that Cremona maps, since they are polynomial, are analytic everywhere. By contrast, symplectic maps of physical interest may have (perhaps complex) singularities; see [2]. It follows that Cremona map approximations can be expected to be valid in a substantial domain $\mathcal{D}$ about the origin only if the singularities of the map being approximated are far from the origin. (Here we must consider a complexified phase space so that "far from the origin" also includes complex directions.) Now consider monomially generated maps. Inspection of (9.5) and (9.6) indicates that they may have singularities. It can be shown that maps arising from integrable polynomial factorization, maps produced by a polynomial generating function of the form (7.4), and maps produced by a Lie completion of the form (8.29), may also have singularities. Suppose, as is very likely, that the map being approximated does not have these particular singularities. Then we may expect the approximation to be valid in a substantial domain $\mathcal{D}$ about the origin only if the singularities of the map being approximated, as well as the singularities of the approximating maps themselves, are far from the origin.

[10]See, however, [26] and [46] for two-dimensional examples.

[11]Note that to construct a symplectic lattice map for $\mathcal{M}$ as given by (5.1), it is necessary to arrange that both the $\exp(:f_1:)\exp(:f_2:)$ and $\mathcal{N}$ factors leave the lattice invariant.

References

[1] Abell, D. T. [1995], *Analytic Properties and Cremona Approximation of Transfer Maps for Hamiltonian Systems*, Ph.D. Thesis, Physics Department, University of Maryland.

[2] Abell, D. T. and Dragt, A. J. [1994], *Taylor Series Maps and Their Domain of Convergence* , In AIP Conference Proceedings 292 (Month, M., Ruggiero, A. and Weng, W. ed.), American Institute of Physics.

[3] Bass, H., Cornell, E. H., and Wright, D. [1982], Bull. Amer. Math. Soc., **7**, 287.

[4] Berg, J. S., Warnock, R. L., Ruth, R. D., and Forest, É. [1994], *Construction of Symplectic Maps for Nonlinear Motion of Particles in Accelerators*, Phys. Rev. E, **49**, 722.

[5] Berz, M. [1992], *Analysis on a Nonarchimedean Extension of the Real Numbers*, Lecture Notes, Studienstiftung Summer School, Budapest. Michigan State University Report MSUCL-933.

[6] Berz, M. [1988], *Differential algebraic treatment of beam dynamics to very high orders including applications to space charge.* In Proc. 1988 Conf. on Computational Accelerator Physics (Eminhizer, C., ed.), AIP Conf. Proc. 177.

[7] Berz, M. [1991], *High-Order Computation and Normal Form Analysis of Repetitive Systems.* Physics of Particle Accelerators (Month, M., ed.), AIP Conf. Proc. 249.

[8] Berz, M. [1988], *High-Order Description of Accelerators using Differential Algebra and First Applications to the SSC.* Proc. 1988 Snowmass Summer Meeting, World Scientific.

[9] Berz, M. [1993], *New features in COSY INFINITY*, in Proc. 1993 Conf. on Computational Accelerator Physics, AIP Conf. Proc.

[10] Berz, M. [1989], Part. Accel., **24**, 109.

[11] Berz, M. [1991], *Symplectic tracking in circular accelerators with high-order maps*, In Proc. Capri Conf. on Nonlinear Problems in Future Particle Accelerators. World Scientific.

[12] Channell, P. J. [1986], *Explicit Integration of Kick Hamiltonians in Three Degrees of Freedom*, Accelerator Theory Note AT-6:ATN-86-6, Los Alamos National Laboratory.

[13] Collatz, L. [1966], *Functional Analysis and Numerical Mathematics*, Academic Press, 177.

[14] Douglas, D. R. and Dragt, A. J. [1983], *Lie Algebraic Methods for Particle Tracking Calculations.* In Proceedings of the 12th International Conference on High Energy Accelerators (Cole, F.T. and Donaldson, R. eds.). Fermilab.

[15] Douglas, D. R., Forest, É., and Servranckx, R. [1985], *A Method to Render Second-Order Beam Optics Symplectic*, IEEE Trans. on Nucl. Sci. NS-**32(5)**, 2279.

[16] Dragt, A. J. [1993], *Lectures on Nonlinear Dynamics and Lie Methods with Applications to Accelerator Physics*, University of Maryland Physics Department.

[17] Dragt, A. J. [1990], Nuclear Instruments and Methods, **A298**, 441.

[18] Dragt, A. J. [1982], *Lectures in Nonlinear Orbit Dynamics*, in Physics of High Energy Particle Accelerators (Carrigan, R. A. et al., eds.), AIP Conf. Proc. **87**.

[19] Dragt, A. J. et al [1994], *MaryLie 3.0 User's Manual*, University of Maryland Physics Department.

[20] Dragt, A. J. and Abell, D. T. [1993], Int. J. Mod. Phys. A, (Proc. Suppl.), **2B**, 1019. World Scientific.

[21] Dragt, A. J. and Finn, J. [1976], J. Math. Phys., **17**, 2215.

[22] Dragt, A. J. and Forest, É. [1983], J. Math. Phys., **24**, 2734.

[23] Dragt, A. J., Gjaja, I., and Abell, D. T. [1993], In Proceedings of the Berlin Workshop on Nonlinear Problems in Accelerator Physics, (Berz, M., Martin, S. and Ziegler, K. eds.), Institute of Physics Conference Series 131. Institute of Physics, Bristol and Philadelphia.

[24] Dragt, A. J., Gjaja, I. M., and Rangarajan, G. [1991], *Kick Factorization of Symplectic Maps*, Proceedings of the IEEE Accelerator Conference, 1621.

[25] Dragt, A. J., Neri, F. et al. [1988], Annual Review of Nuclear and Particle Science, **38**, 455.

[26] Dragt, A. J., Rangarajan, G., et al. [1990], Part. Accel. **28**, 119.

[27] Dür, A. [1989], J. Symbolic Computation, **8**, 327–333.

[28] Earn, D. J. D. [1994], *Symplectic integration without roundoff error*, to appear in Ergodic Concepts in Stellar Dynamics, (Pfenniger, D. and Gurzadyan, V.G., eds.), Lecture Notes in Physics, **430**, Springer-Verlag, Heidelberg.

[29] Earn, D. J. D. and Tremaine, S. [1991], In Dynamics of Disc Galaxies (Sundelius, B., ed.), 137-141. Earn, D. J. D. and Tremaine, S. [1992], Physica D, **56**, 1–22.

[30] Iyanaga, S. and Kawada, Y. eds. [1980], Encyclopedic Dictionary of Mathematics, MIT Press, 359.

[31] Friedland, S. and Milnor, J. [1989], Ergod. Theory and Dyn. Systems, **9**, 67-99.

[32] Gjaja, I. [1994], Part. Accel., **43(3)**, 133–144.

[33] Helmke, U. [1992], J. Pure Appl. Algebra, **80**, 29–45.

[34] Irwin, J. [1995], *A Multi-Kick Factorization Algorithm for Nonlinear Maps*, in Accelerator Physics at the Superconducting Super Collider, (Yan, Y., Naples, J., and Syphers, M., eds.), AIP Conf. Proc. 326, originally written in 1989 as report SSC-228.

[35] Kleiss, R., Schmidt, F., and Zimmermann, F. [1993], *The Use of Truncated Taylor Maps in Dynamic Aperture Studies*, Particle Accelerators, **41**, 117–132.

[36] Kung, J. P. S. [1987], *Canonical Forms for Binary Forms of Even Degree*, in Invariant Theory (Koh, S.S., ed.), Lecture Notes in Math., Springer-Verlag, **1278**, 52–61.

[37] Menyuk, C. R. [1984], Physica D, **11**, 109.

[38] Michelotti, L. [1992], *Comment on the exact evaluation of symplectic maps*, Fermilab preprint.

[39] Moser, J. [1991], *On Quadratic Symplectic Mappings*, Eidgenössische Technische Hochschule preprint.

[40] Rall, L. B. [1981], *Automatic Differentiation: Techniques and Applications.* Springer-Verlag, New York, and references therein.

[41] Rall, L. B. [1986], Math. Magazine **59**, 275.

[42] Rangarajan, G. [1990], *Invariants for Symplectic Maps and Symplectic Completion of Symplectic Jets.* Ph.D. thesis, University of Maryland Physics Department.

[43] Reichstein, B. [1992], Linear Algebra and its Applications, **160**, 1–61.

[44] Reznick, B. [1992], Memoirs of the American Mathematical Society, vol. 96, **463**.

[45] Scovel, C. [1991], Physics Letters A, **159**, 396–400.

[46] Shi, J. [1994], *Integrable Polynomial Factorization for Symplectic Systems*, Physical Review E, **50**, 532.

[47] Shi, J. and Yan, Y. T. [1993], Physical Review E, **48**, 3943.

[48] Testa, F. J. [1973], J. Math. Phys., **14**, 1097.

[49] Wang, S. [1980], J. Algebra, **65**, 453.

[50] Yan, Y. T., Channell, P. S., and Syphers, M. J. [1993], *Performance of an Algorithm for Symplectic Implicit One-Turn Map Tracking*, SSC Laboratory report SSCL-157.

[51] Yan, Y., Sen, T., Chao, A., Bourianoff, G., Dragt, A. J., and Forest, É. [1991], in Nonlinear Problems in Future Particle Accelerators. (Scandale, W. and Turchetti, G., eds.), World Scientific.

Fields Institute Communications
Volume **10**, 1996

Arnold Diffusion in Symplectic Lattice Maps

David J. D. Earn
Physics Department and Racah Institute of Physics
The Hebrew University
Jerusalem 91904 Israel

A. J. Lichtenberg
Department of Electrical Engineering and Computer Sciences
and the Electronics Research Laboratory
University of California
Berkeley, California 94720

Abstract. In numerical experiments with Hamiltonian systems, round-off errors can be avoided by using symplectic lattice maps. This increases the accuracy of long-term studies of individual orbits. However, it is shown here that estimates of the mean local Arnold diffusion rate (in coupled standard maps) do not change when a lattice map is used. This is important because it reinforces the belief that statistical properties of ensembles of orbits are not significantly affected by roundoff error.

1 Introduction

Most numerical studies of Hamiltonian systems involve iterating a symplectic map on a computer using floating-point arithmetic. Although maps take discrete jumps, in most cases roundoff error prevents them from being iterated exactly. The resulting trajectories are different from those of the original system and, moreover, they are not exactly symplectic. The basic reason is that floating-point operations can map many distinct points to the same place even when the exact mapping is one-to-one. This can significantly alter long-term behaviour. In particular, a typical trajectory does not return to its initial point; instead it eventually joins a periodic orbit not including its initial point. Earn and Tremaine ([1991] [1992]; hereafter ET) showed that roundoff errors cause artificial drifting across invariant curves and can even lead to confusion between regularity and chaos.

A *lattice map* is a function that maps a discrete subset of phase space (a lattice) into itself. Any continuous function f on a phase space defines a natural lattice map F on a given lattice via "$F(p,q)$ is the nearest lattice point to $f(p,q)$". Rescaling makes it possible to ensure that all lattice points have integer phase coordinates.

1991 *Mathematics Subject Classification*. Primary: 70K50; Secondary: 65G05.

This means that error-free addition and multiplication of phase coordinates can be done on any computer. Because of this the lattice maps considered below can be iterated exactly without errors of any sort. Rannou [1974] used this method to study area-preserving maps of the plane.

We say that a lattice map is symplectic (or Hamiltonian) if there is a symplectic diffeomorphism that, when applied to points on the lattice, agrees with the lattice map. The natural lattice maps associated with a large class of symplectic maps (and symplectic integration algorithms) are in fact symplectic lattice maps (ET; Scovel [1991]). This is the principal motivation for using lattice maps to study Hamiltonian systems: spurious dissipative behaviour is impossible because the computed system is exactly symplectic.

Using lattice maps rather than ordinary floating-point maps reduces qualitative errors in orbits of Hamiltonian systems (ET). Quantitative errors (in known integrals of motions) are also significantly reduced, even in short trajectory computations (Earn [1994]). Nevertheless, while individual trajectories are better represented with lattice maps, it is not clear that statistical properties of ensembles of orbits are significantly affected by roundoff error.

An important measure of the average behaviour in a particular region of phase space in a nearly integrable Hamiltonian system (with three or more degrees of freedom) is the local Arnold diffusion rate. This is the (typically slow) rate at which a collection of orbits diffuses *across* stochastic layers (as opposed to motion *along* stochastic layers, which does not require three or more degrees of freedom and is typically fast). (See, *e.g.*, Lichtenberg and Lieberman [1992], Chapter 6.)

In this paper we compare the local Arnold diffusion rates computed numerically with floating-point and lattice versions of a map for which the local Arnold diffusion rate has been calculated analytically.

2 Coupled standard maps

The standard map (*e.g.*, Lichtenberg and Lieberman [1992], Chapter 4) is the plane area-preserving map given by

$$\begin{aligned} x_{n+1} &= x_n + y_{n+1}\,, \\ y_{n+1} &= y_n + \tfrac{K}{2\pi}\sin(2\pi x_n)\,, \end{aligned} \tag{2.1}$$

where K is the stochasticity parameter. This map is spatially periodic with period 1 in both x and y. The floating-point version of this map results from iterating (2.1) with floating-point arithmetic. The natural lattice map is obtained by multiplying (2.1) by a (large) integer, say m, and then replacing the sine term with the nearest integer to its value. The integer m gives the number of lattice points per unit in phase space and thus defines the resolution of the lattice map. Floating-point and lattice versions of the standard map were studied in detail by ET.

A simple four-dimensional Hamiltonian map can be constructed by coupling two standard maps,

$$\begin{aligned} u_{n+1} &= u_n + v_{n+1}\,, \\ v_{n+1} &= v_n + \tfrac{K_1}{2\pi}\sin(2\pi u_n) + \tfrac{\mu}{2\pi}\sin 2\pi(u_n + x_n)\,, \\ x_{n+1} &= x_n + y_{n+1}\,, \\ y_{n+1} &= y_n + \tfrac{K_2}{2\pi}\sin(2\pi x_n) + \tfrac{\mu}{2\pi}\sin 2\pi(u_n + x_n)\,. \end{aligned} \tag{2.2}$$

Here K_1 and K_2 are the stochasticity parameters for each standard map and μ is the coupling constant. The two maps are uncoupled for $\mu = 0$. Equation (2.2) can be thought of as a Poincaré return map for a continuous Hamiltonian system with three degrees of freedom. The natural lattice map for (2.2) is derived in the same way as for the ordinary standard map above.

3 Arnold diffusion rate

Wood, Lichtenberg and Lieberman [1990] derived an analytic approximation for the local Arnold diffusion rate in the system of two coupled standard maps [(2.2)] and compared their results with numerical orbit calculations (using floating-point arithmetic). The agreement was good, but it is important to know if it can be improved further by using a lattice map to iterate the orbits.

We choose the three parameters in (2.2) so that the map will display observable Arnold diffusion. Thus we take K_1 large enough to ensure that there are large stochastic regions in u-v space, K_2 small enough to ensure that there are large regular regions in x-y space, and $0 < \mu \ll 1$ so the coupling is weak. Initial conditions are selected in a primary stochastic region of u-v space and a region of x-y space dominated by invariant curves (far from sizable island chains that would distort the diffusion calculation). To achieve the desired property in x-y space we place the initial points as close as possible to the KAM curve that is the last to survive as K is increased in (2.1). Greene [1979] found that the "last KAM curve" has rotation number $\frac{1}{2}(1+\sqrt{5})$, the golden mean, or $1/2\pi$ times this value in our units in which the spatial period of the standard map (2.1) is 1 rather than 2π. [The rotation number or frequency of an invariant curve of the standard map is equal to the average action (y-value) of any orbit on the curve. (See, *e.g.*, Meiss [1992]).]

To estimate the local Arnold diffusion rate, we iterate a sample of trajectories and look at the RMS change in the action y until this change is of order 0.01. Our initial configuration consists of 256 points on a 16^2 grid in a primary stochastic region of u-v space, each associated with the same initial x-y coordinates. The initial (x, y) is chosen by taking $x_0 = 0$ and searching (by bisection) for y_0 such that the average y value after 10^5 iterations of (2.1) agrees with the golden mean to 10 significant figures.

Figure 1 shows the RMS change in y as a function of time n, for four and eight byte floating-point maps and a lattice map with $m = 2^{52}$ points per unit. Wood *et. al.* [1990] predict that $\Delta y_{\text{rms}} \sim n^{1/2}$ for the exact map (2.2) so the curves drawn in Figure 1 should be lines with slope $\frac{1}{2}$. This is in reasonable agreement with all our numerical results; the agreement is no better in the cases in which the lattice map has been used.

4 Discussion

Tennyson, Lichtenberg and Lieberman [1979] investigated Arnold diffusion in a different four-dimensional symplectic map. ET repeated part of their study with a lattice map and found a significantly different global Arnold diffusion rate, which suggested that some of the diffusion seen by Tennyson *et. al.* [1979] might have been due to roundoff errors. In this case the diffusion rate was estimated crudely from the evolution of a single long trajectory; it emphasizes that the long-term qualitative features of individual trajectories can be strongly influenced by numerical errors.

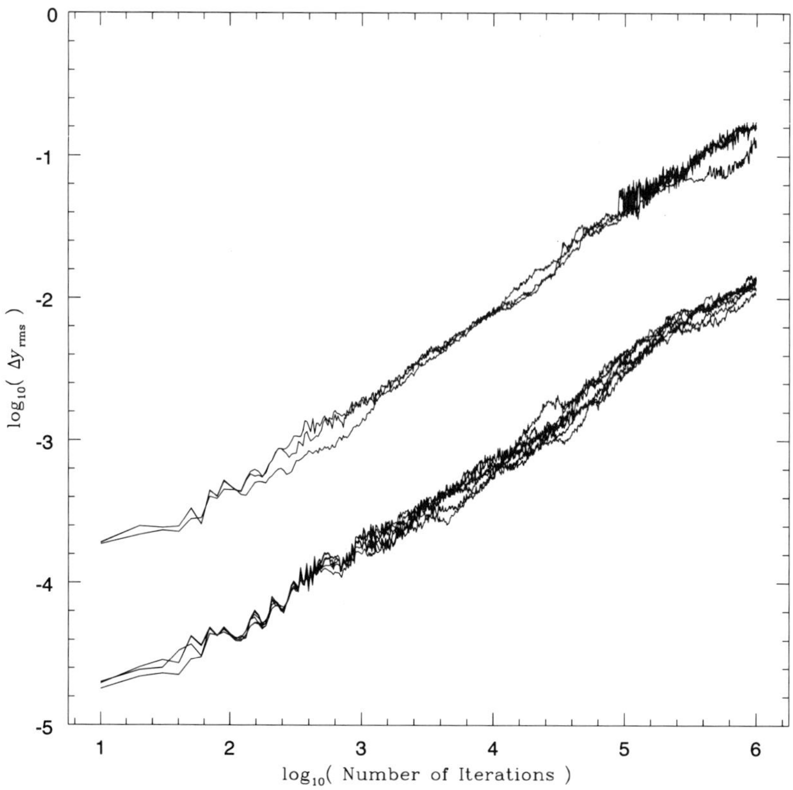

Figure 1 A log-log plot of the RMS action change (due to Arnold diffusion) as a function of time in the system of two coupled standard maps, (2.1). There are 3 curves in the upper group. These all have coupling constant $\mu = 0.001$ and were all obtained using floating-point arithmetic. Two values of K_2 (0.2 and 0.4) were used and the $K_2 = 0.4$ run was done with both 4 byte and 8 byte floating-point arithmetic. There are 6 curves in the lower group. These all have $\mu = 0.0001$. The same values of K_2 and floating-point number size were used as for the upper curves, but this time each run was repeated with a lattice map with $m = 2^{52}$ points per unit. There is nothing to choose from between the various different ways of computing these curves (all are consistent with a slope of 1/2) so this calculation is clearly not seriously affected by roundoff error.

However, precisely because individual trajectories vary so much, it is essential to use a sizable sample to estimate a diffusion rate reliably. The present study has shown that the errors introduced by floating-point arithmetic do not corrupt estimates of the *local* Arnold diffusion rate based on an *ensemble* of trajectories. Strictly, this conclusion is valid only for the specific examples considered here; it remains possible that with sufficiently weak coupling (small enough μ) or with a much more complicated map (requiring many more arithmetic operations per iteration) significant differences may appear between ensemble estimates of Arnold diffusion rates based on ordinary floating-point versus lattice map computations. Nevertheless, the present results are likely to be typical of many symplectic maps and Hamiltonian systems evolved with symplectic integrators, and they strengthen the general belief that statistical properties of ensembles of orbits are not significantly affected by roundoff error.

Acknowledgements

DE thanks Moti Milgrom for his hospitality at the Weizmann Institute, and Scott Tremaine for helpful advice. This research was supported by a Weizmann Institute Exchange Fellowship, a Lady Davis Post-Doctoral Fellowship and a grant from the Albert Einstein Centre for Theoretical Physics.

References

Earn, D. J. D. [1994], *Symplectic integration without roundoff error*, Ergodic Concepts in Stellar Dynamics, (Gurzadyan, V.G. and Pfenniger, D. eds.) Lecture Notes in Physics 430, Springer-Verlag, Heidelberg, 122–130.

Earn, D. J. D. and Tremaine, S. [1991], *Exact solutions for Hamiltonian systems using integer arithmetic*, Dynamics of Disc Galaxies (Sundelius, B. ed.), Göteborg University, Sweden, 137–141.

Earn, D. J. D. and Tremaine, S. [1992], *Exact numerical studies of Hamiltonian maps: Iterating without roundoff error*, Physica D, **56**, 1–22.

Greene, J. M. [1979], *A method for computing the stochastic transition*, J. Math. Phys., **20**, 1183–1201.

Lichtenberg, A. L. and Lieberman, M. A. [1992], Regular and Chaotic Dynamics, Second Edition, Springer-Verlag, New York.

Meiss, J. D. [1992], *Symplectic maps, variational principles, and transport*, Rev. Mod. Phys., **64**, 795–848.

Rannou, F. [1974], *Numerical Study of Discrete Plane Area-preserving Mappings*, Astron. Astrophys., **31**, 289–301.

Scovel, C. [1991], *On Symplectic Lattice Maps*, Phys. Lett. A, **159**, 396–400.

Tennyson, J. L., Lieberman, M. A., and Lichtenberg, A. J. [1979], *Diffusion in near-integrable Hamiltonian systems with three degrees of freedom*, Nonlinear Dynamics and the Beam-Beam Interaction (Month, M. and Herrera, J.C., eds.), AIP Conference Proceedings, **57**, 272-301.

Wood, B. P., Lichtenberg, A. J., and Lieberman, M. A. [1990], *Arnold diffusion in coupled standard maps*, Phys. Rev. A., **42**, 5885-5893.

Fields Institute Communications
Volume 10, 1996

Integrable Hamiltonians From Close Approximations to Invariant Tori

Mikko Kaasalainen
NORDITA
Blegdamsvej 17
DK-2100 Copenhagen, Denmark

James Binney
Theoretical Physics
Keble Road
Oxford OX1 3NP

Abstract. We present a general non-perturbative method for constructing an integrable Hamiltonian H_0 close to a given one H. The existing invariant tori of the original Hamiltonian are accurately approximated by the constructed H_0. In addition to obtaining these tori, we create consistent angle coordinates for them. The constructed Hamiltonian can be used e.g. as a basis for perturbation theory. We illustrate the method by applying it to apparently near-integrable gravitational potentials.

1 Introduction

Many problems in nonlinear dynamics could be conveniently studied if it were possible to fit an integrable Hamiltonian to any given one. For one thing, such a technique would allow the use of Hamiltonian perturbation theory for a much wider class of systems than hitherto has been available. Furthermore, it would be advantageous if the fitted Hamiltonian preserved accurately the 'surviving' invariant phase-space tori of a near-integrable original Hamiltonian ('invariant' is here taken to mean invariant in a given resolution). The computation of these tori is an important subject in many fields ranging from plasma physics and semiclassical quantum theory (Boozer [1982], Martens and Ezra [1987]) to stellar dynamics and accelerator physics, and several schemes for this purpose have been developed, both perturbative and non-perturbative ones (see McGill and Binney [1990] – hereafter Paper I – and Warnock [1991] and references therein). For most problems, non-perturbative methods are the only feasible ones as most often there are no *a priori* known integrable Hamiltonians closely resembling the given one. Of these, the methods introduced in Paper I and (Warnock [1991]) employ a generating function

1991 *Mathematics Subject Classification.* Primary: 58F05; Secondary: 58F07.

periodic in angle coordinates to create a canonical transformation. This approach can be argued to be more efficient and generally usable than previous schemes; however, it has its limitations when the given system is very far from any known integrable one. The method described here uses an additional point transformation to 'bring' some integrable Hamiltonian closer to the original one, when necessary.

In this paper we describe a scheme in which one maps invariant tori in the phase space of a known integrable 'toy' Hamiltonian H_{T} to the phase space of a given 'target' Hamiltonian H with the aid of generating functions and point transformations; we employ the convention that primed variables relate to the tori of H, while the unprimed ones correspond to those of H_{T}. For H_{T}, the transformation between the ordinary phase-space coordinates $(\boldsymbol{x},\boldsymbol{p})$ and the actions and angles $(\mathbf{J},\theta)$ is known. With this approach, one creates invariant tori closely approximating the existing ones of H, and fills the rest of the phase space by creating tori as parts of a global set in which there is a torus corresponding to any given actions $\mathbf{J}'$, thus defining an integrable Hamiltonian $H_0(\mathbf{J}')$. Since the mappings are not obtained perturbatively, H_{T} need not closely resemble H. The scheme consists of three basic steps (Paper I and Binney and Kumar [1993], Kaasalainen and Binney [1994], Kaasalainen [1994] – hereafter Papers II, III and IV): (i) construct the invariant tori, labelled by actions $\mathbf{J}'$, approximating those of H, (ii) construct consistent angle coordinates θ' for the created tori, and (iii) define how to generalize steps (i) and (ii) for values of $\mathbf{J}'$ that do not correspond to invariant tori of H. For step (i), we give a method that is often applicable in step (iii) as well. This is very useful when creating tori for a regular grid of $\mathbf{J}'$-values that may not all correspond to invariant tori of H. One can also use the method of Warnock [1991]; its use is limited to step (i), but it gives very accurate approximations to the invariant tori.

From an integrator's point of view, integrating the motion for the constructed H_0 is very simple: the time development of an orbit on the torus labelled by $\mathbf{J}'$ is given by

$$\theta' = \theta_0' + \omega'(\mathbf{J}')t, \tag{1.1}$$

where ω' is the orbital frequency vector, θ_0' an initial phase and t the time. Thus, because one knows the relationship between the action-angle variables and a point $\boldsymbol{w} = (\boldsymbol{x},\boldsymbol{p})$ in phase-space, integrating the motion is trivial.

The paper is organized as follows. In Section 2 we define our notation and explain how approximate invariant tori are obtained using different toy Hamiltonians H_{T} for topologically different orbit families. In Section 3 we show how the target angle variables and frequencies are obtained, and in Section 4 we describe how a global integrable Hamiltonian can be constructed from the invariant tori and show an example of how resonant islands of the original Hamiltonian can be derived by perturbing the constructed H_0. Section 5 sums up.

2 Mapping the tori

The first goal of our scheme is to seek a generating function $S(\theta,\mathbf{J}')$ that maps the tori of H_{T} into those of H_0. The generating function is given by

$$S(\theta,\mathbf{J}') = \theta\cdot\mathbf{J}' - \mathrm{i}\sum_{\mathbf{n}\neq 0} S_{\mathbf{n}}(\mathbf{J}')\exp(\mathrm{i}\boldsymbol{n}\cdot\theta), \tag{2.1}$$

so that the toy actions $\mathbf{J}$ are related to the target actions $P\mathbf{J}'$ by

$$\mathbf{J} = \frac{\partial S(\theta, \mathbf{J}')}{\partial \theta} = \mathbf{J}' + \sum \mathbf{n}\, S_{\mathbf{n}}(\mathbf{J}')\, \exp(\mathrm{i}n \cdot \theta), \tag{2.2}$$

while the relation between the toy angles θ and the target angles θ' is given by

$$\theta' = \frac{\partial S(\theta, \mathbf{J}')}{\partial \mathbf{J}'} = \theta - \mathrm{i} \sum \frac{\partial S_{\mathbf{n}}(\mathbf{J}')}{\partial \mathbf{J}'} \exp(\mathrm{i}n \cdot \theta). \tag{2.3}$$

The coefficients $S_{\mathbf{n}}$ are found by nonlinear least-squares iteration (Levenberg - Marquardt algorithm), where the objective is to minimize the variation of H, measured for a large set of points θ, along the trial torus labelled by $\mathbf{J}'$ (Paper I). If that torus is an existing invariant torus of H, the value of H on it will not vary after the iteration (except for computational noise), and H_0 will thus coincide with H on it. In addition to $S_{\mathbf{n}}$, the parameters defining the toy Hamiltonians play an important rôle, and they must be optimized together with the coefficients.

In many cases, the generating function approach alone is insufficient. Sometimes the problem is merely that the required generating function has impracticably many significantly non-zero $S_{\mathbf{n}}$. On other occasions the insufficiency is a matter of principle: in general, a generating function of the form (2.1) cannot construct a torus close to the value $J_i' = 0$ for reasons discussed below. These difficulties can be overcome by compounding the transformation (2.2) with suitable point transformations. The basic idea is to map the space of the toy potential into that of the target potential such that the images under this map of toy orbits resemble the target orbits. If $\mathbf{x}$ and $\mathbf{x}'$ are coordinate systems for the toy and target configuration spaces, respectively, the transformation is of the form (Paper III)

$$\mathbf{x} \leftrightarrow \mathbf{x}', \qquad p_i' = \frac{\partial \mathbf{x}}{\partial x_i'} \cdot \mathbf{p}. \tag{2.4}$$

The more the toy and target orbits differ, the more important is the rôle of this point transformation in easing the task of the generating function.

2.1 Single orbit family and H_{T} As an example of a case in which the integrable Hamiltonian H_0 will produce only one family of topologically similar orbits, and consequently only one toy Hamiltonian H_{T} is needed, we study orbits in the axisymmetric Miyamoto and Nagai [1975] potential,

$$\Phi_M(r, z) = -\frac{GM}{\sqrt{r^2 + (a + \sqrt{b^2 + z^2})^2}}, \tag{2.5}$$

where G, M, a and b are potential parameters, $r = \sqrt{x^2 + y^2}$ cylindrical radius, and z the distance from the (x, y)-plane. This potential is generated by a density distribution that approximates the luminous component of a disc galaxy. The motion occurs at a constant z-axis angular momentum $L_z \equiv J_z'$ and can be followed from its projection onto the (r, z)-plane, the meridional plane. The potential parameters are here the same as those used in Paper II. In an $(r, \dot{r})$ surface of section (Figure 1), the invariant curves of the single major-orbit family bend around the point $(r, \dot{r}) \approx (1, 0)$ that corresponds to the closed orbit with zero radial action $J_r' = 0$. A minor-orbit family generates an island centred on the point $(r, \dot{r}) \approx (1.2, 0.45)$ that is associated with the closed orbit of 1:1 resonance between the vertical and radial oscillations. An integrable Hamiltonian does not produce such islands: thus

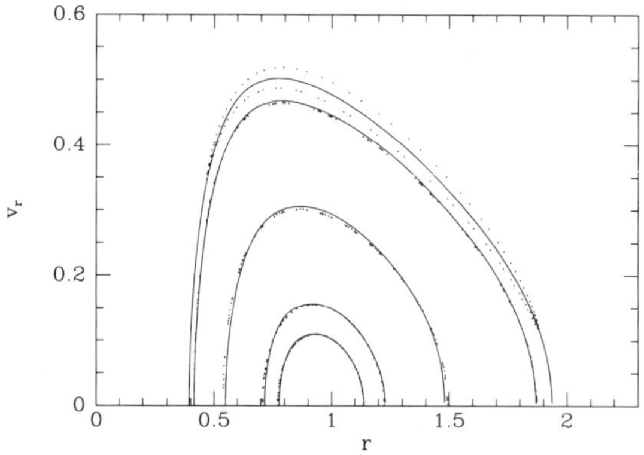

Figure 1 An $(r, \dot{r})$ surface of section for the Miyamoto & Nagai potential with the same H and L_z as in Paper II. Dots are integrated consequents and the full curves sections of numerically generated tori. r is in units of the parameter a of the potential, and v_r is in units of $(GM/a)^{1/2}$.

our H_0 will support only the major family, and no resonant trapping occurs. As the toy Hamiltonian, we use the spherically-symmetric isochrone Hamiltonian

$$H_\mathrm{I} = \frac{1}{2}p^2 - \frac{k}{b + \sqrt{b^2 + (r - r_0)^2}}, \tag{2.6}$$

where k, b and r_0 are parameters.

The solid lines in Figure 1 are sections of tori constructed with our method. The uppermost curve slices right through the resonant island. This illustrates an important advantage of our scheme: it is able to create an invariant torus of H_0 even when there is no corresponding torus of H. The entire island zone can be filled by similar invariant tori of H_0. On these tori, the value of H varies; the constant value of H_0 is the average value of H over the torus.

The two tori in Figure 1 that most closely surround the closed orbit $J_r' = 0$, cannot be obtained by using the generating function (2.1) only. The problem is this. The term periodic in θ has zero mean, and, since $J_r \geq 0$, must vanish whenever $J_r' = 0$. Hence under the transformation (2.2), $J_r' = 0 \Rightarrow J_r = 0$. But $J_r = 0$ in turn implies that $p_r = 0$ and thus that r is constant. The shell of an orbit with $J_r' = 0$ in a non-spherical potential is not confined to a sphere, so $p_r \neq 0$ on it. Thus the torus of such a shell orbit can never be the image under (2.3) of the torus of an orbit with $J_r = 0$ in a spherical potential. Our resolution of this difficulty is to compound the transformation (2.3) with a point transformation which maps a circular orbit in H_I into the closed $J_r' = 0$ orbit in H that has the same value of the third action J_ϑ'. It is best to consider this mapping to be between the four-dimensional phase spaces of the two meridional planes rather than between the full six-dimensional phase spaces. In the meridional planes one orbit forms an arc of a circle that is bisected by the r-axis, and the other forms an arc of a similar but non-circular curve.

Let (r, ϑ) be polar coordinates in the meridional plane of the toy potential, and (r', ϑ') be polar coordinates in the meridional plane of the target potential. We specify a point transformation between these two planes, mapping the two arcs into one another, by writing

$$r = f(\vartheta')r', \qquad \vartheta = g(r')h(\vartheta'), \tag{2.7}$$

where $f(\vartheta')$, $g(r')$ and $h(\vartheta')$ are functions to be determined. On the closed orbit in the toy potential, $r = a$ and $p_r = 0$. Thus along the image $[r'_0(\vartheta'),\, p'_r(\vartheta'),\, p'_\vartheta(\vartheta')]$ of this orbit, found by direct integration of the equations of motion, we have from (2.4)

$$\begin{aligned} f(\vartheta') &= a/r'_0(\vartheta'), \\ p'_r(\vartheta') &= p_\vartheta h \frac{\mathrm{d}g}{\mathrm{d}r'}, \qquad \text{(closed orbit)} \\ p'_\vartheta(\vartheta') &= p_\vartheta g \frac{\mathrm{d}h}{\mathrm{d}\vartheta'}. \end{aligned} \tag{2.8}$$

The last two equations can be considered to be coupled differential equations for g and h. We define $\tilde{g}(\vartheta') \equiv g(r'_0(\vartheta'))$, so that there is only one variable ϑ', and numerically integrate the pair

$$\begin{aligned} \frac{\mathrm{d}\tilde{g}}{\mathrm{d}\vartheta'} &= \frac{p'_r}{hp_\vartheta}\frac{\mathrm{d}r'_0}{\mathrm{d}\vartheta'} \\ \frac{\mathrm{d}h}{\mathrm{d}\vartheta'} &= \frac{p'_\vartheta}{\tilde{g}p_\vartheta}, \end{aligned} \tag{2.9}$$

for which the initial conditions can be chosen to be $\tilde{g}(0) = 1$, $h(0) = 0$. The correct value $J'_\vartheta \equiv (2\pi)^{-1}\oint \boldsymbol{p}'\cdot \mathrm{d}\boldsymbol{x}' - J'_z$ can be found from $\boldsymbol{p}'$ along the numerically integrated closed orbit. To complete the construction of the desired transformation it remains only to express g as a function of r' and to extend the functions f, g and h to all relevant values of (r', ϑ') when $J'_r > 0$. A practical way is to attach suitable polynomials to the endpoints of the ranges defined by the closed orbit. Also, it is helpful to use the variable $\psi \equiv \arcsin(\vartheta'/\vartheta'_{max})$ instead of ϑ' when integrating (2.9) to handle efficiently the condition $p_\vartheta \to 0$ as $\vartheta' \to \vartheta'_{max}$.

The method of mapping a closed orbit in H_T into a closed orbit in H is generally applicable; it can be used, for example, in constructing tori for a minor family of resonantly trapped orbits (Kaasalainen [1995]).

2.2 Two orbit families and H_T:s In this section we study orbits in the two-dimensional logarithmic potential

$$\Phi(x,y) = \frac{1}{2}\ln\Big(x^2 + \frac{y^2}{q^2} + R_\mathrm{c}^2\Big), \tag{2.10}$$

where $R_\mathrm{c} = 0.14$ and $0 < q \le 1$.This represents a nonaxisymmetric planar gravitational potential, such as that of an elongated galaxy (Binney and Tremaine [1987]). A typical $(x, \dot{x})$ surface of section, such as in Figure 2, shows that the consequents define a pattern of invariant curves, consisting of major-family curves and resonant islands.

Two major types of orbits dominate Figure 2: those that have a constant sense of circulation around the origin ('loops') and those that have not ('boxes'); their invariant curves circulate around $(x, \dot{x}) \approx (0.4, 0)$ and $(0, 0)$, respectively. Similar orbits are supported by Stäckel potentials (Binney and Tremaine [1987], de Zeeuw [1985]), whose Hamilton-Jacobi equations separate. In addition to these, a minor family of orbits appears as a pair of islands. These orbits may be considered to be

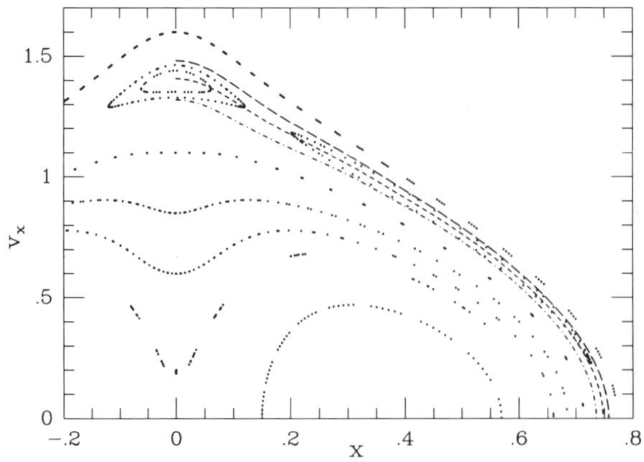

Figure 2 $(x, \dot{x})$ surface of section for the logarithmic potential with $q = 0.8$ and $H = -0.1992$. The dashed lines are curves of numerically constructed tori.

trapped by a 2:3 resonance in a perturbed Stäckel potential. As q is decreased from unity and the potential becomes more elongated, more and more of phase space is occupied by similar families. Our H_0 will be equivalent to a Stäckel one and support only boxes and loops, analogously with the previous section.

By mapping the tori of a known Stäckel Hamiltonian, one could, in principle, construct the invariant tori of both major families from a single toy Hamiltonian H_T. However, general analytic expressions for the transformations $(\mathbf{J}, \boldsymbol{\theta}) \leftrightarrow (\boldsymbol{x}, \boldsymbol{p})$ are not available for the Stäckel tori, so we need two different toy Hamiltonians with orbits that qualitatively resemble boxes and loops; we stress the fact that quantitatively they need not resemble (2.10) closely. For loops, we have used the isochrone Hamiltonian H_I, and for boxes the two-dimensional harmonic oscillator

$$H_\mathrm{H}(\boldsymbol{x}, \boldsymbol{p}) = \frac{1}{2} \sum_{i=1}^{2} (p_i^2 + \omega_i^2 x_i^2), \tag{2.11}$$

where the ω_i are suitable frequencies. In the same way as in Figure 1, one can see three invariant curves (dashed lines) of constructed tori on both sides and in the middle of islands associated with the 2:3 resonance.

Additional point transformations are necessary in this case as well (Paper III). The tori of loop orbits close to $J_r' = 0$ have to be constructed using a point transformation resembling that used for the similar orbits in the meridional plane. Also, the box tori are constructed using point transformations that map the Cartesian coordinates of the harmonic oscillator into confocal elliptic coordinates. This is necessary because most box orbits have narrow waists on the y-axis and are thus physically very different from the rectangular boxes formed by the toy orbits. The parameters describing the transformations are simultaneously optimized with the coefficients $S_\mathbf{n}$.

We have found (Kaasalainen and Binney [1994b]) that, if needed, the constructed tori can be effectively refined by the method of Warnock [1991]. This involves fitting a torus to a numerically integrated orbit by solving for toy actions on an equally-spaced θ-grid, from which the coefficients $S_\mathbf{n}$ can be obtained by

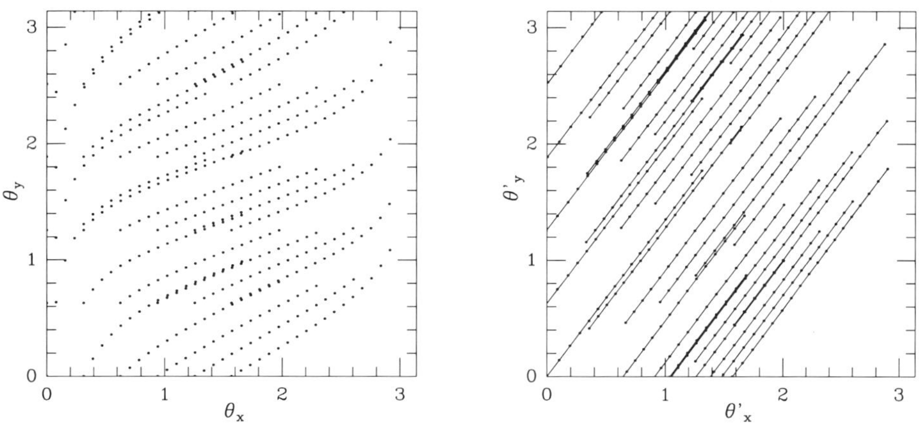

Figure 3 (a) Trajectories in toy angle space obtained by direct integration of the equations of motion from 36 starting points on a box-orbit torus of the logarithmic potential. (b) The same trajectories in target angle space (dots), together with the semi-analytical linear trajectories (1.1) (straight lines).

discrete Fourier transforms. This procedure is numerically stable and converges robustly provided the integrated points are well distributed in θ-space. This is most easily attained when the toy actions do not vary greatly along the orbit. This can in turn be ensured by mapping the toy action-angle variables into the orbit's phase space with a point transformation that has been determined by the least-squares torus-fitting machinery.

3 Recovering the angle variables

To complete the target action-angle coordinate system by deriving the target angles θ' from (2.3), the derivatives $\partial S_{\mathbf{n}}/\partial \mathbf{J}'$ of the coefficients $S_{\mathbf{n}}$ have to be determined. These derivatives are required at constant values of the parameters defining H_{T} and any point transformation. Since we *do* vary these parameters from orbit to orbit, we have to provide a mechanism for determining, torus by torus, the $\partial S_{\mathbf{n}}/\partial \mathbf{J}'$ we would obtain if we did calculate the $S_{\mathbf{n}}$ for fixed parameters.

A convenient method (Paper III) is numerically to integrate the orbit for short periods from initial conditions chosen to lie on the torus $\mathbf{J}'$ and substitute into (2.3) the values $\theta(t)$ taken by the toy angle variables along each orbit section. For the corresponding values of the target angle variables we substitute (1.1). This yields a linear system of equations for the coefficients $\partial S_{\mathbf{n}}/\partial \mathbf{J}'$, the initial phases θ'_0 and ω'. If the created torus approximates an existing invariant torus of H, the motion (1.1) on the torus will thus approximate the time development $\mathbf{w}(t)$ given by H. If the torus lies in a resonant or stochastic region of H, one is free to define how (1.1) projects into phase-space points $\mathbf{w}(t)$: obtaining $\partial S_{\mathbf{n}}/\partial \mathbf{J}'$ from many integrations provides a controlled transition over such a region.

As an example of this procedure, we show in Figure 3a the trajectories in toy-angle space of 36 orbit integrations started from different points on a box torus of (2.10). Figure 3b shows the same trajectories in target-angle space. The dots in Figures 3a,b show the input and output angles θ and θ', respectively, and the straight lines in Figure 3b show the associated trajectories $\theta'(t) = \theta'_0 + \omega' t$.

4 Defining H_0 and applying perturbation theory

A set of constructed tori essentially defines the new integrable Hamiltonian H_0. The H_0 we construct behaves like a separable Hamiltonian in that all its orbits are quasi-periodic and the tori form a global nested sequence at a given value of H_0, smoothly parametrized by either the actions $\mathbf{J}'$ or the orbital frequencies $\omega' = \partial H_0/\partial \mathbf{J}'$. The value of H_0 on the created tori that approximate existing invariant tori of H is defined to be the average value of H (i.e. $H_0 = H$ in principle). By interpolating smoothly in $\mathbf{J}'$ between the tori, we obtain all the quantities needed to define a torus for any given $\mathbf{J}'$ (Kaasalainen [1994]; also, cf. Warnock and Ruth [1991][1992]). Since we are in a position to specify at each grid point not only the function H_0 but its gradients ω' as well, the chosen interpolation scheme for H_0 should utilize all these values consistently.

We define tori for values of $\mathbf{J}'$ that lie between the grid points by independently interpolating the $S_\mathbf{n}$, the $\partial S_\mathbf{n}/\partial \mathbf{J}'$ and the variable parameters $\mathbf{P}$ in H_T and any point transformation – since the derivatives $\partial S_\mathbf{n}/\partial \mathbf{J}'$ are at constant values of the parameters, while the $S_\mathbf{n}$ correspond to varying values of the parameters, the former must be obtained by explicit interpolation rather than by differentiating the latter. Of course, if the parameters $\mathbf{P}$ are kept fixed, S and its derivatives must be interpolated self-consistently rather than independently. In that special case it is also possible to interpolate for S using the function values only, and then take the derivatives: this produces a canonical transformation. However, that transformation will not necessarily be a close approximation [via (1.1) and (2.3)] to the real time development of an orbit on an existing invariant torus of H, if the $\mathbf{J}'$-grid is not dense.

To understand how separate interpolations can be consistent, write the $S_\mathbf{n}$ in the functional form $S_\mathbf{n} = S_\mathbf{n}(\mathbf{J}', \mathbf{P}(\mathbf{J}'))$. Then have

$$\frac{\mathrm{d}S_\mathbf{n}}{\mathrm{d}\mathbf{J}'} = \left(\frac{\partial S_\mathbf{n}}{\partial \mathbf{J}'}\right)_\mathbf{P} + \left(\frac{\partial S_\mathbf{n}}{\partial \mathbf{P}}\right)_{\mathbf{J}'} \cdot \left(\frac{\partial \mathbf{P}}{\partial \mathbf{J}'}\right). \tag{4.1}$$

When $\partial \mathbf{P}/\partial \mathbf{J}'$ is nonzero, there should be many acceptable solutions (via $\partial S_\mathbf{n}/\partial \mathbf{P}$) for the full forms $S_\mathbf{n}(\mathbf{J}', \mathbf{P}(\mathbf{J}'))$ that fulfill consistency between the $S_\mathbf{n}$ and their total derivatives (4.1). The actual solutions are not needed; it is sufficient to know that they can be constructed.

If an integrable Hamiltonian close to a near-integrable one can be constructed in the above manner, is the converse true: can the original Hamiltonian be viewed as a perturbation of the constructed one? The original Hamiltonian can now be written in the form

$$H(\mathbf{J}', \theta') = H_0(\mathbf{J}') + \delta H(\mathbf{J}', \theta'), \tag{4.2}$$

where δH is periodic in the angles θ':

$$\delta H(\mathbf{J}, \theta') = \sum_\mathbf{n} H_\mathbf{n}(\mathbf{J}') \exp(\mathrm{i}\mathbf{n} \cdot \theta'). \tag{4.3}$$

Using the principles of standard secular perturbation theory (e.g. Lichtenberg and Lieberman [1983]), one arrives at a modified form of the usual pendulum-type description of perturbations that gives a good representation of any resonant islands (Paper IV). The standard theory has to be modified mainly because δH often effectively vanishes through much of phase space, whereas the standard treatment

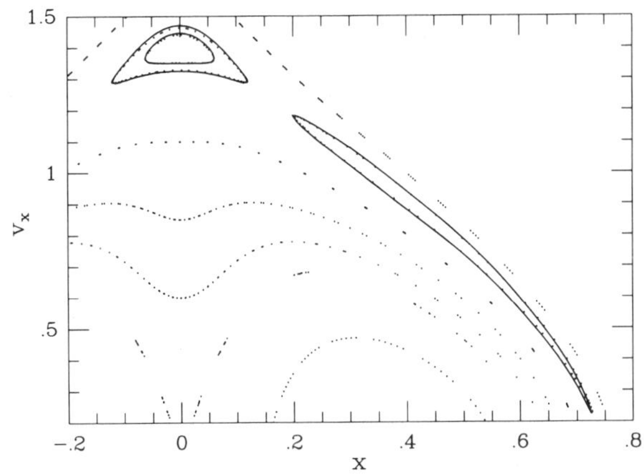

Figure 4 A surface of section as in Figure 2 (the v_x-scale has been slightly expanded); the solid curves show islands obtained by perturbing the numerically-constructed Hamiltonian H_0.

of resonant trapping assumes that δH is significant also far from the resonances. Also, the $H_{\mathbf{n}}(\mathbf{J}')$ sometimes contain antisymmetric components absent from the standard theory, and these have to be included in the formalism. In Figure 4, which presents the same surface of section as Figure 2, we show with solid curves the resonant islands as obtained from perturbation theory – these are in an excellent agreement with the integrated consequents (dots).

5 Conclusions

Techniques have now been developed that enable one to construct an integrable Hamiltonian H_0 that closely fits any given Hamiltonian H. The orbits of most non-trivial Hamiltonians fall into several distinct families. For each significant family one seeks a combination of a point transformation and a generating function (2.1) that maps the invariant tori of a toy Hamiltonian, such as the harmonic oscillator or the isochrone Hamiltonian, into approximately invariant tori of H. Non-linear least-squares minimization of the variation in H on the target torus is generally an effective tool. In difficult cases, the final least-squares fit can be refined by the method of Warnock [1991] once the point transformation has been chosen. Orbit integrations followed by the solution of a set of linear equations enable one to equip each torus with angle variables.

After fitting tori to every family one wishes H_0 to possess, a torus through an arbitrary point in phase space can be found by interpolation in the quantities which define the fitted tori. A value $H_0(\mathbf{J}')$ can, moreover, be assigned to each torus such that the difference $\delta H \equiv H - H_0$ is small everywhere, except where one has chosen not to fit a significant orbit family. Such neglected orbit families can now be treated as comprising orbits resonantly trapped by the perturbation δH to H_0. The perturbation theory required in this step has some novel features because H_0 fits H unusually tightly, especially away from the resonances of H_0.

This machinery promises to be of a sufficiently general nature that with it one should be able to tackle dynamical problems in several different fields of research.

Acknowledgments

MK thanks Osk. Huttunen Foundation, the British Council, and the ORS Awards Scheme for financial support.

References

Binney, J. and Kumar, S. [1993], Mon. Roy. Ast. Soc., **261**, 584. (Paper II).

Binney, J. and Tremaine, S. [1987], *Galactic Dynamics.* Princeton University Press, Princeton.

Boozer, A. [1982], Phys. Fluids, **25**, 520.

de Zeeuw, P. T. [1985], Mon. Roy. Ast. Soc., **216**, 273.

Kaasalainen, M. [1994], Mon. Roy. Ast. Soc., **268**, 1041. (Paper IV).

Kaasalainen, M. [1995], Mon. Roy. Ast. Soc., **275**, 162.

Kaasalainen, M. and Binney, J. [1994a], Mon. Roy. Ast. Soc., **268**, 1033. (Paper III).

Kaasalainen, M. and Binney, J. [1994b], Phys. Rev. Lett., **73**, 2377.

Lichtenberg, A. and Lieberman, M. [1983], *Regular and Stochastic Motion.* Springer, New York.

Martens, C. and Ezra, G. [1987], J. Chem. Phys., **86**, 279.

McGill, C. and Binney, J. [1990], Mon. Roy. Ast. Soc., **244**, 634. (Paper I).

Miyamoto, M. and Nagai R. [1975], Publ. Astr. Soc. Japan, **27**, 533.

Warnock, R. [1991], Phys. Rev. Lett., **66**, 1803.

Warnock, R. and Ruth, R. [1991], Phys. Rev. Lett., **66**, 990; Physica D, **56**, 188 [1992].

Fields Institute Communications
Volume 10, 1996

Exhaustive Search of Symplectic Integrators using Computer Algebra

P.-V. Koseleff
Équipe "Analyse Algébrique", Institut de Mathématiques
Université Pierre et Marie Curie, case 247
4 place Jussieu, F-75252 Paris Cedex 05

Abstract. We find symplectic integrators using universal exponential identities or relations among formal Lie series. We give here general methods to compute such identities in a free Lie algebra. We recover by these methods all the previously known symplectic integrators and some new ones. We list all minimal solutions for integrators of low order. We give some improvement in the case when the Hamiltonian is in form $T(p) + V(q)$. We give also all reversible fourth-order symplectic integrators for the planetary hamiltonian expressed in canonical heliocentric coordinates.

1 Introduction

For very long time integration, there has been recently a development of numerical methods preserving the symplectic structure (see for example [7, 18, 19, 20]), which seem to be more efficient with respect to computational cost.

Symplectic integrators may be seen as the time evolution mapping of a slightly perturbed Hamiltonian, that is to say as a Lie transformation that can be represented either by an exponential, a product of increasing order single exponentials or a proper Lie transformation. Constructing explicit high order symplectic integrators requires the manipulation of formal identities like exponential identities.

In Section 2, we recall first some definitions of the Hamilton formalism. In Section 3, we give some general methods to manipulate formal Lie series and Lie algebra automorphisms. We recall some theorems related to exponential identities and give explicit methods to compute them. They make use the Lyndon basis, which is particularly adapted to this problem. Most of this material has been published already in [8] but is not necessarily known by the reader.

In Section 4, we show how the algorithms described in Section 3 provide symplectic integrators. The idea of such constructions originates in Forest and Ruth

1991 *Mathematics Subject Classification.* Primary: 70H15, 70-08, 17B01; Secondary: 58F05, 58F, 58, 70F10, 70F15, 70F, 70, 17-04, 68-06.

[7], or more recently Yoshida [20]. Our approach in this paper is to combine the use of proper Lie transforms and exponentials. This avoids many unnecessary direct calculations of exponential identities. At the end we propose some improvement in the case when the Hamiltonian is separated into kinetic and potential energies. We give also some fourth-order integrators for the planetary Hamiltonian and show that they are minimals.

All the algorithms described in the present paper have been implemented using AXIOM (NAG) running on IBM-RS/6000-550.

Between the preparation of this paper and its publication, many papers have been published on the subject, especially [13].

1.1 Symplectic Integrators Given a phase space E which can be identified to $\mathbb{R}^{2n}$, a set of variables

$$(q,p) = (q_1, \dots, q_n, p_1, \dots, p_n) = (z_1, \dots, z_{2n}), \tag{1.1}$$

and an Hamiltonian $h = h(p, q, t)$, we consider the system of differential equations

$$\dot{p}_i = -\frac{\partial h}{\partial q_i}, \dot{q}_i = \frac{\partial h}{\partial p_i}, \quad 1 \le i \le n, \tag{1.2}$$

where $\dot{z} = \frac{dz}{dt}$ denotes the total time derivative. Introducing the Poisson bracket

$$\{f,g\} = \sum_{i=1}^{n} \frac{\partial f}{\partial p_i}\frac{\partial g}{\partial q_i} - \frac{\partial g}{\partial p_i}\frac{\partial f}{\partial q_i} \tag{1.3}$$

that turns the set of smooth functions on E onto a Lie algebra, (1.2) becomes

$$\dot{z}_i = \{z_i, h\} = -L_h z_i, \quad 1 \le i \le 2n. \tag{1.4}$$

A transformation on the phase space E is said canonical if it preserves the Poisson brackets. Such transformations are also called symplectic as their Jacobians belong to the symplectic group. One extends the canonical transformations on the functions on the phase space by $Tf(z) = f(T(z))$. Canonical transformations act on the Lie algebra of the Lie operators by $TL_fT^{-1} = L_{Tf}$. Here $L_f : g \mapsto \{f, g\}$. The set of L_f is a Lie algebra with $[L_f, L_g] = L_{\{f,g\}}$.

The time-evolution mapping $S_h(t) : z \mapsto z(t)$ is a canonical transformation. From (1.4), $S_h(t)$ is the solution of the differential equation

$$\frac{d}{dt}S_h(t) = -S_h(t)L_h, S_h(0) = \mathbf{1}. \tag{1.5}$$

If h is not time-dependent we have $S_h(t) = e^{-tL_h}$ and a formal solution of (1.4) is given by its Taylor series, called Lie series

$$z(t) = \sum_{n\ge 0}(-t)^n \frac{L_h^n}{n!} z. \tag{1.6}$$

If h is time-dependent, say for example $h = \sum_{n\ge 0} t^n h_n$, then $S_h(t)z$ may be written as a Lie series. If

$$Z = S_h(t)z = \sum_{n\ge 0} t^n Z_n,$$

we get from (1.4)

$$Z_0 = z, \quad Z_n = -\sum_{p=1}^{n} \tfrac{1}{n} L_{h_p} Z_{n-p}.$$

More generally, see [3], $S_h(t)$ is a series of operators $\sum_{n\geq 0} t^n (S_h)_n$ where

$$(S_h)_0 = \mathbf{1}, (S_h)_n = -\sum_{p=1}^{n} \tfrac{1}{n} L_{h_p} (S_h)_{n-p}.$$

Let us consider an Hamiltonian $h = A + B$, the two time-evolution mappings $S_A(t) = e^{-tL_A}$ and $S_B(t) = e^{-tL_B}$, and a given integer k, one seeks a minimal set of coefficients $c_1, \dots, c_n, d_1, \dots d_n$, such that

$$S^{(n)}(t) = S_A(c_1 t) S_B(d_1 t) \cdots S_A(c_n t) S_B(d_n t) = e^{-tL_h} + o(t^k). \tag{1.7}$$

$S^{(n)}(t)$ is a canonical transformation as composition of canonical transformations. The above expression may be considered as an equality between truncated Lie series. The aim of our paper is to show how one can solve this general problem considering the equation (1.7) as an universal identity between formal transformations on Lie algebra.

2 Lie algebraic formalism

In hamiltonian mechanics, the use of Lie methods or Lie transformations is efficient when it becomes necessary to manipulate Lie polynomials and to express exponential identities like the Baker-Campbell-Hausdorff formula. Our aim in this section is to give general methods for the computation of such identities.

These identities are universal Lie algebraic identities, that is to say they do not depend on the Lie algebra we work in or the Lie bracket we use. We work in free Lie algebras and with formal Lie series, neglecting all the convergence problems that can appear with analytical functions, for example.

We will use the Lyndon basis for the formal computations, but all the identities can be later evaluated in any Lie algebra.

2.1 Definitions X will denote an alphabet, that is to say an ordered set (possibly endless). R is a ring which contains the rational numbers $\mathbb{Q}$. X^* is the free monoid generated by X and is totally ordered with the lexicographic order. $M(X)$ is the free magma generated by X. It contains X and is equipped with a composition law : $(x, y) \mapsto (x, y)$.

$\mathcal{A}(X, R)$ is the associative algebra, that is to say the R-algebra of X^*.

A Lie algebra is an algebra in which the multiplication law $[,]$ is bilinear, alternate and satisfies the Jacobi identity:

$$[a, [b, c]] + [b, [c, a]] + [c, [a, b]] = 0. \tag{2.1}$$

$L(X, R)$ or $L(X)$ is the free Lie algebra on X. It is defined as the quotient of the R-algebra of $M(X)$ by the ideal generated by the elements (u, u) and $(u, (v, w)) + (v, (w, u)) + (w, (u, v))$.

An element of $M(X)$ considered as element of $L(X)$ will be called a Lie monomial. $L_n(X)$ is the free module generated by those of length n. Thus $L(X)$ is graded by the length denoted by $|x|$ for $x \in M(X)$. If $|X| = q < \infty$ we have Witt's formula (see [1, 14, 15]):

$$\sum_{d|n} d \dim L_d(X) = q^n. \tag{2.2}$$

Given a weighted alphabet X in which each letter a has an integer weight $||a||$, we take the weight as graduation for $L(X)$ which is defined as the unique extension of the weight in X. We denote by $\tilde{L}_n(X)$ (resp. $\tilde{\mathcal{A}}_n(X)$) the submodule of $L(X)$ (resp. $\mathcal{A}(X)$) spanned by the elements of weight n.

2.2 Formal Lie series We define the formal Lie series $\tilde{L}(X)$ and $\tilde{\mathcal{A}}(X)$ as

$$\tilde{L}(X) = \prod_{n\geq 0} \tilde{L}_n(X) \quad \text{and} \quad \tilde{\mathcal{A}}(X) = \prod_{n\geq 0} \tilde{\mathcal{A}}_n(X). \tag{2.3}$$

We will write $x \in \tilde{L}(X)$ as a series $\sum_{n\geq 0} x_n$. $\tilde{L}(X)$ is a complete Lie algebra with the Lie bracket

$$([x, y])_n = \sum_{p+q=n} [x_p, y_q]. \tag{2.4}$$

Denoting by $\tilde{L}(X)^+$ (resp. $\tilde{\mathcal{A}}(X)^+$) the ideal of $\tilde{L}(X)$ (resp. $\tilde{\mathcal{A}}(X)$) generated by the elements of positive weight, we can define the exponential and the logarithm as

$$\begin{array}{rcl} \exp : \tilde{\mathcal{A}}(X)^+ & \to & 1 + \tilde{\mathcal{A}}(X)^+ \\ x & \mapsto & \sum_{n\geq 0} \dfrac{x^n}{n!} \end{array} \qquad \begin{array}{rcl} \log : 1 + \tilde{\mathcal{A}}(X)^+ & \to & \tilde{\mathcal{A}}(X)^+ \\ x & \mapsto & -\sum_{n\geq 1} \dfrac{(1-x)^n}{n}. \end{array} \tag{2.5}$$

They are mutually reciprocal functions and we have (see [1], Ch. II, Section 5) the

Theorem 2.1 (Campbell-Hausdorff) *If $x, y \in \tilde{L}(X)^+$ then*

$$\log\left[\exp(x)\exp(y)\right] \in \tilde{L}(X)^+. \tag{2.6}$$

Using the preceding lemma we deduce (see [4, 16]) the

Proposition 2.1 (Factored product expansion) *Given $k \in \tilde{L}(X)^+$, there is a unique series $g \in \tilde{L}(X)^+$ such that*

$$\exp(\textstyle\sum_{n\geq 1} k_n) = \cdots \exp(g_n) \cdots \exp(g_1). \tag{2.7}$$

2.3 Lie series automorphisms For x in $\tilde{L}(X)$ we denote the Lie operator $L_x y = [x, y]$ by $L(x)$ or L_x. From the Jacobi identity (2.1) we have $[L_x, L_y] = L_x L_y - L_y L_x = L_{[x,y]}$. The set of L_x is a Lie algebra that we call the adjoint Lie algebra. For any Lie series automorphisms T, we have by definition $[Tf, Tg] = T[f, g]$. The Lie series automorphisms act on the adjoint Lie algebra by

$$T L_f T^{-1} = L_{Tf}. \tag{2.8}$$

Let us give now some examples of Lie transformations that play an important role in hamiltonian mechanics.

Example 2.2 (The exponential.) Given $x \in \tilde{L}(X)^+$, we consider $\exp(L_x)$ defined as

$$\exp(L_x)y = \sum_{i \geq 0} \frac{L_x^i}{i!} y. \tag{2.9}$$

From the Campbell-Hausdorff theorem (2.1), the set of all $\exp(L_x)$ is a group $\mathbf{G}$ that we will call the Lie transformations group.

Example 2.3 (The Lie transform.) For $w = \sum_{n \geq 1} t^n w_n$, we denote by T_w and T_w^{-1} the solution [3] of

$$\frac{d}{dt} T_w = -T_w L_{\frac{dw}{dt}} \quad \text{and} \quad \frac{d}{dt} T_w^{-1} = L_{\frac{dw}{dt}} T_t^{-1}. \tag{2.10}$$

With the notation of (1.5), we have $T_w^{-1} = S_{\frac{dw}{dt}}$. For $g \in \tilde{L}(X)$, we have (see [2, 3]) $T_w^{-1} g = \sum_{n \geq 0} G_n$ where

$$G_0 = g_0, \quad G_{0,n} = g_n, \quad G_{p,q} = \sum_{k=1}^{p} \tfrac{k}{p} [w_k, G_{p-k,q}], \quad G_n = \sum_{p=0}^{n} G_{p,n-p}. \tag{2.11}$$

We call this transformation the Deprit transform. The composite T of two Lie transforms T_u and T_v satisfies

$$\begin{aligned} \frac{dT}{dt} = \frac{dT_u}{dt} T_v + T_u \frac{dT_v}{dt} &= -T_u L_{\frac{du}{dt}} T_v - T_u T_v L_{\frac{dv}{dt}} && (2.12) \\ &= -T \left(T_v^{-1} L_{\frac{du}{dt}} T_v + L_{\frac{dv}{dt}} \right) && (2.13) \\ &= -T L_{T_v^{-1} \frac{du}{dt} + \frac{dv}{dt}}. && (2.14) \end{aligned}$$

So $T = T_w$ where $\frac{dw}{dt} = T_v^{-1} \frac{du}{dt} + \frac{dv}{dt}$.

Example 2.4 (The Dragt-Finn transform.) The Dragt-Finn transform is the infinite product of exponential maps (see [4]). Given $g = \sum_{n \geq 1} g_n$, we define M_g and M_g^{-1} as

$$M_g = \exp(-L_{g_1}) \cdots \exp(-L_{g_n}) \cdots, \quad M_g^{-1} = \cdots \exp(L_{g_n}) \cdots \exp(L_{g_1}). \tag{2.15}$$

Note that this transformation will be used in this paper as a technical support for proving relations between Lie series.

2.4 Relations between transformations The three above transformations are totally defined by generating series and are connected by the following

Proposition 2.2 ([9]) *Given $w, k, g \in \tilde{L}(X)^+$, there exist*
— *$k' \in \tilde{L}(X)^+$ with $k'_n - w_n \in L(w_1, \ldots w_{n-1})$ such that* $\exp(L_{k'}) = T_w^{-1}$,
— *$g' \in \tilde{L}(X)^+$ with $g'_n - k_n \in L(k_1, \ldots k_{n-1})$ such that* $M_{g'}^{-1} = \exp(L_k)$,
— *$w' \in \tilde{L}(X)^+$ with $w'_n - g_n \in L(g_1, \ldots g_{n-1})$ such that* $T_{w'}^{-1} = M_g^{-1}$.

Proof The third part of the above proposition has been already proved by Finn [6], but not in terms of Lie polynomials. One proves (see [9, 10]) that $T_w = M_g$ if and only if

$$\frac{dw}{dt} = \sum_{n\geq 1} t^{n-1} \left[\sum_{k=1}^{n} k \sum_{\substack{(k+1)m_{k+1}+\cdots \\ +(n-k)m_{n-k}=n-k}} \frac{L_{g_{n-k}}^{m_{n-k}} \cdots L_{g_{k+1}}^{m_{k+1}}}{m_{k+1}! \cdots m_{n-k}!} g_k \right], \tag{2.16}$$

or equivalently

$$w_n = \sum_{k=1}^{n} \frac{k}{n} \sum_{\substack{(k+1)m_{k+1}+\cdots \\ +(n-k)m_{n-k}=n-k}} \frac{L_{g_{n-k}}^{m_{n-k}} \cdots L_{g_{k+1}}^{m_{k+1}}}{m_{k+1}! \cdots m_{n-k}!} g_k = g_n + G_n \tag{2.17}$$

in which $G_n \in L(g_1, \ldots, g_{n-1})$.

Using the Proposition (2.1), one proves the existence of $g = \sum_{n\geq 1} g_n$ such that

$$\exp(\textstyle\sum_{n\geq 1} L_{k_n}) = \cdots \exp(L_{g_n}) \cdots \exp(L_{g_1}), \tag{2.18}$$

in which $g_n = k_n + K_n$ and $K_n \in L(k_1, \ldots, k_{n-1})$. Combining (2.17) and (2.18) we deduce proposition 2.2. □

We deduce in passing that any Lie transformation $T \in \mathbf{G}$ may be expressed as an exponential of a Lie operator or as an infinite product of single exponentials or as a proper Lie transform. The use of a representation depends deeply on the result we look for. For example, if we have to compose transformations, it is much easier to consider Lie transforms because their product is a Lie transform whose generating function appears easily.

Explicit relations up to any given order may be easily computed, using the Lyndon basis. For example, given $w = \sum_{n\geq 1} w_n$, we have at the order 6 $\exp(L_k) = T_w^{-1}$ in which

$$\begin{aligned} k = {} & w_1 + w_2 + w_3 - \tfrac{1}{6}\,[w_1,\, w_2] + w_4 - \tfrac{1}{4}\,[w_1,\, w_3] + w_5 - \tfrac{3}{10}\,[w_1,\, w_4] \\ & - \tfrac{1}{10}\,[w_2,\, w_3] + \tfrac{1}{120}\,[w_1,\, [w_1,\, w_3]] + \tfrac{1}{60}\,[[w_1,\, w_2],\, w_2] \\ & + \tfrac{1}{360}\,[w_1,\, [w_1,\, [w_1,\, w_2]]] + w_6 - \tfrac{1}{3}\,[w_1,\, w_5] - \tfrac{1}{6}\,[w_2,\, w_4] \\ & + \tfrac{1}{60}\,[w_1,\, [w_1,\, w_4]] + \tfrac{1}{30}\,[w_1,\, [w_2,\, w_3]] + \tfrac{1}{24}\,[[w_1,\, w_3],\, w_2] \\ & + \tfrac{1}{240}\,[w_1,\, [w_1,\, [w_1,\, w_3]]] - \tfrac{1}{180}\,[w_1,\, [[w_1,\, w_2],\, w_2]] \end{aligned}$$

3 Search of symplectic integrators

Let $h = A + B$ be an Hamiltonian. For given integers n and k, one looks for a set $c_1, \ldots, c_n, d_1, \ldots d_n$, such that

$$S^{(n)}(t) = S_A(c_1 t) S_B(d_1 t) \cdots S_A(c_n t) S_B(d_n t) = S_h(t) + o(t^k). \tag{3.1}$$

Let $R = \mathbb{Q}[c_1, \ldots, c_n, d_1, \ldots, d_n]$, $\tilde{L}_p(\{A, B\})$ be the submodule of $\tilde{L}(\{A, B\})$ spanned by elements of weight p, where $||A|| = ||B|| = 1$. Using the Witt formula (2.2), the dimension l_p of $\tilde{L}_p(\{A, B\})$ satisfies $\sum_{d|n} d l_d = 2^n$.

Let $\tilde{L}_p\{z, A, B\}$ be the submodule of $\tilde{L}(\{z, A, B\})$, spanned by those of weight p, in which the partial commutative degree in z is 1 (here $||z|| = 0$). Using the Lyndon basis, one proves directly that the dimension of $\tilde{L}_p\{z, A, B\}$ is 2^p.

In order to solve equation (3.1), we have to express exponential identities first. We first prove the algebraic equalities obtained by representing the integrators in several ways.

Proposition 3.1 *Let us consider the following problems:*

1. *Having expressed* $S^{(n)}(t)$ *as* $\exp(-\sum_{p\geq 1} t^p L_{K_p})$, *solve* $K_1 = A + B, K_2 = \cdots = K_k = 0$,
2. *Having expressed* $S^{(n)}(t)$ *as* $\exp(-tL_{G_1})\cdots\exp(-t^p L_{G_p})\cdots$, *solve* $G_1 = A+B, G_2 = \cdots = G_k = 0$,
3. *Having expressed* $S^{(n)}(t)$ *as* $T_{(tW_1+\cdots+t^pW_p+\cdots)}$, *solve* $W_1 = A + B, W_2 = \cdots = W_k = 0$.
4. *Having expressed* $Z = S^{(n)}(t)z$, *solve* $Z_0 = (\exp(-tL_{A+B})z)_0, \ldots, Z_k = (\exp(-tL_{A+B})z)_k$.

The solutions of these four problems are the zeroes of the same polynomial ideal.

Proof Denoting the Lyndon basis $\mathcal{L}_p(\{A, B\})$ by $(x_{p,1}, \ldots, x_{p,l_p})$, we have

$$K_p = \sum_{i=1}^{l_p} K_{p,i}x_{p,i},\ G_p = \sum_{i=1}^{l_p} G_{p,i}x_{p,i},\ W_p = \sum_{i=1}^{l_p} W_{p,i}x_{p,i}. \tag{3.2}$$

For the first three methods, the solutions are the zeroes of the ideals

$$\begin{aligned} \mathcal{I}_K^{(k)} &= (K_{1,1} - 1, K_{1,2} - 1) + (K_{i,j}; 2 \leq i \leq k, 1 \leq j \leq l_i), \\ \mathcal{I}_G^{(k)} &= (G_{1,1} - 1, G_{1,2} - 1) + (G_{i,j}; 2 \leq i \leq k, 1 \leq j \leq l_i), \\ \mathcal{I}_W^{(k)} &= (W_{1,1} - 1, W_{1,2} - 1) + (W_{i,j}; 2 \leq i \leq k, 1 \leq j \leq l_i). \end{aligned} \tag{3.3}$$

We have to bear in mind that S_W is the Deprit transform associated to $\int_0^t W(u)\,du$. We therefore get the relations due to the Proposition 2.2:

$$\begin{aligned} K_p &= \tfrac{1}{p}W_p + R_p^W, & R_p^W &\in L_p(W_1, \ldots, W_{p-1}), \\ \tfrac{1}{p}W_p &= G_p + R_p^G, & R_p^G &\in L_p(G_1, \ldots, G_{p-1}), \\ G_p &= K_p + R_p^K, & R_p^K &\in L_p(K_1, \ldots, K_{p-1}). \end{aligned} \tag{3.4}$$

We first have $K_1 = G_1 = W_1$. For $p > 1$, each R_p^W may be expressed in the Lyndon basis $\mathcal{L}_p(\{A, B\})$ and the coefficients are polynomials. Each monomial contains a $W_{i,j}$ where $i > 1$ and thus belongs to $\mathcal{I}_W^{(k)}$. We therefore deduce that $\mathcal{I}_K^{(k)} \subset \mathcal{I}_W^{(k)}$. On the same way, we deduce $\mathcal{I}_W^{(k)} \subset \mathcal{I}_G^{(k)}$ and $\mathcal{I}_G^{(k)} \subset \mathcal{I}_K^{(k)}$.

We thus have $\mathcal{I}_K^{(k)} = \mathcal{I}_W^{(k)} = \mathcal{I}_G^{(k)}$. Z_p belongs to $\tilde{L}_p\{z, A, B\}$ which basis is denoted by $z_{p,1}, \ldots, z_{p,2^p}$. If $Z_p = \sum_{q=1}^{2^p} Z_{p,q}z_{p,q}$ and $(\exp(-tL_{A+B})z)_p = \sum_{q=1}^{2^p} a_{p,q}z_{p,q}$, the solutions of **4.** are the zeroes of

$$\mathcal{I}_Z^{(k)} \doteq (Z_{i,j} - a_{i,j}, 1 \leq i \leq k, 1 \leq j \leq 2^i) \tag{3.5}$$

From $S^{(n)}(t)z - \exp(-tL_{A+B})z = \left(S^{(n)}(t) - \exp(-tL_{A+B})\right)z = o(t^k)$, we deduce for $p = 1$

$$Z_1 - \{z, A + B\} = -(L_{G_1} - L_{A+B})z = \{z, G_1 - (A + B)\} = 0, \tag{3.6}$$

that is, onto the basis $([z,A],[z,B])$,

$$Z_{1,1}-1=G_{1,1}-1=0,\quad Z_{1,2}-1=G_{1,2}-1=0. \tag{3.7}$$

For $1<p\le k$, we have

$$\begin{aligned} Z_p-\tfrac{(-1)^p}{p!}L^p_{A+B}z &= \left[\sum_{m_1+\cdots+pm_p=p}(-1)^{m_1+\cdots+m_p}\frac{L^{m_1}_{G_1}\cdots L^{m_p}_{G_p}}{m_p!\cdots m_1!}-(-1)^p\frac{L^p_{A+B}}{p!}\right]z\\ &= \tfrac{(-1)^p}{p!}\left(L^p_{G_1}-L^p_{A+B}\right)z-L_{G_p}z+ \qquad (3.8)\\ &\qquad \sum_{\substack{m_1+\cdots+(p-1)m_{p-1}=p\\ m_1<p}}(-1)^{m_1+\cdots+m_{p-1}}\frac{L^{m_1}_{G_1}\cdots L^{m_{p-1}}_{G_{p-1}}}{m_p!\cdots m_1!}z \qquad (3.9)\\ &= 0. \qquad (3.10)\end{aligned}$$

$L^p_{G_1}-L^p_{A+B}$ may be written as

$$\sum_{k_1+\cdots+k_p+l_1+\cdots+l_p=p}\left(G_{1,1}^{k_1+\cdots+k_p}G_{1,2}^{l_1+\cdots+l_p}-1\right)L^{k_1}_A L^{l_1}_B\cdots L^{k_p}_A L^{l_p}_B. \tag{3.11}$$

A coefficient of (3.11) may be expressed as

$$G^k_{1,1}G^l_{1,2}-1=(G^k_{1,1}-1)G^l_{1,2}+(G^l_{1,2}-1)\in(G_{1,1}-1,G_{1,2}-1). \tag{3.12}$$

Each $L^{k_1}_A L^{l_1}_B\cdots L^{k_p}_A L^{l_p}_B$ is a sum of $z_{p,q}$ so any coefficient in $(L^p_{G_1}-L^p_{A+B})z$ belongs to $(G_{1,1}-1,G_{1,2}-1)\subset\mathcal{I}^{(p)}_G$. Other terms in (3.8) and (3.9) have coefficients in $(G_{2,1},\ldots,G_{p,1},\ldots G_{p,l_p})\subset\mathcal{I}^{(p)}_G$. We thus deduce that $\mathcal{I}_Z(p)\subset\mathcal{I}^p_G$.

Suppose we have for each $m<p$, $\mathcal{I}^{(m)}_Z=\mathcal{I}^{(m)}_G$. At the order p, from (3.10), we deduce also that

$$\begin{aligned} -L_{G_p}z &= \overbrace{Z_p-\tfrac{(-1)^p}{p!}L^p_{A+B}z}^{(I)}-\overbrace{\tfrac{(-1)^p}{p!}\left(L^p_{G_1}-L^p_{A+B}\right)z}^{(II)}-\\ &\qquad \underbrace{\sum_{m_1+\cdots+m_{p-1}=p,m_1<n}(-1)^{m_1+\cdots+m_{p-1}}\frac{L^{m_1}_{G_1}\cdots L^{m_{p-1}}_{G_{p-1}}}{m_p!\cdots m_1!}z}_{(III)}. \qquad (3.13)\end{aligned}$$

Each term on the r.h.s. of the above equation, has a decomposition onto the Lyndon basis $z_{p,q}$. Coefficients of (I) belong to $\mathcal{I}^{(p)}_Z$, coefficients of (II) and (III) belong to $(G_{1,1}-1,G_{1,2}-1)\subset\mathcal{I}^{(p-1)}_G\subset\mathcal{I}^{(p-1)}_Z\subset\mathcal{I}^{(p)}_Z$.

We deduce that the coefficients of the $L_{G_p}z$ decomposition onto the $z_{p,q}$ belong to $\mathcal{I}^{(p)}_Z$. As solution of (3.13), the coefficients of G_p onto the $x_{p,q}$ belong to $\mathcal{I}^{(p)}_Z$ We thus have $\mathcal{I}^{(p)}_G\subset\mathcal{I}^{(p)}_Z$ and eventually $\mathcal{I}^{(p)}_G=\mathcal{I}^{(p)}_Z$. □

In order to obtain symplectic integrators, we can use one of these methods which are algebraically equivalent. We will not use the Dragt-Finn representation as its mathematical interpretation in terms of invariants is not clear. Nevertheless, composition of Factored Product transformations is widely used in optics [5].

Direct method: The problem (3.1) may be solved, looking for all z

$$S^{(n)}(t)z = S_A(c_1t)S_B(d_1t)\cdots S_A(c_nt)S_B(d_nt)z = S_h(t)z + o(t^k). \quad (3.14)$$

At each order p, we obtain a system of 2^p polynomial equations.

Invariant function: Problem (3.14) may be solved by expressing $S^{(n)}$ as an exponential and looking for $c_1, \ldots, c_n, d_1, \ldots d_n$ such that

$$S^{(n)}(t) = S_A(c_1t)S_B(d_1t)\cdots S_A(c_nt)S_B(d_nt) = e^{-tL_K} \quad (3.15)$$

with $K = h + o(t^{k-1})$. K is not the Hamiltonian governing the system given by $S^{(n)}$, but an invariant function. At each order p, there is l_p polynomial equations to solve. In order to get these, we have to compute some Baker-Campbell-Hausdorff formulas.

Perturbed Hamiltonian: One can also express $S^{(n)}(t)$ as a Lie transform

$$S_A(c_1t)S_B(d_1t)\cdots S_A(c_nt)S_B(d_nt) = S_W \quad (3.16)$$

where $W = h + o(t^{k-1})$. The condition is obtained by writing

$$\frac{d}{dt}S_W = -S_WL_W = -S_hL_h + o(t^{k-1}). \quad (3.17)$$

W is the Hamiltonian governing the system which time-evolution mapping is $S^{(n)}$. At each order p, there is the same number of equations as previously but that avoids many unnecessary direct calculations of Baker-Campbell-Hausdorff formulas.

3.1 First integrators For a given order k, and for a given method (see Table 1) one seeks a minimal n such that $S^{(n)}(t)$ is a kth-order symplectic integrator. One looks for the zeroes of a polynomial ideal. We use here algebraic methods like Gröbner basis that we compute using AXIOM (when possible) or MACAULAY that works in a ring $\mathbb{Z}/p\mathbb{Z}$ and gives some precious results.

For low orders, these methods furnish symplectic integrators. All the following results have been obtained with algorithms on Lie series and have been implemented t in AXIOM [9]. We then obtain the polynomials that define the variety we look at. For low orders, these can be described [8].

• The solution for $k = 1$ is given by $c_1 = d_1 = 1$ and

$$S_1(t) = S^{(2)}(t) = S_A(t)S_B(t). \quad (3.18)$$

• The solution for $k = 2$ is given by

$$S_2(t) = S^{(3)}(t) = S_A(\tfrac{t}{2})S_B(t)S_A(\tfrac{t}{2}) = S_1(\tfrac{t}{2})S_1^{-1}(-\tfrac{t}{2}). \quad (3.19)$$

This approximant is reversible, that is to say satisfies $S_2^{-1}(t) = S_2(-t)$.

• Solutions for $k = 3$ are

$$\begin{aligned} S_3(t) &= S_A(ct)S_B(ct)S_B(ct)S_A(ct)S_A(\bar{c}t)S_B(\bar{c}t)S_B(\bar{c}t)S_A(\bar{c}t) \\ &= S_2(ct)S_2(\bar{c}t) \end{aligned} \quad (3.20)$$

where $c^2 - \frac{1}{2}\,c + \frac{1}{12} = 0$.

• When $k = 4$, we find two sets of solutions and 5 solutions. The corresponding integrators have 7 factors. If $c^3 - 2c^2 + c = \frac{1}{6}$, we get

$$S_4(t) = S_2(ct)S_2((1-2c)t)S_2(ct). \quad (3.21)$$

These solutions are known and have been given also by Yoshida [20].

If c is a root of $c^2 - \frac{1}{2}c + \frac{1}{6} = 0$, we get two other solutions:

$$S_4(t) = S_2(ct)S_2((c+\bar{c})t)S_2(\bar{c}t). \tag{3.22}$$

• For $k = 5$, one gets exactly 46 solutions as product of 5 approximants S_2.

• For $k = 6$, an exhaustive list is still unknown. If we add the condition of being reversible, we get at most 39 solutions, working in $\mathbb{Z}/31991\mathbb{Z}$. All integrators are products of second-order integrators. The real valued integrators for $k = 2$ or 4 are reversible, which means that $S(-t) = S^{-1}(t)$.

3.2 Reversible Integrators Representing a reversible integrator $S(t)$ by an exponential $\exp(-tL_K)$, we deduce that $K(t) = K(-t)$. Looking for reversible integrators, we can deduce from the Campbell-Hausdorff formula the

Lemma 3.1 ([17]) *If $S_{2k}(t)$ is a reversible symplectic integrator of order $2k$, then*

$$S(t) = S_{2k}\left(\frac{1}{2-\sqrt[2k+1]{2}}t\right) S_{2k}\left(-\frac{\sqrt[2k+1]{2}}{2-\sqrt[2k+1]{2}}t\right) S_{2k}\left(\frac{1}{2-\sqrt[2k+1]{2}}t\right)$$

is a reversible symplectic integrator of order $2k+2$.

This lemma allows us to build reversible symplectic integrators of order $2k$ as products of $2.3^{k-1}+1$ single operators S_A or S_B. With this method we find a 19-factor sixth-order integrator.

One can try to find directly reversible integrators looking for reversible products

$$S_R^{(n)}(t) = S_A(c_nt)S_B(d_nt)\cdots S_A(c_1t)S_B(d_1t)S_A(c_0t)S_B(d_1t)S_A(c_1t)\cdots S_B(d_nt)S_A(c_nt).$$

Denoting by $S^{(n)}(t)$ the operator

$$S_A(c_nt)S_B(d_nt)\cdots S_A(c_1t)S_B(d_1t)S_A(\tfrac{c_0}{2}t), \tag{3.23}$$

we obtain $S_R^{(n)}$ as $S^{(n)}(t){S^{(n)}}^{-1}(-t)$ that we can express as an exponential or a Lie transform. Representing $S_R^{(n)}$ as an exponential e^{-tL_K} has the advantage that $K(t) = K(-t)$. Moreover we have the following lemma resulting from proposition 2.2:

Lemma 3.2 *Let $\mathcal{I}_W^{(k)} = (W_{1,1}-1, W_{1,2}-1, W_{2,1}, \ldots, W_{k,1}, \ldots, W_{k,l_k})$, be the polynomial ideal defining the solutions of $S_R^{(n)}(t) = S_{W(t)} = S_h + o(t^k)$. For each $2p \le k$, we have*

$$\{W_{2p,1}, \ldots, W_{2p,l_{2p}}\} \subset \mathcal{I}_W^{(2p-1)} \quad so \quad \mathcal{I}_W^{(2p)} = \mathcal{I}_W^{(2p-1)}. \tag{3.24}$$

Proof Using proposition 3.1, we get for each $1 \le 2p \le k$,

$$\mathcal{I}_K^{(p)} = \mathcal{I}_W^{(p)} = (K_{1,1}-1, K_{1,2}-1, K_{2,1}, \ldots, K_{p,1}, K_{p,l_p}) \tag{3.25}$$

where the $K_{i,j}$ are the coefficient of K satisfying $\exp(-tK(t)) = S_r^{(n)}$. As S_W is reversible, we have $K_{2p} = 0$, for each p, that is $\mathcal{I}_K^{(2p)} = \mathcal{I}_K^{(2p-1)}$ and $\mathcal{I}_W^{(2p)} = \mathcal{I}_W^{(2p-1)}$. □

This lemma proves that there is no need to consider odd terms of the Hamiltonian obtained with reversible integrators.

• For $k = 4$, one finds 3 reversible integrators obtained with the direct method.

• For $k = 6$, one proves that there is no solutions for $n < 8$. For $n = 8$, one sees, using the Hilbert function implemented in MACAULAY, that the variety of solutions in $\mathbb{Z}/p\mathbb{Z}$ ($p = 31991$) is constituted of 39 points. There are at most 39 algebraic solutions over $\mathbb{Q}$.

Another solution has been proposed by Yoshida [20] consisting in the finding of reversible integrators as reversible product of second-order integrators S_2. We look for

$$S^{(n)}(t) = S_2(c_n t) \cdots S_2(c_1 t) S_2(c_0 t) S_2(c_1 t) \cdots S_2(c_n t) = e^{-tL_{K^{(n)}}}. \tag{3.26}$$

• For $k = 4$, we find the real valued reversible integrator previously found by the direct method or using the Lemma (3.1).

• For $k = 6$, we have four equations with four unknowns $c_0, \ldots, c_3$. The solution is obtained after eliminations with

$$\begin{aligned} P_0(c_0) = {} & c_0^{39} + 4\, c_0^{38} - 18\, c_0^{37} - \tfrac{232}{3}\, c_0^{36} + \tfrac{6469}{45}\, c_0^{35} + \tfrac{8108}{15}\, c_0^{34} - \tfrac{82144}{135}\, c_0^{33} - \\ & \tfrac{239008}{135}\, c_0^{32} + \tfrac{870652}{675}\, c_0^{31} + \tfrac{5898416}{2025}\, c_0^{30} - \tfrac{618824}{675}\, c_0^{29} - \tfrac{5158016}{2025}\, c_0^{28} + \\ & \tfrac{2525372}{30375}\, c_0^{27} + \tfrac{32135888}{30375}\, c_0^{26} - \tfrac{1377776}{10125}\, c_0^{25} - \tfrac{33361568}{91125}\, c_0^{24} + \tfrac{536566}{10125}\, c_0^{23} + \\ & \tfrac{35651416}{455625}\, c_0^{22} - \tfrac{19660868}{1366875}\, c_0^{21} - \tfrac{8051504}{455625}\, c_0^{20} + \tfrac{5636474}{1366875}\, c_0^{19} + \tfrac{11313208}{4100625}\, c_0^{18} - \\ & \tfrac{17674448}{20503125}\, c_0^{17} - \tfrac{8733536}{20503125}\, c_0^{16} + \tfrac{1302268}{6834375}\, c_0^{15} + \tfrac{87632}{2460375}\, c_0^{14} - \tfrac{624184}{20503125}\, c_0^{13} + \\ & \tfrac{288448}{922640625}\, c_0^{12} + \tfrac{3333844}{922640625}\, c_0^{11} - \tfrac{716752}{922640625}\, c_0^{10} - \tfrac{127664}{553584375}\, c_0^{9} + \tfrac{143264}{922640625}\, c_0^{8} - \\ & \tfrac{136499}{4613203125}\, c_0^{7} - \tfrac{19996}{8303765625}\, c_0^{6} + \tfrac{117142}{41518828125}\, c_0^{5} - \tfrac{33848}{41518828125}\, c_0^{4} + \\ & \tfrac{17431}{124556484375}\, c_0^{3} - \tfrac{9668}{622782421875}\, c_0^{2} + \tfrac{656}{622782421875}\, c_0 - \tfrac{64}{1868347265625} = 0, \end{aligned}$$

and $c_1 = P_1(c_0), c_2 = P_2(c_0), c_3 = P_3(c_0)$ where P_1, P_2, P_3 are polynomials of degree 38. P_0 is irreducible over $\mathbb{Q}$ and has only three real roots. All the solutions are reached with this method as there is at most 39 solutions.

• For $k = 8$, Yoshida [20] has found 5 real valued integrators using numerical methods. These integrators involve 31 single integrators S_A or S_B. We proved, using standard basis computed with MACAULAY, that these integrators are not products of 5 fourth-order symplectic integrators.

3.3 Special cases Most of the time, when $h = T(p) + V(q)$, the kinetic energy is just a quadratic form in p. That means that $\{T, V\}$ is of degree one in p, $\{\{T, V\}, V\}$ depends only on q and $\{\{\{T, V\}, V\}, V\} = 0$. We may find symplectic integrators of order 4 or 6 involving less terms.

Unfortunately, there is no integrator of order 4 using less than 7 terms.

As $\{\{T, V\}, V\}$ depends only on q, $V_1 = \alpha V + t^2 \beta \{\{T, V\}, V\}$ depends only on q and t for any α, β and we have

$$e^{-tL_{V_1}} p = p - t \frac{\partial V_1}{\partial q} \quad \text{and} \quad e^{-tL_{V_1}} q = q. \tag{3.27}$$

Denoting $e^{-t(\alpha L_V + \beta t^2 L_{\{\{T,V\},V\}})}$ by $S_{\alpha,\beta}(t)$ we look now for integrators $S^{(n)}$ as product of

$$S_{c_n, z_n}(t) S_T(d_n t) \cdots S_{c_1, z_1}(t) S_T(d_0 t) S_{c_1, z_1}(t) \cdots S_T(d_n t) S_{c_n, z_n}(t) \tag{3.28}$$

or

$$S_T(d_n t)S_{c_n,z_n}(t)\cdots S_T(d_1 t)S_{c_0,z_0}(t)S_T(d_1 t)\cdots S_{c_n,z_n}(t)S_T(d_n t) \qquad (3.29)$$

With this method we found a 5-factor fourth-order integrator and an 9-factor sixth-order integrator (see [8]).

3.4 Decomposition into more terms One can generalize the search of symplectic integrators to the case of 3 operators (see for example Suzuki [17]).

In this part, we show how to find minimal symplectic integrators in the case when $h = A_1+A_2+A_3$, using the algorithms previously described. Such integrators can be useful for planetary problems written in the canonical heliocentric variables of Poincaré (see [12]).

We will give an exhaustive list for orders 1, 2 and 4. It is clear that any permutation on A_1, A_2, A_3 will also give an integrator.

The first-order integrator is

$$S_1(t) = S_{A_1}(t)S_{A_2}(t)S_{A_3}(t). \qquad (3.30)$$

• From $S_2(t) = S_{A_2}(\frac{t}{2})S_{A_3}(t)S_{A_2}(\frac{t}{2})$, we deduce an second-order integrator for $A_1 + (A_2 + A_3)$:

$$\begin{aligned} S_2(t) &= S_{A_1}(\tfrac{t}{2})S_{A_2}(\tfrac{t}{2})S_{A_3}(t)S_{A_2}(\tfrac{t}{2})S_{A_1}(\tfrac{t}{2}). \\ &= S_1(\tfrac{t}{2})S_1^{-1}(-\tfrac{t}{2}) \end{aligned} \qquad (3.31)$$

• Looking for reversible fourth-order integrators, we could deduce from the 7-factor fourth-order integrator S_4 for $A_1 + A_3$, that

$$S_{A_3}(d_2 t)S_4(c_1 t)S_{A_3}(d_1 t)S_4(c_0 t)S_{A_3}(d_1 t)S_4(c_1 t)S_{A_3}(d_2 t) \qquad (3.32)$$

is a 25-factor fourth-order integrator for $h = (A_1 + A_2) + A_3$.

Using the Lemma 3.1, one obtains a 13-factor fourth-order symplectic integrator as a product of 3 reversible second-order integrators:

$$\begin{aligned} &S_{A_1}(\tfrac{a}{2}t)S_{A_2}(\tfrac{a}{2}t)S_{A_3}(at)S_{A_2}(\tfrac{a}{2}t)S_{A_1}(\tfrac{a+b}{2}t)S_{A_2}(\tfrac{b}{2}t)S_{A_3}(bt)S_{A_2}(\tfrac{b}{2}t) \\ &\qquad S_{A_1}(\tfrac{a+b}{2}t)S_{A_2}(\tfrac{a}{2}t)S_{A_3}(at)S_{A_2}(\tfrac{a}{2}t)S_{A_1}(\tfrac{a}{2}t). \end{aligned} \qquad (3.33)$$

where $a = \frac{1}{2-\sqrt[3]{2}}, b = -\frac{\sqrt[3]{2}}{2-\sqrt[3]{2}}$.

Let us prove now that these integrators are minimal regards to the number of factors. For a given order k, a kth-order integrator is an operator of length m:

$$S^{(m)}(t) = S_{A_1}(x_1 t)S_{A_2}(x_2 t)S_{A_3}(x_3 t)\cdots S_{A_m}(x_m t), \qquad (3.34)$$

where $i - 1 = m \bmod 3$. A p-factor kth-order integrator is a sequence $(x_1, \ldots, x_m)$ in which $x_1 x_m \neq 0$ and $m - p$ of the x_i's are equals to zero. In such a sequence, there is no 2 consecutive zeroes, otherwise its length would be $m - 2$. For the integrator (3.31), we have

$$x_1 = \tfrac{1}{2}, x_2 = \tfrac{1}{2}, x_3 = 1, x_4 = 0, x_5 = \tfrac{1}{2}, x_6 = 0, x_7 = \tfrac{1}{2}.$$

For given m and k the set of sequence $(x_1, \ldots, x_m)$ satisfying

$$S_{A_1}(x_1 t)S_{A_2}(x_2 t)S_{A_3}(x_3 t)\cdots S_{A_m}(x_m t) = S_h(t) + o(t^k), \qquad (3.35)$$

is an algebraic variety. For a given order k, let us denote by M_k the minimal integer such that each minimal kth-order integrator (up to a permutation of (A_1, A_2, A_3))

may be written as $S^{(m)}$ where $m \leq M_k$. Each kth-order integrator (up to a permutation of A_1, A_2, A_3) will be a sequence that is solution of

$$S^{(m)}(x_1, \ldots, x_m) = S_{A_1}(x_1 t) \cdots S_{A_m}(x_m t) = S_{A_1+A_2+A_3}(t) + o(t^k), \qquad (3.36)$$

with $m \leq M_k$.

The second-order example shows that $M_2 \geq 7$. Let S_2 be any 5-factor second-order integrator. Its length is at most 9. If its length is 9 then we have $x_2 = x_4 = x_6 = x_8 = 0$. By transposing A_2 and A_3, the sequence $(x_1, x_3, x_5, x_7, x_9)$ gives also a 5-factor integrator.

If its length is 8, then there are 3 zeroes in the subsequence $x_2, \ldots, x_7$, because $x_1 x_8 \neq 0$. There is only two possibilities: $x_2 = x_4 = x_6 = 0$ or $x_3 = x_5 = x_7 = 0$. In the first case, by transposing A_2 and A_3, the sequence $(x_1, x_3, x_5, x_7, 0, x_8)$ gives a 6-factor integrator. In the second case, the same transposition gives also a 5-factor integrator with the sequence $(x_1, 0, x_2, x_4, x_6, x_8)$. It proves that any 5-factor operator (e.g. second-order integrator) (up to permutations) has a length less than 7.

- There is no 5-factor second-order integrator of length 5.
- Looking for integrators of length 6, one finds

$$S_2(t) = S_{A_1}(ct) S_{A_2}(ct) S_{A_3}(ct) S_{A_1}(\bar{c}t) S_{A_2}(\bar{c}t) S_{A_3}(\bar{c}t) = S_1(ct) S_1(\bar{c}t) \qquad (3.37)$$

where c is a complex root of $c^2 - c + \frac{1}{2}$.

- Looking for all possible solution involving 7 variables, we have $x_1 x_7 \neq 0$ so for a 5-factor integrator we must have $x_2 x_3 x_5 x_6 = 0$. We thus find

$$x_1 = x_7 = \tfrac{1}{2}, x_4 = 0, (x_2 - \tfrac{1}{2}) x_2 = 0, x_2 = x_3 - \tfrac{1}{2} = \tfrac{1}{2} - x_6 = 1 - x_5. \qquad (3.38)$$

If $x_2 = 0$ or $x_2 = 1$, we find the integrator (3.31). That proves that any minimal real second-order integrator has 5 factors and length 7.

Let S be a 7-factor operator. One can suppose that the 2 first factors are S_{A_1} and S_{A_2}. Let m be its length and express S as

$$S(t) = S_{A_1}(x_1 t) S_{A_2}(x_2 t) S_{A_3}(x_3 t) \cdots S_{A_m}(x_m t) \qquad (3.39)$$

in which $x_1 x_2 x_m \neq 0$. The subsequence $(x_1, \ldots, x_{m-1})$ has a length $m - 2$ and corresponds to a 5-factor operator. We thus deduce that $m - 2 \leq 8$ because x_{m-1} could be zero. So any 7-factor operator may be written as an operator of length $m \leq 10$.

Suppose now that S has length 10 and that $x_1 x_2 x_{10} \neq 0$. There are 3 zeroes in the subsequence $x_3, \ldots, x_9$ and there are not consecutive. The only solutions are

a) $x_3 = x_5 = x_7 = 0$, *b*) $x_3 = x_5 = x_8 = 0$, *c*) $x_3 = x_5 = x_9 = 0$,

d) $x_3 = x_6 = x_8 = 0$, *e*) $x_3 = x_6 = x_9 = 0$, *f*) $x_3 = x_7 = x_9 = 0$,

g) $x_4 = x_6 = x_8 = 0$, *h*) $x_4 = x_6 = x_9 = 0$, *i*) $x_4 = x_7 = x_9 = 0$,

j) $x_5 = x_7 = x_9 = 0$.

Let us suppose that $S(t)S^{-1}(-t)$ is a reversible fourth-order integrator. If we suppose now that $A_1 = 0$, then $S(t)S^{-1}(t)$ is still a fourth-order integrator for $A_2 + A_3$. It implies that we must have at least 2 factors S_{A_2} and 2 factors S_{A_3} in S. So the only cases to consider are *a*), *g*), *j*).

In the first case, suppose that $A_2 = 0$, then we obtain

$$S'(t) = S_{A_1}((x_1 + x_4)t)S_{A_3}((x_6 + x_9)t)S_{A_1}(x_{10}t).$$

In the second case, suppose that $A_3 = 0$, then we get

$$S'(t) = S_{A_1}(x_1 t)S_{A_2}((x_2 + x_5)t)S_{A_1}((x_7 + x_{10})t).$$

In the third case, suppose that $A_1 = 0$, then we obtain

$$S'(t) = S_{A_2}(x_2 t)S_{A_3}((x_3 + x_6)t)S_{A_2}(x_8 t).$$

One of those case would imply that there is 5-factor fourth-order integrator which is impossible.

It shows that any minimal reversible fourth-order integrator S_4 may be found by looking for a 7 factor operator $S(t)$ of maximal length 9, such that $S_4 = S(t)S^{-1}(-t)$.

If we look for all possible reversible 13 factor fourth-order integrators we get a zero dimensional algebraic variety of degree 12 and c_0 satisfies

$$(c_0^9 - 2\,c_0^8 + c_0^7 + \tfrac{2}{3}\,c_0^6 - c_0^5 + \tfrac{2}{3}\,c_0^3 - \tfrac{5}{9}\,c_0^2 + \tfrac{2}{9}\,c_0 - \tfrac{1}{27})(c_0^3 + c_0^2 - c_0 + \tfrac{1}{3}) = 0.$$

We therefore get 2 sets of solutions:

$$\begin{aligned}
&c_0 = 2e_1 = 2d_1,\\
&c_1 = e_2 = d_2 = 27\,c_0^8 - \tfrac{81}{2}\,c_0^7 + \tfrac{9}{2}\,c_0^6 + \tfrac{45}{2}\,c_0^5 - 15\,c_0^4 - 9\,c_0^3 + \tfrac{27}{2}\,c_0^2 - \tfrac{15}{2}\,c_0 + 2,\\
&c_2 = d_3 = e_3 = -c_1 - \tfrac{1}{2}c_0 - \tfrac{1}{2},\\
&c_0^9 - 2\,c_0^8 + c_0^7 + \tfrac{2}{3}\,c_0^6 - c_0^5 + \tfrac{2}{3}\,c_0^3 - \tfrac{5}{9}\,c_0^2 + \tfrac{2}{9}\,c_0 - \tfrac{1}{27} = 0.
\end{aligned}$$

and

$$\begin{aligned}
&c_1 = e_2 = 0,\\
&d_2 = d_3 = e_3 = \tfrac{1}{2}c_2 = -\tfrac{1}{4}\,c_0 + \tfrac{1}{4},\\
&d_1 = \tfrac{1}{2}c_0,\\
&e_1 = \tfrac{1}{4}c_0 + \tfrac{1}{4},\\
&c_0^3 + c_0^2 - c_0 + \tfrac{1}{3} = 0.
\end{aligned}$$

The first set gives 17-factor integrators while the second set gives 13-factor integrators as product of four S_{A_1}, six S_{A_2} and three S_{A_3}. As we can exchange A_1, A_2 and A_3, we shall take for A_2 the part for which the time-evolution mapping has the lowest cost.

3.5 Planetary Hamiltonian Wisdom and Holman [18] have used a symplectic integrator for their integrations of the solar system which allowed them to use a longer step size. They used an expression of the Planetary Hamiltonian in term of Jacobi coordinates for which the Hamiltonian is splitted in two parts. Here we prefer to use the canonical heliocentric variables of Poincaré which provides a more elegant and symmetrical formulation of the hamiltonian [12], which is then expressed in three integrable parts: H_0, T_1, U_1.

H_0 corresponds to the sum of n disjoint Keplerian problems. T_1 is the perturbation depending only on the actions and U_1 depends only on the positions. We are thus led to search for symplectic integrators for Hamiltonians which decompose in three integrable parts $H = A + B + C$.

Let us consider O the center of mass of $n+1$ bodies of masses $m_0, \dots, m_n$ in gravitational interaction. Let u_i be the coordinates with respect to O and $\Delta_{i,j} = \|u_i - u_j\|$, the Hamiltonian becomes

$$H = T + U = \tfrac{1}{2}\sum_{i=0}^{n} m_i \|\dot{u}_i\|^2 - G \sum_{0 \le i < j \le n} \tfrac{m_i m_j}{\Delta_{i,j}}. \tag{3.40}$$

Let $\tilde{u}_i = m_i \dot{u}_i$, we obtain in canonical coordinates

$$T = \tfrac{1}{2}\sum_{i=0}^{n} \frac{\|\tilde{u}_i\|}{m_i}. \tag{3.41}$$

Let us consider now the heliocentric coordinates: $r_0 = u_0, r_i = u_i - u_0$. In order to have canonical variables, we take

$$\tilde{r}_0 = \sum_{i=0}^{n} u_i = 0, \tilde{r}_i = \tilde{u}_i, \quad 1 \le i \le n \tag{3.42}$$

or

$$\tilde{u}_0 = -\sum_{i=1}^{n} \tilde{r}_i, \tilde{u}_i = \tilde{r}_i, \quad 1 \le i \le n. \tag{3.43}$$

The kinetic energy becomes

$$T = \tfrac{1}{2}\sum_{i=1}^{n} \frac{\|\tilde{r}_i\|^2}{m_i} + \tfrac{1}{2}\frac{\|\sum_{i=1}^{n} \tilde{r}_i\|^2}{m_0} \tag{3.44}$$

$$= \tfrac{1}{2}\sum_{i=1}^{n} \|\tilde{r}_i\|^2 \left[\tfrac{1}{m_i} + \tfrac{1}{m_0}\right] + \sum_{0<i<j} \frac{\tilde{r}_i . \tilde{r}_j}{m_0} \tag{3.45}$$

and

$$U = -G\sum_{i=1}^{n} \frac{m_0 m_i}{r_i} - G \sum_{0<i<j\le n} \frac{m_i m_j}{\Delta_{i,j}}. \tag{3.46}$$

One can write $H = H_0 + H_1$ with $H_0 = T_0 + U_0$, $H_1 = T_1 + U_1$ where H_0 is the Hamiltonian of n disjoint two body problems: the planet of mass $\frac{m_0 m_i}{m_0 + m_i}$ around the sun of mass $m_0 + m_i$. H_1 may be considered as an interactive perturbation. We thus have

$$T_0 = \tfrac{1}{2}\sum_{i=1}^{n} \|\tilde{r}_i\|^2 \left[\tfrac{1}{m_i} + \tfrac{1}{m_0}\right], U_0 = -G\sum_{i=1}^{n} \frac{m_0 m_i}{r_i} \tag{3.47}$$

$$T_1 = \sum_{0<i<j} \frac{\tilde{r}_i . \tilde{r}_j}{m_0}, U_1 = -G \sum_{0<i<j\le n} \frac{m_i m_j}{\Delta_{i,j}}. \tag{3.48}$$

H_1 is particularly simple as the kinetic energy and the potential energy depend on coordinates and momenta respectively.

Writing $H = H_0 + T_1 + U_1$ where T_1 is very easy to integrate, we can try to use the integrator defined in (3.32). That is what we will do in a near future at the Bureau des Longitudes.

4 Conclusion

We showed in this paper that there are exactly 5 7-factor fourth-order symplectic integrators. Three of them are known (see [7, 20]). There are exactly 46 11-factor fifth-order symplectic We showed, that there are exactly 39 15-factor reversible sixth-order symplectic integrators. All of them are reversible products of second-order integrators. Three of them were known [20]. In the case when $h = A + B + C$, we show that minimal second-order integrator have length 5 and reversible fourth-order integrator have length 13.

References

[1] Bourbaki, N. [1972], *Groupes et algèbres de Lie.* Éléments de Mathématiques. Hermann, Paris.

[2] Cary, J. R. [1981], *Lie transform perturbation theory for Hamiltonian systems*, Phys. Rep., **79**, 129–159.

[3] Deprit, A. [1969], *Canonical transformations depending on a small parameter*, Cel. Mech., **1**, 12–30.

[4] Dragt, A. J. and Finn, J. M. [1976], *Lie series and invariant functions for analytic symplectic maps*, J. Math. Phys., **17**.

[5] Dragt, A. J. and Healy, L. M. [1988], *Concatenation of Lie algebraic maps*, In Lie Methods in Optics II, volume 352 of Lec. Notes in Physics, Springer Verlag.

[6] Finn, J. M. [1984], *Lie series: a perspective*, In Local and Global Methods of Nonlinear Dynamics, volume 252 of Lec Notes in Physics, Springer Verlag.

[7] Forest, E. and Ruth, D. [1990], *Fourth-order symplectic integration*, Physica D, **43**, 105–117.

[8] Koseleff, P.-V. [1993], *Relations among formal Lie series and construction of symplectic integrators*, In AAECC'10 Proceedings, volume 673 of Lect. Not. Comp. Sci.

[9] Koseleff, P.-V. [1993], *Thèse de troisième cycle 1993*, École Polytechnique, January 1993.

[10] Koseleff, P.-V. [1994a], *Comparison between Deprit and Dragt-Finn perturbation methods*, Cel. Mech.

[11] Koseleff, P.-V. [1994b], *Formal Calculus for Lie Methods in Hamiltonian Mechanics*, Ph.D. Thesis, English translation, Lawrence Berkeley Laboratory, University of California, LBID-2030, Rev. UC-405.

[12] Laskar, J. [1991], *Analytical framework in Poincaré variables for the motion of the solar system*, In Predictability, Stability and Chaos in N-Body Dynamical Systems (Roy, A., ed.), NATO ASI. Plenum Press.

[13] MacLachlan, R.I. [1995], *On the numerical integration of ordinary differential equations by symmetric composition methods*, SIAM J. Sci. Comp., **16**, 151–168.

[14] Michel, J. [1974], *Bases des algèbres de lie libres, Étude des coefficients de la formule de Campbell-Hausdorff*, Thèse, Orsay.

[15] Perrin, D. [1983], *Combinatorics On Words*, Chapter 5. M. Lothaire, Factorization of free monoids. Addison-Wesley.

[16] Steinberg, S. [1985], *Lie series, Lie transformations, and their applications*, In Lie Methods in Optics, volume 250 of Lec. Notes in Physics, Springer Verlag.

[17] Suzuki, M. [1990], *Fractal decomposition of exponential operators with applications to many-body theories and monte carlo simulations*, Phys. Lett. A, **146**, 319–323.

[18] Wisdom, J. and Holman, M. [1991], *Symplectic maps for the n-body problem*, Astronom. J, **102**, 1528–1538.

[19] Yoshida, H. [1990], *Conserved quantities of symplectic integrators for Hamiltonian systems*, Physica D.

[20] Yoshida, H. [1990], *Construction of higher order symplectic integrators*, Phys. Lett. A, **150**, 262–268.

Fields Institute Communications
Volume **10**, 1996

Conserving Algorithms for the N Dimensional Rigid Body

D. Lewis
Mathematics Board
University of California at Santa Cruz
Santa Cruz CA 95064

J.C. Simo*
Division of Applied Mechanics
Stanford University
Stanford, CA 94305

1 Introduction

Recently, (Moser and Veselov [1991]) and (Lewis and Simo [1994]) proposed techniques for constructing conserving algorithms for the rigid body. The former authors concern themselves with the particular case of the N dimensional rigid body whereas the latter authors address the general problem of numerical integration of Hamiltonian systems on arbitrary Lie groups. The goal of this paper is to specialize the results in (Lewis and Simo [1994]) to the case of the N dimensional rigid body and to draw a detailed comparison between the two algorithms. In doing so, we arrive at new results which further develop our earlier treatment. In particular, we obtain an explicit estimate of the critical time step below which existence of an algorithmic solution is ensured and establish a variational formulation of the algorithm using the discrete Lagrangian of Moser and Veselov.

Although concerned with the same application, the emphasis in the treatment of this application in the current work differs significantly from that of the Moser-Veselov treatment. While Moser and Veselov concern themselves primarily with issues pertaining to the integrability and the Hamiltonian structure of the discrete rigid body, leading to questions about matrix factorization, we emphasize the design and analysis of effective and computationally efficient algorithms for the *numerical simulation* of Hamiltonian systems, including the rigid body. In particular, we are interested in numerical schemes that preserve exactly the Hamiltonian and the momentum map. We expect that the conservation of these quantities is relevant to the capture of key qualitative features of the long term dynamics. The remarkable

1991 *Mathematics Subject Classification.* Primary: 65L05, 70E15; Secondary: 65L70, 70H05, 70H33.
*Deceased.

feature that our exact energy-momentum schemes are also symplectic appears to be related to the complete integrability of the N dimensional rigid body.

The key points of the current work can be summarized as follows. We construct an implicit, second order accurate equivariant exact energy-momentum conserving symplectic algorithm for the N dimensional rigid body using the results given in Lewis and Simo [1994]. We prove the solvability of the implicit equations for our scheme and provide an estimate of the critical time step below which existence of a solution is guaranteed. We show that our scheme can be interpreted as the Euler-Lagrange equations of a function defined by an infinite series. This formulation is the point of departure in the analysis of Moser and Veselov. Our algorithm differs from that of Moser and Veselov in the following aspects: Our scheme automatically preserves the true momentum map of the Hamiltonian dynamics for the N rigid body, while the update map of Moser and Veselov preserves an *algorithmic* momentum map which coincides with the true spatial momentum in the limit as the time-step approaches zero; the 'algorithmic' trajectory produced by the Moser-Veselov update map must be corrected if momentum conservation is desired. In addition, our algorithm is a single-step algorithm, while the Moser-Veselov algorithm is a multi-step scheme. These differences between two algorithms with a common variational characterization are due to different choices for the momentum equation supplementing the variational condition for the configurations. In theory, there is a one-to-one mapping between exact discrete time trajectories computed using our scheme and exact trajectories computed using the Moser-Veselov algorithm. Essentially, this map corresponds to shifting the body momentum by one time step. It is only in this sense that the two algorithms are equivalent.

Remark 1.1 The existence of a one-to-one mapping relating the discrete-time trajectories of the two algorithms does not imply that the two algorithms are identical. A classical example is provided by the relation between the mid-point rule and the trapezoidal rule. A well-known result going back to (Dahlquist [1975]) shows that the trajectories of the two algorithms are related by a one-to-one mapping. (See Hairer et al. [1991], page 330.) However, the mid-point rule is B-stable and symplectic, while the trapezoidal rule is neither. In theory, the trajectories of one scheme can be reconstructed from the trajectories of the other; in practice, finite-precision arithmetic will typically lead to incompatible results.

Remark 1.2 In the presentation given below an effort has been made to minimize the use of the notation and terminology of differential geometry and geometric mechanics. Our intention is to make the presentation accessible to a wide audience. We expect that geometrically inclined readers will recognize the relevant constructs. For a more formal approach, the interested reader is directed to (Lewis and Simo [1994]).

2 Design of energy-momentum and symplectic algorithms for $SO(N)$

The problem of interest here is the unreduced Euler equations for the N dimensional rigid body, obtained by appending the evolution equation for the configuration of the dynamical system to the reduced Euler equations in the form generalized by (Arnold [1966][1978]). The system of ODE's with phase space $P = SO(N) \times so(N)^*$ reads as follows

$$\dot{\Lambda} = \Lambda\Omega \qquad \text{and} \qquad \dot{\Pi} = [\Pi, \Omega], \tag{2.1}$$

where $\Lambda \in SO(N)$ is the generalized attitude matrix, $\Omega \in so(N)$ is the skew-matrix of generalized body angular velocities and $\Pi \in so(N)^* \approx so(N)$ is the skew-matrix of generalized body momenta, defined in terms of the Euler tensor E as

$$\Pi = E\Omega + \Omega E = 2\,\mathrm{skew}\,[\Omega E]. \tag{2.2}$$

Recall that E is a symmetric positive definite $N \times N$ matrix determined by the inertia properties of the reference configuration. The bracket pairing $[\cdot\,,\,\cdot]$ is the usual matrix commutator.

The classic paper of (Arnold [1966]) explains in detail the Hamiltonian nature of equations (2.1) supplemented by (2.2). Here we simply recall that the Hamiltonian associated to (2.1) can be written using relation (2.2) as

$$H(\Lambda, \Pi) = \tfrac{1}{2} \langle \Pi, \Pi \rangle_{E^{-1}} := \tfrac{1}{4} \mathrm{tr}\left[\Pi^T \Omega\right]. \tag{2.3}$$

In addition to conserving the Hamiltonian, the dynamics of the rigid body conserve quantities determined by the rotational invariance of the Hamiltonian (2.3); these quantities comprise an element of $so(N)^*$. Specifically, the momentum map $\boldsymbol{J} : P \to so(N)^*$ defined as

$$\boldsymbol{J}(\Lambda, \Pi) = \Lambda \Pi \Lambda^T =: \pi \tag{2.4}$$

is conserved by equations (2.1). These conservation laws can be used to prove the remarkable and far from trivial fact that equations (2.1) and (2.2) define a completely integrable system. See (Manakov [1976] and Ratiu [1980]) for proofs of the integrability of the N dimensional free rigid body with the metric determined by (2.2). For more general metrics on $so(N)$, i.e. a more general map relating the body velocity Ω and the body momentum Π than that determined by the Euler tensor (2.2), the equations (2.1) may not be integrable.

Remark 2.1 We note that the equations (2.1) can be written in the alternative but entirely equivalent form

$$\dot{\Lambda} = \Lambda\Omega \qquad \text{and} \qquad \dot{\pi} = 0, \tag{2.5}$$

where Ω is now specified by combining (2.2) and (2.4) to obtain

$$E\Omega + \Omega E = \Lambda^T \pi \Lambda. \tag{2.6}$$

The algorithmic counterpart of the equations (2.5) and (2.6) is used in the actual implementation of the algorithm described below.

The general algorithm design strategy described in (Lewis and Simo [1994]) is to consider a general class of exact momentum conserving algorithms, then successively enforce equivariance, energy conservation and finally, if there is "enough room", enforce the symplectic character of the flow. This program can be fully carried out for the unreduced Euler equations (2.1) for $SO(N)$. A key feature of the algorithms derived in Lewis and Simo [1994] and here is the formulation of computationally intensive steps in the algorithm, such as the iterative solution of implicit equations, on the vector space $so(N)$, rather than the nonlinear manifold $SO(N)$. To do so, we make use of the Cayley transform cay : $so(N) \to SO(N)$, given by

$$\mathrm{cay}\,[\Theta] := \left(\mathbf{1} + \tfrac{1}{2}\Theta\right)\left(\mathbf{1} - \tfrac{1}{2}\Theta\right)^{-1}. \tag{2.7}$$

Recall the following two properties of the Cayley transform, which will be used repeatedly in what follows:

1. *Equivariance.* $Q\mathrm{cay}\,[\Theta]Q^T = \mathrm{cay}\left[Q\Theta Q^T\right]$ for all $\Theta \in so(N)$ and $Q \in SO(N)$.
2. *Lie algebra fixed point property.* $[\mathrm{cay}\,[\Theta], \Theta] = 0$, i.e. $\mathrm{cay}\,[\Theta]\,\Theta\,\mathrm{cay}\,[\Theta]^{-1} = \Theta$, for all $\Theta \in so(N)$. (The bracket pairing $[\ ,\ ]$ denotes the standard matrix commutator.)

These properties can easily be verified using (2.7). Also note that $\mathrm{cay}\,[\Theta]^{-1} = \mathrm{cay}\,[-\Theta]$. Use of the Cayley transform and the algebra $so(N)$ results in an efficient, easily implemented algorithm. No coordinate charts are required — the vector space $so(N)$ can be globally identified with $\mathbb{R}^{N(N-1)/2}$. As we shall see in section 3, we can take advantage of familiar analytic results for functions on vector spaces when carrying out the estimates required to establish existence of solutions to the implicit equation in our algorithm.

The algorithm design can naturally be described in a series of steps, each step implementing another conservation law.

Step 1. Exact momentum conserving algorithms. Consider the family of algorithms

$$\left.\begin{array}{rcl} \Lambda_{n+1} & = & \Lambda_n \mathrm{cay}\,[\Theta_n] \\ \pi_{n+1} & = & \pi_n \end{array}\right\}, \tag{2.8}$$

where Θ_n is an as yet unspecified element of $so(N)$. These algorithms are formulated in terms of the spatial momentum, corresponding to the form (2.5) and (2.6) of the unreduced Euler equations. By design, the algorithms (2.8) produce trajectories lying in the level set of spatial momentum π_0 in P. In practice, we need only compute the configuration update; the spatial momentum update is trivial. Note that $\mathrm{cay}\,[\Theta] \in SO(N)$ implies that $\Lambda_{n+1} \in SO(N)$ for *any* choice of $\Theta \in so(N)$ if $\Lambda_n \in SO(N)$.

Remark 2.2 Let $\Pi_n = \Lambda_n^T \pi_n \Lambda_n$ denote the body angular momentum associated to the configuration Λ_n and spatial momentum π_n. The analog of algorithm (2.8) in terms of the body momenta is

$$\left.\begin{array}{rcl} \Lambda_{n+1} & = & \Lambda_n \mathrm{cay}\,[\Theta_n] \\ \Pi_{n+1} & = & \mathrm{cay}\,[-\Theta_n]\,\Pi_n\,\mathrm{cay}\,[\Theta_n] \end{array}\right\}. \tag{2.9}$$

Using the properties of the Cayley transform for $N = 3$, (2.9) can be written in the form

$$\left.\begin{array}{rcl} \Lambda_{n+1} - \Lambda_n & = & \Lambda_{n+\frac{1}{2}} \Theta_n \\ \Pi_{n+1} - \Pi_n & = & \left[\Pi_{n+\frac{1}{2}}, \Theta_n\right] \end{array}\right\}, \tag{2.10}$$

on $SO(3)$, where

$$\Lambda_{n+\frac{1}{2}} = \tfrac{1}{2}(\Lambda_n + \Lambda_{n+1}) \qquad \text{and} \qquad \Pi_{n+\frac{1}{2}} = \tfrac{1}{2}(\Pi_n + \Pi_{n+1}). \tag{2.11}$$

The expression (2.10) is reminiscent of a mid-point rule approximation to the unreduced Euler equations (2.1) in the body representation. In fact, the midpoint rule can be obtained from (2.10) by setting Θ_n equal to Δt times the midpoint body angular velocity $\Omega_{n+\frac{1}{2}}$, defined by Euler's formula

$$\Pi_{n+\frac{1}{2}} = E\Omega_{n+\frac{1}{2}} + \Omega_{n+\frac{1}{2}}E. \tag{2.12}$$

Note that the matrix $\Lambda_{N+\frac{1}{2}}$ is typically not an element of $SO(N)$. In (2.10) we make use of the fact that $SO(N)$ can be viewed as a submanifold of the vector space $L(N)$ to express (2.9) in a familiar form.

Step 2. Enforcement of equivariance. The Euler equations (2.1) are equivariant with respect to the spatial (left) action of $SO(N)$ on the phase space P, i.e. the Hamiltonian flow determined by (2.1) commutes with the left group action of $SO(N)$ on P, given by

$$Q \cdot (\Lambda, \Pi) = (Q\Lambda, \Pi) \text{ or, equivalently, } Q \cdot (\Lambda, \pi) = (Q\Lambda, Q\pi Q^T) \tag{2.13}$$

for any $Q \in SO(N)$. Equivariance of the algorithmic version (2.8) of (2.1) mapping (Λ_n, π_n) to $(\Lambda_{n+1}, \pi_{n+1})$ corresponds to the condition that the scheme map $Q \cdot (\Lambda_n, \pi_n)$ to $Q \cdot (\Lambda_{n+1}, \pi_{n+1})$. The equivariance property of the Cayley transform and conservation of spatial momentum imply that the algorithm (2.8) is equivariant if and only if the relative rotation vector Θ_n between the configurations Λ_n and Λ_{n+1} equals the relative rotation vector between $Q\Lambda_n$ and $Q\Lambda_{n+1}$ for all $(\Lambda_n, \pi_n) \in P$ and $Q \in SO(N)$. Equations (2.2), (2.12) and (2.13) suggest the use of the body angular velocity Ω as the building block for the relative rotation vector. Specifically, if Θ_n is a function of the body angular velocities Ω_n and Ω_{n+1} alone, then (2.8) yields an equivariant single-step scheme.

Step 3. Conservation of energy. Note that for the free rigid body the incremental change in the energy (2.3) satisfies

$$H_{n+1} - H_n = \left\langle \Pi_{n+\frac{1}{2}}, \Pi_{n+1} - \Pi_n \right\rangle_{E^{-1}}. \tag{2.14}$$

Using (2.9) and (2.12), it follows that

$$H_{n+1} - H_n = \left\langle \Omega_{n+\frac{1}{2}}, \operatorname{cay}[-\theta_n]\Pi_n \operatorname{cay}[\theta_n] - \Pi_n \right\rangle, \tag{2.15}$$

where the pairing $\langle\ ,\ \rangle$ is given by $\langle \Omega, \Xi \rangle = \frac{1}{2}\operatorname{tr}\left[\Omega^T \Xi\right]$. Regrouping terms, we obtain

$$H_{n+1} - H_n = \left\langle \Pi_n, \operatorname{cay}[\theta_n]\Omega_{n+\frac{1}{2}} \operatorname{cay}[-\theta_n] - \Omega_{n+\frac{1}{2}} \right\rangle, \tag{2.16}$$

Thus a sufficient condition for energy conservation is

$$\left[\Theta_n, \Omega_{n+\frac{1}{2}}\right] = 0. \tag{2.17}$$

In particular, the midpoint rule obtained by setting $\Theta_n = \Delta t\, \Omega_{n+\frac{1}{2}}$ is an energy and momentum conserving algorithm.

Step 4. Conservation of the symplectic form. We shall not rederive the condition for conservation of the symplectic form by algorithms of the class (2.8) here. The reader is referred to (Lewis and Simo [1994]) for the derivation of the general form of the condition for algorithms on arbitrary Lie groups with general Hamiltonians. This general expression involves various geometric structures, such as the co-adjoint action of the Lie group G on the dual of the Lie algebra and the infinitesimal action of the Lie algebra on its dual. Rather than discuss these structures in detail, we derive the particular expression for the symplectic condition for the scheme (2.8) from the general symplectic condition in the appendix. The resulting condition can be expressed purely in terms of left and right matrix multiplication. Define $\boldsymbol{m} : so(N) \to so(N)^* \approx so(N)$ by $\boldsymbol{m} := L_E + R_E$, where L_E (respectively R_E) denotes left (respectively right) multiplication by the matrix E. Given an element Θ of $so(N)$, define the linear map $\beta_\Theta : so(N) \to so(N)$ by

$$\beta_\Theta \Xi := 2\,\Xi - \tfrac{1}{2}\,\Theta\,\Xi\,\Theta \qquad \text{for all } \Xi \in so(N). \tag{2.18}$$

The algorithm (2.8) is symplectic if there exists a differentiable map $\sigma : so(N) \to so(N)$ such that the residual equation $\sigma(\Theta_n) = \Delta t\, \Omega_{n+\frac{1}{2}}$ has a solution for each n and

$$2\,\text{skew}\left[\boldsymbol{m}\,\nabla\sigma(\Theta_n)\,\beta_{\Theta_n}\right]\Xi = \left[\beta_{\Theta_n}^{-1}\left[\boldsymbol{m}\,\sigma(\Theta_n), \Theta_n\right], \Xi\right] \tag{2.19}$$

for all $\Xi \in so(N)$. Here the notation skew$[\,\cdot\,]$ denotes the projection onto its skew-symmetric component of a linear map of a vector space into its dual; e.g., if $M : so(N) \to so(N)^*$, then $2\,\langle \Theta, \text{skew}\,[M]\,\Xi\rangle = \langle \Theta, M\Xi\rangle - \langle \Xi, M\Theta\rangle$ for all Θ, $\Xi \in so(N)$.

Theorem 2.3 *The algorithm*

$$\boxed{\begin{aligned} \Lambda_{n+1} &= \Lambda_n \text{cay}\,[\Theta_n] \\ \pi_{n+1} &= \pi_n \\ \Delta t\,\Omega_{n+\frac{1}{2}} &= \text{skew}\,[\text{cay}\,[\Theta_n]], \end{aligned}} \tag{2.20}$$

where $\Omega_{n+\frac{1}{2}}$ is given by (2.11) and (2.12), exactly conserves the energy, momentum and symplectic form and is equivariant and second order accurate.

Proof Clearly (2.20) is of the form (2.8) and hence exactly conserves the spatial momentum. The Lie algebra fixed point property of the Cayley transform and (2.20_3) imply that

$$\left[\Theta_n, \Omega_{n+\frac{1}{2}}\right] = \tfrac{1}{\Delta t}\left[\Theta_n, \text{skew}\,[\text{cay}\,[\Theta_n]]\right] = \tfrac{1}{\Delta t}\text{skew}\,[[\Theta_n, \text{cay}\,[\Theta_n]]] = 0. \tag{2.21}$$

Hence the energy conservation condition (2.17) derived in Step 2 holds for the scheme (2.20). Equivariance of the map $\sigma : so(N) \to so(N)$ given by

$$\sigma(\Theta) := \text{skew}\,[\text{cay}\,[\Theta]] = \left(\mathbf{1} - \tfrac{1}{2}\Theta\right)^{-1} - \left(\mathbf{1} + \tfrac{1}{2}\Theta\right)^{-1}, \tag{2.22}$$

and hence of the algorithm (2.20), is a direct consequence of the equivariance of the Cayley transform.

We now turn to the symplectic condition (2.19). We need to show that the map σ satisfies (2.19). (The existence of solutions to the equation $\sigma(\Theta_n) = \Delta t\,\Omega_{n+\frac{1}{2}}$ is addressed in section 3.) To simplify the relevant expressions, we introduce the map

$\Upsilon : so(N) \to S(N)$, where $S(N)$ denotes the space of N by N symmetric matrices, given by $\Upsilon(\Theta) := \operatorname{cay}[\Theta] - \sigma(\Theta)$. The identity $\nabla\sigma(\Theta) = (L_{\Upsilon(\Theta)} + R_{\Upsilon(\Theta)})\beta_\Theta^{-1}$ implies that

$$\begin{aligned} 2\,\mathrm{skew}\,[\boldsymbol{m}\,\nabla\sigma(\Theta)\,\beta_\Theta]\Xi &= 2\,\mathrm{skew}\left[(L_E + R_E)\left(L_{\Upsilon(\Theta)} + R_{\Upsilon(\Theta)}\right)\right]\Xi \\ &= [[E, \Upsilon(\Theta)], \Xi]\,. \end{aligned} \tag{2.23}$$

The right-hand side of equation (2.19) is simplified as follows: Using the identities

$$\Upsilon(\Theta) = \mathbf{1} + \tfrac{1}{2}\sigma(\Theta)\Theta \qquad \text{and} \qquad \sigma(\Theta) - \Theta = \tfrac{1}{4}\Theta\sigma(\Theta)\Theta, \tag{2.24}$$

we find that

$$[\boldsymbol{m}\sigma(\Theta), \Theta] = 2\,\mathrm{skew}\,[[E, \Theta]\,\sigma(\Theta)] = \beta_\Theta\,[E, \Upsilon(\Theta)]\,. \tag{2.25}$$

Thus, $\beta_\Theta^{-1}\,[\boldsymbol{m}\sigma(\Theta), \Theta] = [E, \Upsilon(\Theta)]$ and (2.19) holds.

The fact that (2.20) is second order can be verified using (2.10). Spatial momentum is exactly conserved; it is treated as a constant parameter in an actual implementation of the algorithm (2.20). (See Section 5.) Hence only the configuration update need be considered in the order analysis. If we let $\mathcal{F}_t : SO(N) \times so(N)^* \to SO(N)$ denote the configuration component of the flow at time t determined by the unreduced Euler equations (2.1), then the difference between the true configuration update $\mathcal{F}_{\Delta t}(\Lambda_n, \Pi_n)$ and the algorithmic configuration update (2.20_1) determines the order of the algorithm. The Taylor expansion of $\mathcal{F}_{\Delta t}(\Lambda_n, \Pi_n)$ at $\Delta t = 0$ is

$$\mathcal{F}_{\Delta t}(\Lambda_n, \Pi_n) = \Lambda_n\left(\mathbf{1} + \Delta t\,\Omega_n + \tfrac{(\Delta t)^2}{2}\left(\Omega_n^2 + \dot{\Omega}_n\right)\right) + \mathcal{O}\left(\Delta t^3\right), \tag{2.26}$$

where $\dot{\Omega}_n = \boldsymbol{m}^{-1}\dot{\Pi}_n = \boldsymbol{m}^{-1}\left[\Pi_n, \Omega_n\right]$ is determined by (2.1) and (2.2). We now derive the second order expansion for the configuration update Λ_{n+1} determined by (2.20_1). The Cayley transform has the second order expansion

$$\operatorname{cay}[\Theta] = \mathbf{1} + \Theta + \tfrac{1}{2}\Theta^2 + \mathcal{O}\left(|\Theta|^3\right). \tag{2.27}$$

Taking the skew component of (2.27), we see that

$$\Theta_n = \sigma^{-1}(\Delta t\,\Omega_{n+\frac{1}{2}}) = \Delta t\,\Omega_{n+\frac{1}{2}} + \mathcal{O}\left(\Delta t^3\right). \tag{2.28}$$

To make the expansion of the relative rotation matrix Θ_n in terms of Δt more explicit, we note that (2.10_2) and (2.12) imply that the midpoint body velocity $\Omega_{n+\frac{1}{2}}$ satisfies

$$\Omega_{n+\frac{1}{2}} = \boldsymbol{m}^{-1}\Pi_{n+\frac{1}{2}} = \boldsymbol{m}^{-1}\left(\Pi_n + \tfrac{1}{2}\left[\Pi_{n+\frac{1}{2}}, \Theta_n\right]\right) = \Omega_n + \tfrac{\Delta t}{2}\dot{\Omega}_n + \mathcal{O}\left(\Delta t^2\right). \tag{2.29}$$

Combining (2.27), (2.28), and (2.29), we obtain

$$\Lambda_{n+1} = \Lambda_n\operatorname{cay}[\Theta_n] = \Lambda_n\left(\mathbf{1} + \Delta t\,\Omega_n + \tfrac{(\Delta t)^2}{2}\left(\Omega_n^2 + \dot{\Omega}_n\right)\right) + \mathcal{O}\left(\Delta t^3\right). \tag{2.30}$$

Thus $\mathcal{F}_{\Delta t}(\Lambda_n, \Pi_n) - \Lambda_{n+1} = \mathcal{O}\left(\Delta t^3\right)$. Since the spatial momentum is preserved by both the true flow and the algorithmic flow, it follows that (2.20) is a second order algorithm. □

Remark 2.4 The algorithm (2.20) for $N = 3$ is identical to the algorithm described in (Lewis and Simo [1994]). The Cayley transform for $N = 3$ can be written in the form

$$\operatorname{cay}[\Theta] = \mathbf{1} + \frac{2}{1 + \left|\frac{1}{2}\Theta\right|^2}\left(\tfrac{1}{2}\Theta + \left(\tfrac{1}{2}\Theta\right)^2\right). \tag{2.31}$$

Thus the residual equation (2.20_3) takes the form

$$\frac{1}{1 + \left|\frac{1}{2}\Theta_n\right|^2}\Theta_n = \operatorname{skew}\left[\operatorname{cay}\left[\Theta_n\right]\right] = \Delta t\, \Omega_{n+\frac{1}{2}}. \tag{2.32}$$

Equation (2.32) implies that $\Theta_n = \kappa\, \Delta t\, \Omega_{n+\frac{1}{2}}$, where

$$\kappa = \frac{2}{1 \pm \sqrt{1 - \left|\Delta t\, \Omega_{n+\frac{1}{2}}\right|^2}}. \tag{2.33}$$

Consistency of the scheme imposes the condition that κ approach one as Δt approaches zero, which leads to the smaller value of κ. Thus (2.20_3) is satisfied by

$$\Theta_n = \frac{2\,\Delta t}{1 + \sqrt{1 - \left|\Delta t\, \Omega_{n+\frac{1}{2}}\right|^2}}\Omega_{n+\frac{1}{2}}, \tag{2.34}$$

which is the expression derived in Lewis and Simo [1994].

3 Existence of solutions to the algorithmic problem

The configuration update $\Lambda_n \to \Lambda_{n+1}$ determined by the algorithm (2.20) is found by solving the implicit equation (2.20_3) for Θ_n and then substituting Θ_n into the explicit equation (2.20_1) for Λ_{n+1}. If the equation (2.20_3) fails to have a solution, then the configuration update is undefined and the algorithm breaks down. Thus, as always when using implicit schemes, it is highly desirable to find conditions guaranteeing the existence of a solution to the implicit equation. The following theorem provides such a condition in terms of a bound on the time-step Δt. Note that if the bound is initially satisfied, it is satisfied for the entire algorithmic trajectory.

Theorem 3.1 *The residual equation (2.20_3) for trajectories in $so(N)$ with spatial momentum π has a solution if Δt satisfies*

$$\Delta t \leq \frac{m_{\min}}{|\pi|}, \tag{3.1}$$

where $m_{\min}$ denotes the minimum eigenvalue of the map $\boldsymbol{m} = L_E + R_E$ determined by the Euler tensor E.

Remark 3.2 The eigenvalue $m_{\min}$ is bounded below by $2\,e_{\min}$, where $e_{\min}$ denotes the minimum eigenvalue of the Euler tensor.

As the first step in proof of Theorem 2, we reformulate the implicit residual equation (2.20_3) as a fixed point condition.

Lemma 3.3 *The residual equation* (2.20$_3$) *determining* Θ_n *is equivalent to the condition that* Θ_n *be a fixed point of the quadratic map* $\varphi_n : so(N) \to so(N)$ *given by*

$$\varphi_n(\Theta) := \Delta t\, \boldsymbol{m}^{-1}\left(\left(\mathbf{1} - \tfrac{1}{2}\Theta\right)\Pi_n\left(\mathbf{1} + \tfrac{1}{2}\Theta\right)\right), \tag{3.2}$$

where $\Pi_n := \Lambda_n^T \pi_n \Lambda_n$ *is the body momentum at the* n*-th time-step.*

Proof Given $\Theta \in so(N)$, we define the map $A_\Theta : so(N)^* \to so(N)^*$ by

$$A_\Theta \Pi := \operatorname{cay}[-\Theta]\,\Pi\,\operatorname{cay}[\Theta]. \tag{3.3}$$

The identity $\Pi_{n+1} = A_{\Theta_n}\Pi_n$ implies that

$$\Pi_{n+\frac{1}{2}} = \tfrac{1}{2}\left(\mathrm{id} + A_{\Theta_n}\right)\Pi_n. \tag{3.4}$$

Equation (2.20$_3$) and invertibility of $\mathrm{id} + A_{\Theta_n}$ imply that

$$\begin{aligned} \tfrac{\Delta\Theta}{2}\Pi_n &= \Delta t\left(\mathrm{id} + A_{\Theta_n}\right)^{-1}\Pi_{n+\frac{1}{2}} \\ &= \left(\mathrm{id} + A_{\Theta_n}\right)^{-1}\boldsymbol{m}\sigma(\Theta) \\ &= \operatorname{skew}\left[\operatorname{cay}[\Theta_n]\,E\right]. \end{aligned} \tag{3.5}$$

Multiplying both sides of (3.5) by $\left(\mathbf{1} + \frac{1}{2}\Theta_n\right)$ on the right and $\left(\mathbf{1} - \frac{1}{2}\Theta_n\right)$ on the left, we find that

$$\Delta t\left(\mathbf{1} - \tfrac{1}{2}\Theta_n\right)\Pi_n\left(\mathbf{1} + \tfrac{1}{2}\Theta_n\right) = 2\operatorname{skew}[\Theta_n E] = \boldsymbol{m}\,\Theta_n. \tag{3.6}$$

□

Proof of Theorem 2. Using the standard identification of $so(N)$ with $\mathbb{R}^d$, where $d = \frac{1}{2}N(N-1)$, we can identify the map φ_n with a map on $\mathbb{R}^d$. (To reduce notational clutter, we shall use the notation φ_n for this map as well.) The Brouwer fixed point theorem states that a smooth map taking a closed ball in $\mathbb{R}^d$ into itself has a fixed point. Hence if there exists r such that $|\Theta| \le r$ implies $|\varphi_n(\Theta)| \le r$, then the map φ_n has a fixed point and the residual equation at the n-th time-step has a solution. We shall show that for Δt satisfying (3.1) r can be taken equal to two.

We first estimate $|\varphi_n(\Theta)|$. Using the identities

$$\langle[\Pi, \Theta], \Pi\rangle = 0 = \langle[\Pi, \Theta], \Theta\Pi\Theta\rangle \tag{3.7}$$

and

$$\tfrac{1}{2}\left|[\Pi, \Theta]\right|^2 = |\Pi\Theta|^2 + \langle\Pi, \Theta\Pi\Theta\rangle, \tag{3.8}$$

we see that

$$\left|\left(\mathbf{1} - \tfrac{1}{2}\Theta\right)\Pi\left(\mathbf{1} + \tfrac{1}{2}\Theta\right)\right|^2 = |\Pi|^2 + \tfrac{1}{2}|\Pi\Theta|^2 + \tfrac{1}{16}|\Theta\Pi\Theta|^2 = \langle\Pi, S_\Theta\Pi S_\Theta\rangle, \tag{3.9}$$

where $S_\Theta := \mathbf{1} - \frac{1}{4}\Theta^2$. The map $L_{S_\Theta}R_{S_\Theta}$ taking Π to $S_\Theta\Pi S_\Theta$ is a self-adjoint linear map on $so(N)$. Moreover, multiplication by a symmetric matrix is a self-adjoint map on $L(N)$; left and right multiplication by a symmetric matrix have the same real eigenvalues, with transposed eigenvectors, on $L(N)$. Thus we can bound separately left and right multiplication with respect to the extension of the inner product $\langle\ ,\ \rangle$ to $L(N)$, rather than directly estimating the norm of $L_{S_\Theta}R_{S_\Theta}$ on $so(N)$. For any $M \in L(N)$ we have

$$\langle M, MS_\Theta\rangle = |M|^2 + \tfrac{1}{4}|M\Theta|^2 . \tag{3.10}$$

The pairing $|M\Theta|^2 = -\langle M^T M, \Theta^2\rangle$ attains its maximum over M of unit norm at $M = \Theta/|\Theta|$. Thus

$$k_{\max}(\Theta) := \max_{|M|=1}\langle M, S_\Theta M\rangle = \max_{|M|=1}\langle M, MS_\Theta\rangle = 1 + \frac{|\Theta^2|^2}{4|\Theta|^2}. \tag{3.11}$$

Combining (3.9) and (3.11) yields the estimate

$$\left|\left(\mathbf{1} - \tfrac{1}{2}\Theta\right)\Pi\left(\mathbf{1} + \tfrac{1}{2}\Theta\right)\right| \le k_{\max}(\Theta)\,|\Pi| . \tag{3.12}$$

Hence, since $|\Pi_n| = |\pi|$,

$$|\varphi_n(\Theta)| = \left|\Delta t\, \boldsymbol{m}^{-1}\left(\mathbf{1} - \tfrac{1}{2}\Theta\right)\Pi_n\left(\mathbf{1} + \tfrac{1}{2}\Theta\right)\right| \le \frac{\Delta t\,|\pi|\,k_{\max}(\Theta)}{m_{\min}}. \tag{3.13}$$

We now estimate $k_{\max}(\Theta)$. We shall show that the eigenvalues of the symmetric linear transformation $f_\Theta(\Pi) := -\text{skew}\left[\Theta^2\Pi\right]$ on $so(N)$ are bounded above by $|\Theta|^2$ and thus

$$|\Theta^2|^2 = \langle f_\Theta(\Theta), \Theta\rangle \le |\Theta|^4 . \tag{3.14}$$

The identity $\langle\Theta^2\boldsymbol{x}, \boldsymbol{x}\rangle = -|\Theta\boldsymbol{x}|^2$ implies that Θ^2 is negative semi-definite. Let μ denote the minimum eigenvalue of Θ^2; since all non-zero eigenvalues of Θ^2 are double, we have the estimate $\mu \ge \frac{1}{2}\text{tr}\left[\Theta^2\right] = -|\Theta|^2$. Hence for any eigenvector Π of f_Θ, with eigenvalue λ, and any $\boldsymbol{x} \in \mathbb{R}^N$

$$|\Theta|^2|\Pi\boldsymbol{x}|^2 \ge \langle f_\Theta(\Pi)\boldsymbol{x}, \Pi\boldsymbol{x}\rangle = \lambda|\Pi\boldsymbol{x}|^2 . \tag{3.15}$$

Thus $|\Theta|^2 \ge \lambda$, $|\Theta|^4 \ge |\Theta^2|^2$, and

$$k_{\max}(\Theta) \le 1 + \left|\tfrac{1}{2}\Theta\right|^2 . \tag{3.16}$$

Combining (3.13) and (3.16), we obtain

$$|\varphi_n(\Theta)| \le \frac{\Delta t\,|\pi|\left(1 + |\Theta/2|^2\right)}{m_{\min}}. \tag{3.17}$$

If $|\Theta| \le r$, then $|\varphi_n(\Theta)| \le r$ if

$$\Delta t \le \frac{r\,m_{\min}}{(1 + (r/2)^2)\,|\pi|}. \tag{3.18}$$

The right-hand side of (3.18) attains its maximum value $m_{\min}/|\pi|$ at $r = 2$. This value of r yields the bound (3.1). □

Remark 3.4 If Θ has exactly one pair of nonzero entries, then $S_\Theta\Theta S_\Theta = (1 + |\Theta/2|^2)^2\Theta$. Hence the estimate $\langle\Pi, S_\Theta\Pi S_\Theta\rangle \le (k_{\max}(\Theta)|\Pi|)^2$ given by (3.16) is sharp.

Remark 3.5 For practical numerical implementations of the algorithms, we may need an initial guess for the fixed point of φ_n. If we take as our initial guess the rescaled current body angular velocity, $\Theta = \Delta t\, \boldsymbol{m}^{-1}\Pi_n$, then

$$|\Theta| \leq \frac{\Delta t\ |\pi|}{m_{\min}} \leq \frac{r}{1+(r/2)^2} \leq r, \tag{3.19}$$

i.e. this initial guess lies inside the ball of radius r, if Δt satisfies (3.18).

Remark 3.6 If, for some reason, tight control on the size of the update step is desired, the inequality (3.18) can be used to determine an appropriate value of Δt.

The approach used here to estimate the maximal time-step for which the update for the N dimensional rigid body is known to exist can be applied to an arbitrary natural Hamiltonian system on a Lie group. The implicit residual equation can easily be expressed as a fixed point condition on the Lie algebra of the Lie group of interest. Thus the computationally intensive step of the scheme is formulated as a relatively well-understood problem on a relatively small vector space. We shall briefly comment on this aspect of the general problem here, referring the interested reader to (Lewis and Simo [1994]) for a more detailed discussion of the treatment of general Hamiltonian systems on Lie groups.

The most general class of systems considered in (Lewis and Simo [1994]) consists of natural Hamiltonian systems, i.e. systems with Hamiltonian of the form 'kinetic plus potential energy' on an arbitrary Lie group G. At the n-th time-step we denote the current configuration by q_n, the current spatial momentum by π_n, and the current body momentum by $\Pi_n = \mathrm{Ad}^*_{q_n}\pi_n$. The algorithms considered in (Lewis and Simo [1994]) are of the form

$$\left.\begin{array}{rcl} q_{n+1} & = & q_n\tau(\Theta_n) \\ \pi_{n+1} & = & \pi_n + \Delta t\,\mathcal{F}(q_n,\Theta_n) \\ \Theta_n & = & \Theta(\Delta t, \Pi_n, \Pi_{n+1}) \end{array}\right\}, \tag{3.20}$$

where $\tau : \mathcal{G} \to G$ is the 'algorithmic exponential map' (which is typically *not* the true exponential map of G), $\mathcal{F} : G \times \mathcal{G} \to \mathcal{G}^*$ is the algorithmic force, and $\Theta : \mathbb{R}\times\mathcal{G}^*\times\mathcal{G}^* \to \mathcal{G}$ determines the residual equation (3.20_3). For the N dimensional rigid body $\tau = \mathrm{cay}\,[\,\cdot\,]$, $\mathcal{F} = 0$, and $\Theta(\Delta t, \Pi_n, \Pi_{n+1}) = \Psi\left(\frac{\Delta t}{2}\boldsymbol{m}^{-1}(\Pi_n + \Pi_{n+1})\right)$, where Ψ is a local inverse of σ.

Combining the three equations of (3.20), we find that

$$\Theta_n = \varphi_n(\Theta_n) := \Theta(\Delta t, \Pi_n, \mathrm{Ad}^*_{\tau(\Theta_n)}(\Pi_n + \Delta t\,\mathrm{Ad}^*_{q_n}\mathcal{F}(q_n,\Theta_n))). \tag{3.21}$$

Thus the residual equation (3.20_3) is, in fact, a fixed point condition on the Lie algebra $\mathcal{G}$. Any finite dimensional Lie algebra can be identified with $\mathbb{R}^d$, where d is the dimension of the Lie group G. Thus, if Δt and r can be found such that the map φ_n maps the sphere of radius r in $\mathbb{R}^d$ into itself and the initial approximation $\Delta t\,\Omega_n$ to Θ_n is contained in this sphere, then the Brouwer fixed point theorem guarantees that a solution to the equation (3.20_3) exists.

Remark 3.7 We note that the conservation properties of the algorithm (3.20) do not depend on the choice of time-step. Thus the step size Δt can be adjusted as desired at each time-step.

Remark 3.8 The formulation of the residual equation on the vector space $\mathcal{G}$ involves neither the use of coordinate charts nor the embedding of the (typically nonlinear) configuration manifold G into a higher dimensional vector space.

4 Variational formulation of the algorithm and comparison to other algorithms

At least three recent papers have proposed algorithms for the discrete time integration of the rigid body equations. The first of these is by (Moser and Veselov [1991]); this work contains a derivation of a discrete time version of the Euler equations for the N dimensional free rigid body. The second paper, by (Lewis and Simo [1994]), contains an explicit construction for $N = 3$ of the algorithm presented in the current work, as well as algorithm design criteria for more general systems. Finally, (McLaughlin and Scovel [1995]) apply integration techniques recently developed by (Reich [1993] [1994]) to several classes of mechanical systems, including the N dimensional rigid body. While the Moser-Veselov algorithm and our algorithm share several common features, there are significant differences between them. On the other hand, the McLaughlin-Scovel algorithm is formally equivalent to ours, although implemented differently.

The algorithmic problem (2.20) can be expressed in terms of the body momenta $\Pi_n = \Lambda_n^T \pi_n \Lambda_n$ as

$$\left.\begin{array}{rcl} \Lambda_{n+1} & = & \Lambda_n Q_n \\ \Pi_{n+1} & = & Q_n^T \Pi_n Q_n \\ \Delta t\, \Pi_n & = & 2\,\mathrm{skew}\,[Q_n E] \end{array}\right\}, \tag{4.1}$$

where $Q_n = \mathrm{cay}\,[\Theta_n] = \Lambda_n^T \Lambda_{n+1}$ is the primary unknown in the problem and equation (4.1_3) is the alternate expression (3.5) for the residual equation. Equation (4.1_2) simply expresses the conservation of spatial momentum in terms of the body momenta. Equation (4.1_3) can be derived from (2.20) by noting that (2.2) and (2.20_3) imply that

$$\begin{aligned} \Pi_{n+\frac{1}{2}} &= 2\,\mathrm{skew}\left[\Omega_{n+\frac{1}{2}} E\right] \\ &= \tfrac{2}{\Delta t}\mathrm{skew}\,[\mathrm{skew}\,[Q_n]E] = \tfrac{1}{\Delta t}\,(\mathrm{id} + A_{\Theta_n})\,\mathrm{skew}\,[Q_n E]. \end{aligned} \tag{4.2}$$

Equivalence of (4.1) and (2.20) follows from (3.4) and the invertibility of $\mathrm{id} + A_{\Theta_n}$.

We shall show that equations (4.1), and thus the algorithm (2.20), are equivalent to a constrained minimization problem for a discrete time trajectory supplemented by the condition that the natural relationship $\pi_n = \Lambda_n \Pi_n \Lambda_n^T$ between spatial and body momenta hold along the trajectory. Using this formulation of the algorithm, we can compare (2.20) to the algorithm obtained by Moser and Veselov, who make use of the identical constrained minimization problem, but impose a different relationship between the spatial and body momenta in the construction of their update map.

The equations (4.1_1) through (4.1_3) can be viewed in part as a critical point condition on sequences in $SO(N)$ corresponding to the configuration components of discrete time trajectories of the Euler equations (2.1). Equation (4.1_3) implies that

$$\begin{aligned}\Delta t\,\pi_n &= \Lambda_n Q_n E \Lambda_n^T - \Lambda_n E(\Lambda_n Q_n)^T \\ &= \Lambda_{n+1} E \Lambda_n^T - \Lambda_n E \Lambda_{n+1}^T. \end{aligned} \tag{4.3}$$

Equation (4.1_2) is the body form of the statement of conservation of spatial momentum. In particular, $\pi_n = \pi_{n-1}$; i.e.

$$\Lambda_{n+1} E \Lambda_n^T - \Lambda_n E \Lambda_{n+1}^T = \Lambda_n E \Lambda_{n-1}^T - \Lambda_{n-1} E \Lambda_n^T. \tag{4.4}$$

Rearranging terms yields

$$(\Lambda_{n+1} + \Lambda_{n-1}) E \Lambda_n^T = \Lambda_n E (\Lambda_{n+1} + \Lambda_{n-1})^T, \tag{4.5}$$

i.e.

$$\text{skew}\left[(\Lambda_{n+1} + \Lambda_{n-1}) E \Lambda_n^T\right] = 0. \tag{4.6}$$

We now show that equation (4.6) has a natural interpretation as a critical point condition. Any tangent vector $\delta\Lambda_n \in T_{\Lambda_n} SO(N)$ satisfies $\delta\Lambda_n = \Xi\Lambda_n$ for some skew symmetric matrix Ξ. (Recall that the Lie algebra $so(N)$ of $SO(N)$ is the space of skew symmetric $N \times N$ matrices.) Thus (4.6) is equivalent to the statement that

$$\text{tr}\left[(\Lambda_{n+1} + \Lambda_{n-1}) E \delta\Lambda_n^T\right] = 0 \tag{4.7}$$

for all $\delta\Lambda_n \in T_{\Lambda_n} SO(N)$. If we define a function $\mathcal{S}$ on the space of sequences in $SO(N)$ by

$$\mathcal{S}(\{\Lambda_n\}) := \sum_{n=0}^{\infty} \text{tr}\left[\Lambda_{n+1} E \Lambda_n^T\right], \tag{4.8}$$

then equation (4.7) is equivalent to the condition that the partial derivative of $\mathcal{S}$ with respect to Λ_n equal zero. Thus trajectories determined by the algorithm (2.20) correspond to critical points of $\mathcal{S}$. In summary, if a sequence $\{\Lambda_n\}$ is a critical point of $\mathcal{S}$, then the sequence $\{(\Lambda_n, \pi_n)\}$, where the (constant) π_n's are determined by (4.3), is equal to the trajectory of the algorithm (2.20) with initial conditions (Λ_0, π_0).

It is possible to reverse the preceding argument and derive the algorithm (4.1) (equivalently (2.20)) starting from the variational principle based on the functional (4.8). However, the criticality condition (4.7) alone does not uniquely determine the algorithm. The variational problem associated to the function $\mathcal{S}$ given in (4.8) is the point of departure taken in (Moser and Veselov [1991]), but the Moser-Veselov algorithm is not identical to (4.1). In particular, the update map used in the Moser-Veselov algorithm does *not* automatically conserve the spatial momentum of the system: the 'algorithmic' configuration and momentum determined by the update map must be corrected to obtain configuration-momentum pairs satisfying the conservation laws of the rigid body.

We now take as our starting point the condition that a sequence $\{\Lambda_n\}$ associated to an initial configuration Λ_0 and a time-step Δt be a critical point of $\mathcal{S}$; this condition must be supplemented by an appropriate definition of the momentum to obtain an algorithm on the phase space P. Consistency of the algorithm imposes the condition that the body velocity Ω determined by (2.6) satisfy the relation

$$\Omega = \lim_{\Delta t \to 0} \frac{Q_n}{\Delta t}, \tag{4.9}$$

where $Q_n := \Lambda_n^T \Lambda_{n+1}$. Dropping the limit in (4.9) and applying the map (2.2) taking body velocities to body momenta, we obtain

$$\Pi = \tfrac{2}{\Delta t} \operatorname{skew}[Q_n E]. \tag{4.10}$$

For the moment, we do not specify the indices of Ω and Π in (4.9) and (4.10). Assigning the 'backward' index n to Ω and Π, corresponding to the right-handed limit in (4.9), leads to the algorithm (4.1), while, as we shall show below, the 'forward' assignment $n+1$ corresponding to the left-handed limit results in the Moser-Veselov scheme. The choice of index in the momentum equation (4.10) must be decided by additional considerations. For us, the guiding consideration is conservation of momentum. Spatial momentum is conserved only if the configuration Λ_n and the body momentum Π_n are both 'rotated' by the same matrix Q_n, as in (4.1_1) and (4.1_2). As was shown above, the sequence $\{\Lambda_n\}$ is a critical point of $\mathcal{S}$ if and only if

$$\operatorname{skew}[Q_{n+1}E] = Q_n^T \operatorname{skew}[Q_n E] Q_n \tag{4.11}$$

for all n. If we set $\Pi_n = \frac{2}{\Delta t} \operatorname{skew}[Q_n E]$, then (4.11) is equivalent to (4.1_2). Thus the critical point condition (4.11), the finite Δt interpretation (4.10) of the consistency equation (4.9), and conservation of spatial momentum completely determine the scheme (4.1).

We now briefly discuss the scheme proposed by Moser and Veselov [1991] and its relation to the scheme (4.1). Moser and Veselov introduce the update map

$$\left.\begin{aligned} \Lambda'_{n+1} &= \Lambda'_n Q'_n \\ \Pi'_{n+1} &= (Q'_{n-1})^T \Pi'_n Q'_{n-1} \\ \Delta t\, \Pi'_{n+1} &= 2 \operatorname{skew}[Q'_n E] \end{aligned}\right\}. \tag{4.12}$$

of the algorithmic configuration-momentum pair (Λ'_n, Π'_n). (The variables Λ'_n and Π'_n correspond to the variables X_n and M_n in the notation of Moser and Veselov.) The momentum equations (4.12_2) and (4.12_3) are consistent with the choice of index $n+1$ in (4.10). As a result, the quantities Λ'_n and Π'_n do *not* satisfy the relationship $(\Lambda'_n)^T \Pi'_n \Lambda'_n = constant$, i.e. conservation of spatial momentum. Rather, (4.12) satisfies the staggered conservation law $(\Lambda'_n)^T \Pi'_{n+1} \Lambda'_n = constant$, which combines data from two consecutive time-steps. The algorithm (4.12) is a multi-step method involving an explicit update of the algorithmic momentum as a function of the two previous configurations Λ'_{n-1} and Λ'_n and the previous momentum Π'_n using (4.12_2), followed by a configuration update (4.12_1) using the solution Q'_n of the equation (4.12_3), which depends on the momentum Π'_{n+1}. In comparison, the algorithm (2.20) (equivalently (4.1)) is a single-step method, i.e. the update depends only on the data from the previous time-step. The update specified by (4.1) is performed by *first* solving (4.1_3) for Θ_n as a function of Π_n, *then* 'rotating' the configuration and body momentum using $Q_n = \operatorname{cay}[\Theta_n]$. Note that while the algorithms (4.1) and (4.12) share several common features, they are not the same.

Remark 4.1 A discrete time trajectory conserving spatial momentum can be constructed from a trajectory generated by (4.12) by any of the following means:

1. Shift the algorithmic configuration-momentum pairs, i.e. replace the n-th step (Λ'_n, Π'_n) by the 'n-th step' (Λ'_{n+1}, Π'_n). Note that this correction cannot be performed time-step by time-step, but can easily be applied to a previously computed trajectory.
2. Discard the algorithmic body momentum update Π'_n and represent the phase space in its spatial trivialization, with (constant) spatial momentum $\pi'_n = \pi'_0 = \Lambda'_0 \Pi'_0 (\Lambda'_0)^T$. Equivalently, in the body trivialization of the phase space, replace the algorithmic body momentum Π'_n with the 'true' body momentum $(\Lambda'_n)^T \pi'_n \Lambda'_n$.
3. Discard the algorithmic configuration update Λ'_n and compute a 'true' configuration update Λ'_n using the relationship $\Pi'_{n+1} = \Lambda'_n \pi'_n (\Lambda'_n)^T$.

Remark 4.2 (McLaughlin and Scovel [1995]) also consider the N dimensional rigid body. The algorithm presented in McLaughlin and Scovel is formally equivalent to (4.1), but the relative rotation matrix Q_n is found as the solution to a constrained problem on $L(N)$, rather than as the Cayley transform of the relative rotation vector, which is an element of the Lie algebra $so(N)$. Thus McLaughlin and Scovel formulate the residual equation as a constrained equation on the N^2 dimensional vector space $L(N)$, rather than an unconstrained equation on the $N(N-1)/2$ dimensional vector space $so(N)$. We note that the McLaughlin-Scovel version of the algorithm is a *formal* algorithm, in the sense that conditions for solvability of the constrained residual equation on $L(N)$ are not given. We reiterate that while our algorithm, and hence the formally equivalent McLaughlin-Scovel algorithm, and the Moser-Veselov algorithm model the same discrete time dynamical system, they differ in several essential regards.

5 Implementation

The algorithm can easily be implemented using the representation of $so(N)$ as $\mathbb{R}^d$, where $d = \frac{1}{2}N(N-1)$. It is straightforward to express the maps $\boldsymbol{m} = L_E + R_E$ and $R_{(1+\frac{1}{2}\Theta)} L_{(1-\frac{1}{2}\Theta)}$ as linear maps on $\mathbb{R}^d$ and solve the residual equation there. To illustrate this approach, we present a simple program for implementing the algorithm (2.20). This implementation makes use of the alternative form (3.5) of the residual equation for Θ. For the sake of convenience and conciseness, we implement the algorithm in the high-level programming language *Mathematica*. However, it is by no means necessary to use a symbolic programming language such as *Mathematica* to implement the algorithm (2.20). The list manipulations carried out in the maps given here are easily implemented and efficiently executed in most modern programming languages. The implicit equation, which is a quadratic, unconstrained equation on $\mathbb{R}^d$, can be solved using standard routines for finding roots or fixed points.

In the *Mathematica* program given below, a function determining the residual equation on $\mathbb{R}^d$ is automatically determined as the initial step in the simulation. The functions defined below can be briefly described as follows:

- The maps `upper`, `skew`, and `cayley` construct upper triangular, skew, and special orthogonal $N \times N$ matrices from an d-vector, where $d = \frac{1}{2}N(N-1)$.

The d-vector is partitioned to form the upper triangle of an upper triangular matrix, which is then skew-symmetrized, as shown below:

$$\mathtt{skew}[\boldsymbol{x}] = \begin{pmatrix} 0 & x_1 & \cdots & x_{N-1} \\ -x_1 & 0 & \cdots & x_{2N-3} \\ \vdots & \vdots & \ddots & \vdots \\ -x_{N-1} & -x_{2N-3} & \cdots & 0 \end{pmatrix}. \tag{5.1}$$

The map **unskew** is the inverse of the map **skew**; it maps an $N \times N$ skew matrix to a d-vector.

- The maps **sym** and **minv** are used to construct the inverse to the metric associated to the Euler tensor; the Euler tensor itself is entered as a pair of vectors: an N-vector containing the diagonal entries of the Euler tensor and a d-vector containing the off-diagonal entries.

$$\mathtt{sym}[\boldsymbol{x}, \mathbf{y}] = \begin{pmatrix} x_1 & y_1 & \cdots & y_{N-1} \\ y_1 & x_2 & \cdots & y_{2N-3} \\ \vdots & \vdots & \ddots & \vdots \\ y_{N-1} & y_{2N-3} & \cdots & x_N \end{pmatrix}. \tag{5.2}$$

- The map **step** maps a configuration matrix $\Lambda_n \in SO(N)$ to the updated configuration matrix Λ_{n+1}. The map $\mathtt{rhs}\,[\Lambda_n]$ corresponds to the map φ_n; the initial guess for Θ_n is the rescaled current body angular velocity $\Delta t\, \Lambda_n^T \pi \Lambda_n = \mathtt{rhs}\,[\Lambda_n]\,[\mathbf{0}]$.
- Finally, the map **int** computes the trajectory in the configuration space $SO(N)$ with initial condition Λ (**lambda** $\in SO(N)$), spatial momentum π (**pi** $\in \mathbb{R}^d$), Euler tensor E (**e** = {**ed, eo**}, **ed** $\in \mathbb{R}^N$, **eo** $\in \mathbb{R}^d$), total simulation time range t (**t** $\in \mathbb{R}$), and time-step Δt (**dt** $\in \mathbb{R}$). The final optional argument **options** of **int** allows the user to invoke the options of the built-in *Mathematica* function **FindRoot**, which solves the implicit equation $\Theta = \mathtt{rhs}\,[\Lambda_n]\,[\Theta]$, i.e. $\Theta = \varphi_n(\Theta)$, for the relative rotation vector Θ.

The precise definitions of the *Mathematica* routines are as follows:

```
upper[x_] := Table[Join[Array[0&, k], Array[x[[#]]&, # - k,
   Sum[# - i, {i, k - 1}] + 1]], {k, #}]&[(1 + Sqrt[1 + 8 Length[x]])/2]

skew[x_] := (# - Transpose[#])&[upper[x]]

unskew[s_] := Join@@Array[Drop[s[[#]], #]&, Length[s]]

sym[x_, y_] := DiagonalMatrix[x] + (# + Transpose[#])&[upper[y]]

cayley[x_] := Inverse[#1 - #2].(#1 + #2)&\
   [IdentityMatrix[Length[#]], #]&[skew[x/2]]

minv[{ed_, eo_}] :=
   Block[{y, n = Length[eo]},
      Inverse[Array[Coefficient[Evaluate[Expand[unskew[#1.#2 + #2.#1]]&\
      [sym[ed, eo], skew[Array[y, n]]]], y[#]]&, n]]
      ]

step[lambda_, rhs_, n_,  options___] :=
   Block[{theta},
      lambda.cayley[#1]/.FindRoot[#1 == rhs[#1], ##2, options]&@@\
      Prepend[Transpose[{#, rhs[Array[0&, n]]}], #]&[Array[theta, n]]
      ]
```

```
int[lambda_, pi_, e_, t_, dt_, options___] :=
  Block[{n = Length[pi], tm = dt minv[e], sm = skew[pi], rhs},
    rhs[lam_] = tm.unskew[Transpose[#].sm.#&[lam + lam.skew[#/2]]]&;
    NestList[step[#, rhs[#], n, options]&, N[lambda], Round[t/dt]]
    ]
```

Remark 5.1 We note that the identification of a skew matrix with a vector used here does *not* coincide with the usual identification for $N = 3$, which makes use of the cross product. In the representation used here **skew** $[(x, y, z)]\,(u, v, w) = (-z, y, -x) \times (u, v, w)$. The entries of the Euler tensor are analogously permuted with respect to the standard ordering. Straightforward modifications of the routines **upper** and **unskew** are sufficient to adapt the code to a preferred representation of skew (respectively symmetric) N by N matrices as d-vectors (respectively as N-vector, d-vector pairs).

Acknowledgements.

Support for this research was provided by NSF Grant DMS-9122708 and FRC grant 503128-09523 through U.C. Santa Cruz and AFOSR Contract No. 2-DJA-826 through Stanford University. This support is gratefully acknowledged.

References

Arnold, V. I. [1966], *Sur la geometrie differentielle des groupes de Lie de dimension infinie et ses applications a l'hydrodynamique des fluids parfaits*, Ann. Inst. Fourier Grenoble, **16(1)**, 319–361.

Arnold, V. I. [1978], *Mathematical Methods of Classical Mechanics*, Springer-Verlag, New York.

Dahlquist, G. [1975], *Error analysis for a class of methods for stiff nonlinear initial boundary value problems*, Lecture Notes in Mathematics **506**, 60–74.

Haire, E. and Wanner, G. [1991], *Solving ordinary differential equations II. Stiff and algebraic problems*, Springer-Verlag, New York.

Lewis, D. and Simo, J. C. [1994], *Conserving algorithms for the dynamics of Hamiltonian systems on Lie groups*, J. Nonlinear Sci. **4**, 253–299.

McLachlan, R. I. and Scovel, C. [1995], *Equivariant constrained symplectic integration*, J. Nonlinear Sci., **5**, 233–256.

Manakov. S. V. [1976], *Note on the integration of Euler's equations of the dynamics of an n-dimensional rigid body*, Funkts. Anal. Prilozh., **10(4)**, 93–94. English translation: Funct. Anal. Appl. **10**, 328–329.

Moser, J. and Veselov, A. P. [1991], *Discrete versions of some classical integrable systems and factorization of matrix polynomials*, Comm. Math. Phys., **139**, 217–243.

Ratiu, T. [1980], *The motion of the free n-dimensional rigid body*, Indiana Univ. Math. J., **29(4)**, 609–629.

Reich, S. [1993], *Symplectic integration of constrained Hamiltonian systems by Runge-Kutta methods*, Technical report 93–13, Univ. British Columbia.

Reich, S. [1994], *Momentum preserving symplectic integrators*, Phys. D., **76**, No. 4, 375-383.

Appendix: Derivation of the symplectic condition on $SO(N)$

We specialize the symplectic conditions derived in Lewis and Simo [1994] to the case of the rotation groups $SO(N)$. The general condition that a generalized midpoint algorithm, i.e. an algorithm of the form (3.20) with $\Theta_n = \Delta t\, \Omega_{n+\frac{1}{2}}$, with no forcing be symplectic is that

$$\begin{aligned} &\operatorname{skew}\left[\boldsymbol{m}\,\boldsymbol{H}(\Theta)^{-1}\left(\mathrm{id} + \mathrm{Ad}_{\tau(\Theta)^{-1}}\right)\right] \\ &\qquad = \mathrm{ad}^*\left(\left(\left(\mathrm{id} + \mathrm{Ad}^*_{\tau(\Theta)^{-1}}\right)^{-1} - \tfrac{1}{2}\,\mathrm{id}\right)\boldsymbol{m}\,\Theta\right). \end{aligned} \tag{5.3}$$

Here $\boldsymbol{H}(\Theta) : so(N) \to so(N)$ is given by $\boldsymbol{H}(\Theta) := T_{\tau(\Theta)} L_{\tau(\Theta)^{-1}} T_\Theta \tau$, $\boldsymbol{m} : so(N) \to so(N)^*$ denotes the mass matrix determined by the kinetic energy of the system, and $\mathrm{Ad} : G \times \mathcal{G} \to \mathcal{G}$, $\mathrm{Ad}^* : G \times \mathcal{G}^* \to \mathcal{G}^*$, and $\mathrm{ad}^* : \mathcal{G} \times \mathcal{G}^* \to \mathcal{G}^*$ denote the adjoint, coadjoint, and infinitesimal coadjoint actions. $\tau : \mathcal{G} \to G$ is the 'algorithmic exponential' map, satisfying $\tau(0) = \mathbf{1}$.

For the rotation groups $SO(N)$, we consider maps τ given by the composition of the Cayley transform with some invertible map $\Psi : so(N) \to so(N)$ that fixes zero. In specializing (5.3) to $SO(N)$ we shall make use of the alternative expressions

$$\mathrm{cay}\,[\Theta] = 2\left(\mathbf{1} - \tfrac{1}{2}\Theta\right)^{-1} - \mathbf{1} = \left(\mathbf{1} - \tfrac{1}{2}\Theta\right)^{-1}\Theta + \mathbf{1} \tag{5.4}$$

for the Cayley transform (2.7). The Cayley transform has the linearization

$$\begin{aligned} T_\Theta \mathrm{cay} &= \tfrac{1}{2}\left(\mathrm{id} + L_{\mathrm{cay}[\Theta]}\right) R_{(\mathbf{1}-\Theta/2)^{-1}} \\ &= \left(\left(\mathrm{id} - \tfrac{1}{2}R_\Theta\right)\left(\mathrm{id} - \tfrac{1}{2}L_\Theta\right)\right)^{-1} \\ &= \left(L_{\mathrm{cay}[\Theta]} + R_{\mathrm{cay}[\Theta]}\right)\beta_\Theta^{-1}, \end{aligned} \tag{5.5}$$

where, as before, $\beta_\Theta := 2\,\mathrm{id} - \frac{1}{2}\,L_\Theta\,R_\Theta$.

Lemma 5.2 *Let $\tau : so(N) \to SO(N)$ be given by $\tau(\Theta) := \mathrm{cay}\,[\Psi(\Theta)]$ for some invertible map $\Psi : so(N) \to so(N)$. The generalized mid-point rule associated to τ is symplectic if and only if*

$$2\,\mathrm{skew}\left[\boldsymbol{m}(\nabla\Psi(\Theta))^{-1}\beta_{\Psi(\Theta)}\right] = \left[\beta^{-1}_{\Psi(\Theta)}\left[\boldsymbol{m}\,\Theta, \Psi(\Theta)\right], \cdot\,\right]. \tag{5.6}$$

for all $\Theta \in so(N)$. Equivalently, setting $\sigma := \Psi^{-1}$, the symplectic condition can be expressed in the form

$$2\,\mathrm{skew}\left[\boldsymbol{m}\,\nabla\sigma(\Theta)\,\beta_\Theta\right] = \left[\beta_\Theta^{-1}\left[\boldsymbol{m}\,\sigma(\Theta), \Theta\right], \cdot\,\right] \tag{5.7}$$

for all $\Theta \in so(N)$.

Proof We can simplify the symplectic condition (5.3) to (5.6) using the identities (5.4) for the Cayley transformation. We first consider the left-hand side of (5.3). Recall the map $A_\Theta : so(N)^* \to so(N)^*$ given by $A_\Theta \Pi = \mathrm{cay}\,[-\Theta]\,\Pi\,\mathrm{cay}\,[\Theta] = \mathrm{Ad}^*_{\mathrm{cay}[\Theta]^{-1}}\Pi$. For $\tau(\Theta) = \mathrm{cay}\,[\Psi(\Theta)]$,

$$\begin{aligned} \boldsymbol{H}(\Theta)^{-1} &= \left(T_{\mathrm{cay}[\Psi(\Theta)]} L_{\mathrm{cay}[-\Psi(\Theta)]}\, T_{\Psi(\Theta)}\mathrm{cay}\, \nabla\Psi(\Theta)\right)^{-1} \\ &= \left(\nabla\Psi(\Theta)\right)^{-1}\left(T_{\mathrm{cay}[\Psi(\Theta)]} L_{\mathrm{cay}[-\Psi(\Theta)]} T_{\Psi(\Theta)}\mathrm{cay}\right)^{-1}. \end{aligned} \tag{5.8}$$

Hence the identity

$$\begin{aligned}
&\left(T_{\mathrm{cay}[\Theta]} L_{\mathrm{cay}[-\Theta]} T_\Theta \mathrm{cay}\right)^{-1} (\mathrm{id} + A_\Theta) \\
&= 2\, R_{\left(1-\frac{1}{2}\Theta\right)} \left(\mathrm{id} + L_{\mathrm{cay}[-\Theta]}\right)^{-1} \left(\mathrm{id} + L_{\mathrm{cay}[-\Theta]} + L_{\mathrm{cay}[-\Theta]} \left(R_{\mathrm{cay}[\Theta]} - \mathrm{id}\right)\right) \\
&= 2 \left(R_{\left(1-\frac{1}{2}\Theta\right)} + \left(\mathrm{id} + L_{\mathrm{cay}[\Theta]}\right)^{-1} R_\Theta \right) \\
&= \beta_\Theta \qquad (5.9)
\end{aligned}$$

for all $\Theta \in so(N)$ implies that

$$\operatorname{skew}\left[\boldsymbol{m}\, \boldsymbol{H}(\Theta)^{-1} \left(\mathrm{id} + \mathrm{Ad}_{\tau(\Theta)^{-1}}\right)\right] = \operatorname{skew}\left[\boldsymbol{m}\, (\nabla \Psi(\Theta))^{-1} \beta_{\Psi(\Theta)}\right]. \qquad (5.10)$$

The right-hand side of (5.3) can be simplified using the identity

$$\begin{aligned}
&(\mathrm{id} + A_\Theta)^{-1} \qquad (5.11) \\
&\quad = \left(\mathrm{id} + L_{\mathrm{cay}[\Theta]} R_{\mathrm{cay}[\Theta]}\right)^{-1} \\
&\quad = \left(L_{\left(1-\frac{1}{2}\Theta\right)} R_{\left(1+\frac{1}{2}\Theta\right)} + L_{\left(1+\frac{1}{2}\Theta\right)} R_{\left(1-\frac{1}{2}\Theta\right)} \right)^{-1} L_{\left(1-\frac{1}{2}\Theta\right)} R_{\left(1+\frac{1}{2}\Theta\right)} \\
&\quad = \tfrac{1}{2} \left(\mathrm{id} + \beta_\Theta^{-1} \left(R_\Theta - L_\Theta\right)\right) \qquad (5.12)
\end{aligned}$$

for any $\Theta \in so(N)$, which implies that

$$\left(\left(\mathrm{id} + A_{\Psi(\Theta)}\right)^{-1} - \tfrac{1}{2}\,\mathrm{id}\right) \boldsymbol{m}\, \Theta = \tfrac{1}{2}\, \beta_{\Psi(\Theta)}^{-1} \left[\boldsymbol{m}\Theta, \Psi(\Theta)\right]. \qquad (5.13)$$

Combining (5.10) and (5.13), we see that the symplectic condition (5.3) for the algorithm (2.20) is equivalent to (5.6). The alternate form (5.7) for the symplectic condition follows directly from the identity $(\nabla\Psi(\Theta))^{-1} = \nabla(\Psi^{-1})(\Psi(\Theta))$. □

Fields Institute Communications
Volume **10**, 1996

More on Symplectic Correctors

Robert I. McLachlan
Mathematics Department
Massey University
Palmerston North
New Zealand

Abstract. We give a different point of view of the symplectic correctors of Wisdom, Holman, and Touma and study the condition that errors committed by the kernel can be corrected. Because of this requirement there are no outstandingly good new methods of this form, although any fourth-order kernel can be corrected for near-integrable problems. For systems that are not near-integrable we find that correctors bring no clear advantage.

1 Symplectic correctors

Wisdom, Holman, and Touma [6] (hereafter WHT) have introduced and studied in depth the concept of a *symplectic corrector*, that is, a numerical integrator defined by an operator of the form

$$e^C e^K e^{-C} \tag{1.1}$$

in which the kernel e^K is symplectic, and the corrector e^C may or may not be. This is a useful form when many time steps can be taken without output, as then one only computes $e^C(e^K)^n e^{-C}$. Symplectic methods of this type were mentioned but not explored by Suzuki [5]; general methods of the form (1.1) were first introduced by Butcher [1] who found that for a fifth-order Runge-Kutta method, the kernel need only have five stages instead of the minimum of six needed for general Runge-Kuttas. However, the improvements are much more dramatic when the vector field is near-integrable and the kernel is a composition method, that is, the vector field to be integrated is of the form $A+B$ where A and B can be integrated exactly and $A = \mathcal{O}(1)$, $B = \mathcal{O}(\epsilon)$. In this case WHT show that for all n the dominant $\mathcal{O}(\epsilon t^n)$ errors can be eliminated by the corrector. WHT develop in depth one possible

1991 *Mathematics Subject Classification.* Primary: 65L05; Secondary 70–08.
Key words and phrases. Initial value problems, composition methods, operator splitting, symplectic integrators, Hamiltonian systems.

approach to the design of correctors and kernels for these problems; but they also comment that

> We may also develop composition integrators which are intended from the beginning to be used with a corrector. Presumably, [this] is the most economical path to high order, since the kernel must satisfy a much reduced set of constraints, only that the error terms be correctable, not that they be zero.

In this note we explore this suggestion and also give a simple alternative derivation of symplectic correctors. WHT develop an "ideal" corrector and then approximate it by composition, the corrector to be used with various kernels; here we formulate and study the kernel and the corrector simultaneously. Regarding their hope expressed above, our conclusions are pessimistic: the "correctability constraint" does indeed restrict the choice of kernel, and seems to impose much the same order and stability limits as exist with ordinary composition methods. Our notation follows that of WHT, and in fact this is best read as a footnote to that paper. For most of the paper we have the near-integrable case in mind; the last section addresses the general case with regard to the above suggestion of WHT.

For the near-integrable case, the most important errors are those due to $\mathcal{L}_A^m B$ $(=[A,[A,\ldots,[A,B]\ldots]])$. These can be studied by considering the test problem $A=\partial_x$, $B=f(x)\partial_y$, for which the *only* nonzero Lie brackets, and hence errors, are due to $\mathcal{L}_A^m B$. A method which has order n on this problem will have no errors due to $\mathcal{L}_A^m B$, $m<n$, when applied to arbitrary A and B. For the test problem, a composition method of the form

$$\ldots e^{b_2 tB} e^{a_2 tA} e^{b_1 tB} e^{a_1 tA}, \tag{1.2}$$

where t is the time step, gives

$$\begin{aligned} x \mapsto x' &= x + t\textstyle\sum a_i \\ y \mapsto y' &= y + t\textstyle\sum b_i f(c_i) \end{aligned}$$

where $c_i=\sum_{j=1}^i a_i$; the true solution is

$$\begin{aligned} x' &= x+t \\ y' &= y+\textstyle\int_0^1 f(x+\alpha t)\,d\alpha \end{aligned}$$

So we have order n if the c_i are the nodes and the b_i the weights of a quadrature formula of order n on $[0,1]$. This may in turn be expressed by the quadrature conditions

$$\sum b_i c_i^m = \frac{1}{m+1}, \qquad m=0,\ldots,n-1 \tag{1.3}$$

obtained with the trial functions $f(x)=x^m$.

Now turn to corrector methods in which e^C and e^K both take the form (1.2). Let k be the number of nodes (i.e., the number of applications of e^{tB} in e^C). Fixing the kernel is the same as fixing some of the nodes and weights in the quadrature; if e^C contributes a set of weights $\{b_i : i=1,\ldots,k\}$ and nodes $\{c_i\}$ then e^{-C} contributes another set $\{-b_i\}$ and $\{c_i+1\}$, assuming the kernel is consistent. The necessity of using negative weights indicates that the correction process is rather ill-conditioned. A reasonable assumption is that the quadrature be symmetric about $x=\frac{1}{2}$; so e^C must contain pairs $\pm b_i$ at nodes $\pm c_i$. The quadrature conditions for

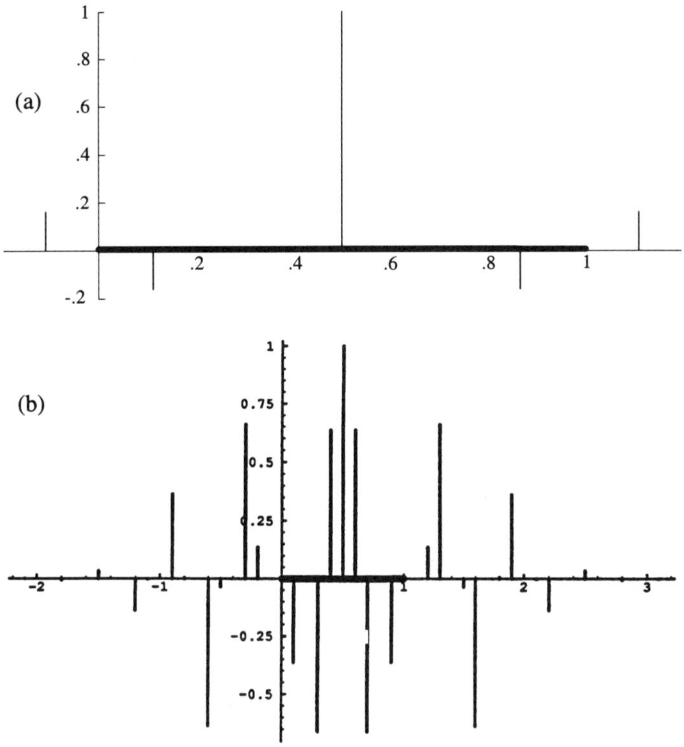

Figure 1 Placement of weights b_i and nodes c_i for two second-order methods. (a), $\mathcal{L}_A^n B$ errors removed for $n < 4$ (the left two nodes form the corrector e^C); (b), for $n < 16$. The bold line marks the quadrature domain $[0, 1]$.

odd m are then satisfied automatically. Figure 1(a) illustrates the symmetry of the nodes and weights.

2 The leapfrog kernel

The corrector equations for any kernel can be written down immediately from (1.3). Take $e^K = e^{\frac{1}{2}tA}e^{tB}e^{\frac{1}{2}tA}$ which corresponds to a single node at $x = \frac{1}{2}$ with weight 1, so the quadrature condition for $m = 0$ is already satisfied. The contributions to (1.3) from the nodes at $\pm c_i$ cancel, leaving

$$\left(\frac{1}{2}\right)^m + \sum_{i=1}^{k/2} b_i\Big((1-c_i)^m - (1+c_i)^m\Big) = \frac{1}{m+1}, \quad m = 2, 4, 6, \ldots n-2. \tag{2.1}$$

If $n = k + 2$ and the c_i are given, distinct numbers, then these equations have a unique solution for the b_i. Furthermore, an exploration of the determining equations (2.1) suggests that no choice of nodes c_i can increase the order n beyond $k+2$, i.e., there is an order barrier for this type of quadrature. Since the efficiency of the corrector is not crucial we have not explored this point further.

The simplest, fourth-order corrector has $k = 2$ and $c_1 b_1 = -\frac{1}{48}$. This quadrature is shown in Figure 1(a) for $c_1 = \frac{1}{8}$. For higher orders, we follow WHT and take

$c_i = ic$, $i = 1, \dots, k/2$, obtaining a family of eighth-order correctors with $k = 6$:

$$\begin{aligned} b_1 &= -\frac{155+3822\,c^2+60480\,c^4}{1935360\,c^5} \\ b_2 &= \frac{124+2352\,c^2+12096\,c^4}{1935360\,c^5} \\ b_3 &= -\frac{31+294\,c^2+1344\,c^4}{1935360\,c^5} \end{aligned} \tag{2.2}$$

As c decreases, $\max|b_i|$ increases ($b_1 \approx -(0.15/c)^5$ as $c \to 0$), so some experimentation is needed to find the best value of c. The nodes c_i may be taken in any order when reconstructing the a_i in (2); taking $(-c_1, c_1, -c_2, c_2 \dots)$ gives the corrector

$$\dots e^{-c_1 tA} e^{b_1 tB} e^{2c_1 tA} e^{-b_1 tB} e^{-c_1 tA} \tag{2.3}$$

as also used by WHT. The maximum step in A is $2ic$, so the high-order correctors of this family take large steps. Figure 1(b) illustrates the weights and placement of nodes for the sixteenth-order corrector with $k = 14$ and $c = 0.3$.

3 Hybrid methods

In [4], methods are designed which eliminate the errors due to $\mathcal{L}_A^m B$ without using a corrector. One such class is based on Gaussian quadrature. Initial trials indicate that these are roughly as efficient as the corrector methods of the previous section, even in the limit of large numbers of steps between corrections. However, they also suggest a hybrid approach which may ameliorate some of the problems of the correctors used in WHT: use the accurate, stable methods of [4] to eliminate $\mathcal{L}_A^m B$ for small m, and leave the errors for larger m to the corrector.

For example, take the kernel

$$e^{\alpha tA} e^{\frac{1}{2}tB} e^{(1-2\alpha)tA} e^{\frac{1}{2}tB} e^{\alpha tA}$$

where $\alpha = (3 - \sqrt{3})/6$. The first $\mathcal{O}(\epsilon)$ error is now at $\mathcal{O}(t^5)$, instead of $\mathcal{O}(t^3)$ for the leapfrog kernel. As before, one finds an eighth-order corrector with $k = 6$:

$$\begin{aligned} b_1 &= \frac{40+819c^2}{2177280c^5} \\ b_2 &= -\frac{32+504c^2}{2177280c^5} \\ b_3 &= \frac{8+63c^2}{2177280c^5} \end{aligned}$$

These are much smaller in magnitude than with the leapfrog kernel; and for large c they are $\mathcal{O}(c^{-3})$ instead of $\mathcal{O}(c^{-1})$. Figure 2 shows the largest weight for each kernel for a sixteenth-order corrector. This approach seems promising but a definite conclusion will depend on pinpointing the cause of the problems discussed in WHT.

4 Higher orders

The corrector idea offers a possibility for avoiding some of the problems of high-order composition methods: that some a_i and some b_i must be negative for order more than 2 [2], and that the minimum number of stages needed rise rapidly with order—3 applications of e^{tB} are needed for fourth order, 7 for sixth order, and 15 for eighth order. To this end we studied fourth-order symplectic correctors.

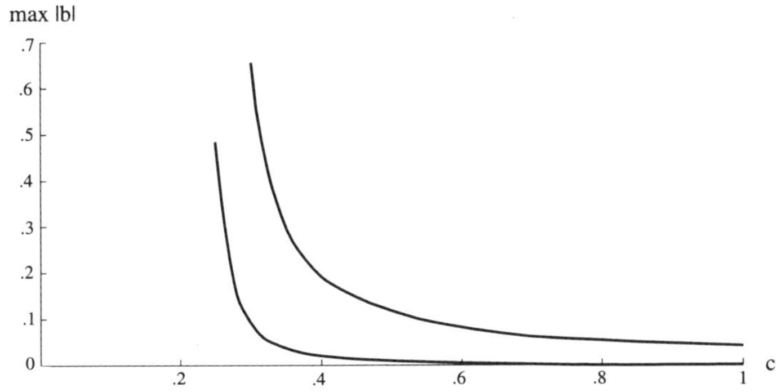

Figure 2 Largest weight ($= \max_i |b_i|$) for two families of sixteenth-order correctors. The upper curve is for a leapfrog kernel, the lower curve for the two-stage kernel of section 3.

We take a symmetric kernel to avoid destroying the symmetry of the quadrature:

$$\begin{aligned} e^K &= e^{\alpha_r tA} e^{\beta_r tB} \dots e^{\alpha_0 tA} \dots e^{\beta_r tB} e^{\alpha_r tA} \\ \Rightarrow K &= t(A+B) + t^3 (p_1 \mathcal{L}_A^2 B + p_2 \mathcal{L}_B^2 A) + \mathcal{O}(t^5) \end{aligned}$$

and the same symmetric corrector as before, with nodes $\pm a_i$ (the notation following WHT):

$$\begin{aligned} e^{X(a,b)} &= e^{atA} e^{btB} e^{-atA} \\ \Rightarrow X(a,b) &= \exp(\mathcal{L}_{atA}) btB; \end{aligned}$$

$$\begin{aligned} e^{Z(a,b)} &= e^{X(a,b)} e^{X(-a,-b)} \\ \Rightarrow Z(a,b) &= 2\sinh(\mathcal{L}_{atA}) btB - t^3 2ab^2 L_B^2 A + \mathcal{O}(t^4); \end{aligned}$$

$$\begin{aligned} e^C &= \prod_{i=1}^{k/2} e^{Z(a_i,b_i)} \\ \Rightarrow C &= 2t^2 \Big(\sum_{i=1}^{k/2} a_i b_i \Big) \mathcal{L}_A B - 2t^3 \Big(\sum_{i=1}^{k/2} a_i b_i^2 \Big) \mathcal{L}_B^2 A + \mathcal{O}(t^4). \end{aligned}$$

For any kernel the errors corresponding to $t^{m+1}\mathcal{L}_A^m B$ can be eliminated as before, using the quadrature conditions (3). But now we must also eliminate all terms at order 3 and 4—the order 4 terms are not trivial because the method (1) cannot itself be symmetric. We have

$$e^C e^K e^{-C} = e^W,$$

$$\begin{aligned} W &= K + \mathcal{L}_C K + \mathcal{O}(t^5) \\ &= t(A+B) + t^3(p_1 \mathcal{L}_A^2 B + p_2 \mathcal{L}_B^2 A) + t^3 2 \Big(\sum_{i=1}^{k/2} a_i b_i \Big) \mathcal{L}_{A+B} \mathcal{L}_B A \\ &\quad + t^4 2 \Big(\sum_{i=1}^{k/2} a_i b_i^2 \Big) \mathcal{L}_{A+B} \mathcal{L}_B^2 A + \mathcal{O}(t^5) \end{aligned} \tag{4.1}$$

so, because the corrector cannot control the two third-order error terms independently, we obtain the *third-order correctability condition* on the kernel:

$$p_1 + p_2 = 0$$

For example, if the kernel is itself third order then $p_1 = p_2 = 0$ and correctability is ensured. More generally, an application of the Baker-Campbell-Hausdorff formula to e^K gives

$$4(p_1+p_2) = -\frac{1}{3} + \sum_{i=1}^{r}\left[\beta_i\Big(\alpha_0 + 2\sum_{j=1}^{i-1}\alpha_j\Big)^2 + \alpha_i\Big(2\sum_{j=1}^{i}\beta_j\Big)^2\right] = 0 \qquad (4.2)$$

Although we have only one equation instead of the two required for a third-order kernel, it seems to impose much the same restrictions. With $r=1$ (for a two-stage kernel), we must have $\beta_1 = \frac{1}{2}$, $\alpha_1 = (1-\alpha_0)/2$, and (7) simplifies to

$$3\alpha_0^2 - 3\alpha_0 + 1 = 0,$$

which has no real solutions. For the simplest kernel, take $r = 2$ and $\alpha_2 = 0$, $\alpha_1 = (1-\alpha_0)/2$, $\beta_2 = \frac{1}{2} - \beta_1$ to get

$$\alpha_0^2\beta_1 - 2\alpha_0\beta_1^2 + \beta_1(2\beta_1 - 1) + \frac{1}{6} = 0$$

which has real solutions for $\beta_1 < 0$ and $\beta_1 \geq \beta^* = 1/(2-2^{1/3})$. The well-known three-stage fourth-order method corresponds to $\beta_1 = (1-\beta^*)/2$.

So including a corrector *does* allow new kernels, but there is no clear reason to use them over kernels which already have the required order [3]. One does get slightly shorter steps (e.g., the largest step when $\beta_1 = -\frac{1}{2}$ is $\alpha_0 = (\sqrt{93}-3)/6 \approx 1.11$, compared to -1.70 for the fourth-order kernel), but this does not seem to balance the practical disadvantage of having to use a corrector. We conclude that the only application for symplectic correctors is the near-integrable case, for which extra, dominant errors can be eliminated by the corrector.

Having chosen a kernel one needs a corrector which leads to a method with overall fourth-order accuracy. From (6), the only new determining equation is $\sum_{i=1}^{k/2} a_i b_i^2 = 0$. WHT solved this equation by taking $\pm a_i$, $\pm b_i$ pairs. This does not affect the rest of the determining equations; it merely corresponds to a "doubling up" of the quadrature nodes. Although it seems wasteful to use half the unknowns to satisfy one equation, an algebraic search for various values of k did not uncover any better solutions. One piece of the corrector then takes the form

$$e^{Z(a,b)}e^{Z(-a,-b)} = e^{atA}e^{btB}e^{-2atA}e^{-2btB}e^{2atA}e^{btB}e^{-atA}.$$

For example, three such pieces ($k = 12$) make an eighth-order corrector; taking the fourth-order kernel and $c_i = i/3$ gives

$$\begin{aligned} b_1 &= 0.11365774925958663923, \\ b_2 &= -0.07195437705518615839, \\ b_3 &= 0.01008366828359522585 \end{aligned}$$

(the general, analytic solution is rather lengthy).

The above analysis can be extended to higher orders. For example, for a method with no $\mathcal{O}(\epsilon)$ or $\mathcal{O}(\epsilon^2)$ terms up to sixth order in t, one finds two correctability constraints on the three terms in question ($\mathcal{L}_A^4 B$, $\mathcal{L}_B\mathcal{L}_A^3 B$, and $\mathcal{L}_A\mathcal{L}_B\mathcal{L}_A^2 B$). Again they are satisfied if the three terms are already zero in the kernel.

5 Extensions

Suppose the vector field cannot be split into two solvable parts. Then each of these methods corresponds naturally to a method of the same order involving an arbitrary first-order method and its inverse [3]. The only new case to be dealt with is the near-integrable one: when the vector field can be split as $A+B$ with $B=\mathcal{O}(\epsilon)$, but the vector field B cannot be solved exactly. One might have to approximate its flow by, say, $\prod e^{tX_i}$ if $B=\sum X_i$. So suppose we have some approximation of the flow of B,

$$e^{B^+}, \quad B^+ = tB + t^2B_2 + t^3B_3 + \dots,$$

and its adjoint

$$e^{B^-}, \quad B^- = tB - t^2B_2 + t^3B_3 - \dots,$$

where $B_n = \mathcal{O}(\epsilon^n)$.

If the method is second order, the only new lower-order error term is that due to B_2. This can be eliminated by replacing each e^{tbB}, e^{tbB} pair in the method (1), (2) by the pair e^{tbB^+}, e^{tbB^-}. Any unpaired term (such as e^{tB} in the leapfrog kernel) must be replaced by $e^{\frac{1}{2}tB^+}e^{\frac{1}{2}tB^-}$. Furthermore, such a replacement introduces no new $\mathcal{O}(\epsilon)$ terms at any order in t.

This approach is not sufficient at higher order. The $\mathcal{O}(\epsilon^3t^3B_3)$ term, for example, would be independent of the corrector, leading to a new constraint on the kernel. But there is a simple solution: just sequentially replace $e^{atA}e^{btB}e^{ctA}\dots$ by $e^{atA}e^{atB^+}e^{(b-a)tB^-}e^{(b-a)tA}e^{(c-(b-a))tA}\dots$ in the kernel and corrector separately. For example, the piece of the corrector in Eq. (5) becomes

$$\dots\varphi(-c_1t)\varphi^*((c_1+b_1)t)\varphi((c_1-b_1)t)\varphi^*(-c_1t)$$

where $\varphi(x)=e^{xA}e^{xB^+}$ and the adjoint $\varphi^*(x)=e^{xB^-}e^{xA}$. Now from the general theory in [3], the overall order of the method is preserved; and the errors incurred in replacing B by B^+ or B^- are all $\mathcal{O}(\epsilon^2)$, so the new method still removes all $\mathcal{O}(\epsilon t^n)$ terms to the same order in t.

6 Methods formed from symmetric stages

The existence of correctability constraints is most easily seen when the kernel and corrector are both composed of the same symmetric method—the standard approach for reaching high order [7]. Let $S(t)$ be symmetric, i.e., $S(t)S(-t)=1$, and take

$$e^K = \prod_{i=1}^{r} S(w_it), \quad e^C = \prod_{i=1}^{k} S(z_it)$$

Suppose

$$S(t) = \exp(tX_1 + t^3X_3 + t^5X_5 + \dots);$$

for the overall method there is one determining equation for each independent Lie bracket of the X_n. Take e^K symmetric. The equations for the errors corresponding to the X_n themselves are, for odd n,

$$\delta_{n1} = \sum_{i=1}^{r} w_i^n + \sum_{i=1}^{k} z_i^n + \sum_{i=1}^{k}(-z_i)^n = \sum_{i=1}^{r} w_i^n$$

—i.e., the corrector makes no contribution. Thus these equations are included among the correctability conditions.

There are in fact no more at sixth order. To see this we have to actually construct a corrector. Solving the above equations provides a K with

$$K = tX_1 + t^5 k_1 \mathcal{L}_X^2 X_3 + \mathcal{O}(t^7)$$

where k_1 is given in [7]. To avoid even-order errors, one finds that C must have no terms at order t, t^3, or t^5. Those due to X_1, X_3, and X_5 are avoided by taking $e^C = e^{D(t)}e^{D(-t)}$; that due to $\mathcal{L}_X^2 X_3$ by taking $e^{D(t)} = e^{E(t)}e^{E(-t)}$ where now

$$e^{E(t)} = \prod_{i=1}^{k/4} S(z_i t).$$

This gives

$$C = t^4 c_1 \mathcal{L}_X X_3 + \mathcal{O}(t^6), \quad c_1 = 4\sum_{i=2}^{k/4} \left[z_i^3 \Big(\sum_{j=1}^{i-1} z_j \Big) - z_i \Big(\sum_{j=1}^{i-1} z_j^3 \Big) \right].$$

Finally, the overall method has the expansion $K + \mathcal{L}_C K + \mathcal{O}(t^9)$, so the term $\mathcal{L}_X^2 X_3$ *can* be corrected, if the corrector is chosen so that

$$c_1 = k_1.$$

From the antisymmetry of c_1 we see that this equation has a solution for any k_1; for example, $k = 8$ (leading to two unknowns z_1, z_2) is sufficient.

So the shortest sixth-order symplectic corrector has $r = 5$ stages, as opposed to $r = 7$ for uncorrected methods. The determining equations for the kernel have two real solutions (e.g., $w_1 \approx 1.452$, $w_2 \approx -2.151$, $w_3 \approx 2.397$, with corrector $z_1 = \frac{1}{4}$, $z_2 \approx -0.127$) which are close to Yoshida's methods "B" and "C" [7]. Both of these are much less accurate than his method "A"—unfortunately this third method just fails to exist as a corrector. One can buy a little more freedom by going to $r = 7$, e.g., $w_1 = w_2 \approx 0.431$, $w_3 \approx -0.834$, $w_4 \approx 0.944$ has slightly smaller coefficients that Yoshida's method A. Overall, it seems that dropping the determining equation $k_1 = 0$ from the kernel does not gain much. Extending the above derivation to $\mathcal{O}(t^8)$ shows that in addition to the X_7 term, $\mathcal{L}_{X_3}^2 X_1$ cannot be corrected, so the kernel requires at least $r = 9$ stages.

A corrector method essentially allows one to leave the secular part of the error to be dealt with later. One's intuition is that errors which can be made secular by the small change of coordinates corresponding to e^C are not in general the most important ones. Our fairly thorough exploration of the determining equations at fourth and sixth order suggests that, indeed, the correctable portion of the error is neither large enough, nor a sufficiently significant constraint on the efficiency of the kernel, to make leaving it for later worth while.

7 Acknowledgements

I'd like to thank Jack Wisdom for his patient, extensive discussions of [6], and Jürgen Moser for inviting me to the Forschungsinstitut für Mathematik where this work was carried out.

References

Butcher, J. C. [1968], *The effective order of Runge-Kutta methods*, Lecture Notes in Mathematics, Springer, **109**, 133–139.

Goldman, D., and Kaper, T. J. [1996], *Nth-order operator splitting schemes and nonreversible systems*, SIAM J. Numer. Anal. **33**, 349.

McLachlan, R. I. [1995], *On the numerical integration of ordinary differential equations by symmetric composition methods*, SIAM J. Sci. Comp., **161**, 151-168.

McLachlan, R. I. [1995], *Composition methods in the presence of small parameters*, BIT, **35**, 258-268.

Suzuki, M., [1993], *Improved Trotter-like formula*, Phys. Lett. A, **180**, 232–234.

Wisdom, J., Holman, M. and Touma, J., *Symplectic correctors*, these proceedings.

Yoshida, H. [1990], *Construction of higher order symplectic integrators*, Phys. Lett. A, **150**, 262–268.

Fields Institute Communications
Volume **10**, 1996

A Survey of Open Problems in Symplectic Integration

Robert I. McLachlan
Mathematics Department
Massey University
Palmerston North
New Zealand

Clint Scovel
Computer Research Group, C–3, MS–B265
Los Alamos National Laboratory,
Los Alamos, NM USA 87545

Dedicated to Feng Kang

Abstract. We present a personal view of the most important open problems and areas for research in symplectic integration and in the more general study of geometric structure in numerical methods.

1 Introduction

In the past few years there has been a substantial amount of research on symplectic integration. The subject is only part of a program concerned with numerically preserving a system's inherent geometrical structures, such as volume preservation, reversibility, differential constraints on PDEs such as those involving *div* and *curl*, and integral invariants. In many cases one requires a numerical method to stay in the smallest possible appropriate group of phase space maps. It is not the authors' opinion that symplecticity, for example, automatically makes a numerical method superior to all others, but it is our opinion that it should be taken seriously and that a conscious, informed decision be made in that regard.

We present here a survey of open problems in symplectic integration, including other problems from the larger program. This is not intended as a review of symplectic integration (see [36, 121, 122]) and is naturally derived from the authors' own research interests. Many of the problems mentioned here call for numerical experimentation, some for application of suggested but untested methods, some for new methods, and some for theorems. We hope that this survey helps to communicate the outstanding problems in this rapidly growing field to the research community.

1991 *Mathematics Subject Classification.* Primary: 65L05; Secondary: 70-08, 58F05, 65-02.

In most of the paper we have in mind that one is trying to solve initial value problems for known, fixed ODEs; PDEs are considered separately in Section 11 and other computational tasks in Section 14. The flavor of this subject is a mixture of numerical analysis and geometric mechanics, but a necessary aim of all this work is to produce algorithms. Coordinates must be used and all functions must be computable; coordinate-free geometry can be at most an intermediate step.

2 Composition methods

2.1 High-order methods It is now a standard practice in structure-preserving integration to compose one or more low-order methods (usually first or second order) with appropriately chosen time steps to obtain a higher-order method. If one has a differential equation $\dot{x} = X(x) = A(x) + B(x)$ such that $\dot{x} = A$ and $\dot{x} = B$ can both be solved analytically, with flows $\exp(tA)$ and $\exp(tB)$, then one may use the compositions

$$\ldots \exp(a_2 tA) \exp(b_1 tB) \exp(a_1 tA) \tag{2.1}$$

or

$$\ldots S(w_2 t) S(w_1 t) \tag{2.2}$$

where $S(t) = \exp(\frac{1}{2}tA)\exp(tB)\exp(\frac{1}{2}tA)$ is the leapfrog method. The time-step map is then in the Lie group whose Lie algebra is generated by the vector fields A and B. If they were Hamiltonian vector fields, for instance, this group would be a subgroup of the group of symplectic maps. If their flows had identical integrals such as angular momentum, then so would the composition. In particular, for an invariant of the full system to be preserved by this method, each component of the decomposition must preserve that invariant. Some invariants appear to be easier than others to preserve under decomposition. For example, symplecticity is easily preserved, but integrability and energy conservation are destroyed in general.

To analyze such methods, one may expand the compositions (2.1, 2.2) as $\exp(\sum_{i=1}^{\infty} t^i X_i)$ where $X_i \in L^n(A, B)$, the vector space spanned by the n-fold Lie brackets of A and B. This space is a subspace of the free Lie algebra generated by A and B [12, 111] whose dimension is known. There are thus $\sum_{n=1}^{p} \dim L^n(A, B)$ *determining equations* which must be satisfied for a method to have order at least p. It is traditional to compare different solutions of the determining equations based on an *effective error constant*—some measure of the first term in the local truncation error, adjusted for the complexity of the method. In this section we compare methods only in this limited sense.

Thus, this avenue leads in primarily algebraic rather than geometric directions. We also need to study curves through the identity in Lie groups, here with parameter t, rather than points and one-parameter subgroups.

Various symmetries can be imposed on (2.1) and (2.2) which turn out to reduce the number of determining equations, i.e., which in the language of Runge-Kutta methods act as simplifying assumptions. A family of maps $\varphi(t)$ is said to be *symmetric* (also self-adjoint, reversible, or revertible) if $\varphi(t)\varphi(-t) = 1$. Note that in (2.2), the individual stages $S(t)$ are themselves symmetric. Roughly, it has been found that the higher the order, the more symmetry it is worth imposing. Some borderline cases remain [84]:

- Is there a symmetric sixth-order method which is better than those composed of symmetric steps?

Now consider the case of "simple mechanical systems," which are Hamiltonian with the standard symplectic structure, and have Hamiltonians of the form $H = H_A(p) + H_B(q)$ where H_A is quadratic in p. (Composition methods based on this splitting are instances of "Runge-Kutta-Nystrøm" methods.) Let A be the Hamiltonian vector field corresponding to the Hamiltonian H_A, etc. One can check that now $[B,[B,[B,A]]]$ (which we abbreviate as $B^{(3)}$) is identically zero, so the determining equations corresponding to this term, and to all its Lie brackets, may be dropped. This has been shown to lead to a definite advantage for nonsymmetric fourth-order and for symmetric sixth-order methods. Some eighth-order methods of this type have been found by Okunbor [97].

- Is there a symmetric eighth-order Runge-Kutta-Nyström method which is better than general methods of type (2.2)?

- Describe the Lie algebra generated by the vector fields A and B when $B^{(3)} = 0$.

- Describe the Lie algebra generated by the Hamiltonian functions p^2 and $H_B(q)$.

There are no differences between these two cases $B^{(3)} = 0$ and $H = p^2 + H_B(q)$ for $n \leq 7$, so the popular Runge-Kutta-Nyström methods of order 7 or less [84] apply in the more general case $B^{(3)} = 0$, which has some applications in nonlinear wave equations [83]. However, one can check that the eighth-order term $[B^{(2)},[B^{(2)},B^{(1)}]]$ is zero under the $p^2 + H_B(q)$ splitting, but is not explicitly a bracket of $B^{(3)}$.

There are some other cases of particular As and Bs which are advantageous to consider separately. For example, if $B \ll A$ so that the vector field is near-integrable, then the Lie brackets of A and B are of differing orders—e.g., $[B,[B,A]] \ll [A,[A,B]]$ and can be profitably ignored [85]. In a related case, Saha and Tremaine [115] and Wisdom, Holman, and Touma [142] take advantage of the relationships between the particular A and B occurring in solar system dynamics. On the other hand, the separable case with arbitrary $A = A(p)$ and $B = B(q)$ has been shown *not* to be a special case, in the sense that a composition method which works (has a certain order) in this case will also work for all A and B [1]. Therefore, the same is true for the case that A and B have linear flows (i.e., $\exp(tA) = 1+tA$), as they do in the separable case and elsewhere [36].

- Are there other useful special cases, subspaces of particular Lie algebras, or kinds of compositions, which lead to better methods?

2.2 Existence of splittings What is special about the class of Hamiltonian systems—or of all vector fields—which possess splittings $X = \sum_i X_i$ for which the pieces can be integrated exactly? Even many non-canonical systems can be split in this way, such as those Lie-Poisson systems in which the Hamiltonian only couples variables within the Lie algebra's abelian subalgebras [82]. This class includes, for example, the moment algebra with separable energy function, the rigid body, and the sine-bracket truncation of the 2D Euler equations (see Section 11.1).

For most splittings used in practice, the pieces X_i are either linear or have linear flows. One can leave this class and ask,

- which Hamiltonian vector fields X can be written as $X = \sum_{i=1}^{n} X_i$ where each X_i is completely integrable?

Such a situation is surely not generic, but there are many such vector fields to call on, for both canonical [105] and noncanonical [29, 39] systems. In this problem one is supposed to *start* with X, not with the X_i. A recent example arose in celestial mechanics, where the first post-Newtonian correction, previously supposed to be not splittable, has been neatly split [116].

Naturally one does not want n to be too large. Consider, for example, "Euler equations." These are Poisson systems with Lie-Poisson structure, i.e., linear brackets, and quadratic Hamiltonians. In coordinates, $\dot{x}_i = \sum_{jkl} c_{ij}^k x_k a_{jl} x_l$. The Hamiltonian $\sum_{jl} \frac{1}{2} a_{jl} x_j x_l$ may be diagonalized and each term integrated [109], and this is often the best approach. But in one case, the $N \times N$ sine-bracket truncation of the 2D Euler fluid equations discussed in Section 11.1, it leads to an $\mathcal{O}(N^5)$ algorithm, as opposed to the $\mathcal{O}(N^3 \log N)$ possible with a different splitting [82].

Canonical systems with polynomial Hamiltonians are another important case. Channell ([19]) splits them into monomials, but Feng and Wang ([36]) split them into polynomials with linear flows. These different approaches have not been compared.

Of course $\exp(tX_i)$ should not be too difficult to evaluate—there is a big difference between proving a system integrable and actually computing its solution. For example, the Hamiltonian for the motion of n point vortices is $H = \sum_{i<j} V(x_i, x_j)$, where vortex i is located at $x_i \in \mathbb{R}^2$. The two-vortex interaction term $V(x_1, x_2) = -\frac{1}{4\pi} \log |x_1 - x_2|$ has a simple flow (in translating coordinates, the vortices rotate about their center of vorticity [114]), leading to an explicit symplectic integrator of any order for this problem. It is not at all clear that the integrability of the three-vortex problem can be exploited here. Another example is due to Wisdom and Holman [141], who successfully integrated the n-planet solar system with such a method. In Jacobi coordinates, they wrote $H = K(q,p) + V(q)$ where K represents n Sun-planet two-body Kepler problems (all mutually commuting, and easily solved) and V represents the $n(n-1)/2$ planet-planet interaction terms, whose flow is a shear. Because it preserves the near-integrability of the problem, this splitting is much more accurate than the $H_A(p) + H_B(q)$ splitting. However, it is only accurate for solar system-like configurations; it is not uniformly accurate in the planet-planet distances. We therefore pose the problem:

- Develop a symplectic integrator for the Newtonian n-body problem which is uniformly accurate near two-body collisions.

A promising approach is to use a partition of unity on the Hamiltonian so that each close pair is integrated exactly, with smooth transitions in the splitting as pairs break up.

2.3 Symplecticity for free Take a Hamiltonian differential equation $\dot{x} = X(x)$. Let $\varphi(t) = 1+tX$ be Euler's method, which is not symplectic. It has been noticed independently by Iserles and by Sanz-Serna (unpublished), that $\varphi(t/2)\varphi^{-1}(-t/2)$ is the midpoint rule, $x' = x + tX((x+x')/2)$, which explains the good long-time behavior of the nonsymplectic composition $\varphi^{-1}(-t/2)\varphi(t/2)$, which is the second-order Adams-Moulton method: $x' = x + t(X(x) + X(x'))/2$. Namely, it is conjugate to the midpoint rule by the map $\varphi(t/2)$, and thus preserves the perturbed symplectic form $\varphi(t/2)^*\omega$, where ω is the standard symplectic 2-form. Generalizing this cute example would give a huge new class of symplectic integrators.

A direct calculation shows that

$$\varphi(t)^*\omega = \varphi(-t)^*\omega,$$

is a necessary and sufficient condition for $\varphi(t/2)\varphi^{-1}(-t/2)$ to be symplectic. That is, $\varphi(t)^*\omega$ must be even in t. When $\varphi(t)$ is Euler's method, $\varphi(t)^*\omega$ is even because it is quadratic and the linear terms vanish when the vector field is Hamiltonian. Other methods based on this idea have not been constructed.

- Give a general construction of methods $\varphi(t)$ such that $\varphi(t)^*\omega$ is even in t.

- Construct integrators which are conjugate to symplectic maps for each t.

3 Multiderivative methods

For some systems, such as those of molecular dynamics, the marginal cost of evaluating the *derivative* of the forces is small [126]. Therefore one should

- Investigate all possible ways of including extra derivative information in a symplectic integrator.

In fact, the earliest symplectic integrators, based on the Taylor series solution of the Hamiltonian-Jacobi equation [34, 20], *did* use derivatives, but these methods are not considered competitive today. An exception is the Stofer ansatz [133, 122], which achieves fourth order in one stage using only the second derivative of the Hamiltonian. There are also explicit multiderivative symplectic schemes available for some systems [86, 68]. Unfortunately, *linear* multiderivative symplectic schemes (those linear in $d^n x/dt^n$) do not exist [50].

4 No-derivative methods

Sometimes the Hamiltonian is so complicated that one does not wish to calculate its derivative by hand, or even by a symbolic manipulator.

- How efficient can symplectic integrators be, given only the Hamiltonian function?

One approach is to form the vector field using automatic differentiation, which is now a mature technique [37, 64]. Another is to use a fixed lattice in phase space to define an interpolated piecewise-polynomial Hamiltonian. A symplectic map $f(x)$ can be defined by taking the lattice points nearest x and using the vector field corresponding to that piece of the interpolating Hamiltonian. (Or can one finite-difference the Hamiltonian about x?—see Section 12.2.) There is no cost associated with making the interpolation error as small as desired, but care must be taken to make the $f(x)$ as smooth as possible, so that, e.g., the map will preserve invariant tori (see Section 12.1).

5 Implicit methods

Some vector fields cannot be split. Then one has the generating function and Gauss-Legendre Runge-Kutta methods. Note that as the order increases, the latter do not suffer combinatorial explosion as composition methods do—so they can sometimes be more efficient anyway [84]. Several authors [61, 117] have considered different symplectic Runge-Kuttas, but these have not been tested yet. If one is willing to form and invert Jacobians then Runge-Kuttas with real roots may be superior.

- Find the best symplectic Runge-Kuttas at each order.

When solving the implicit equations, the authors have always been content to extrapolate x^{n+1} from previous time steps, and then solve $x^{n+1} = x^n + f(x^n, x^{n+1})$, say, by iteration to roundoff error—probably unbeatable as $\Delta t \to 0$, but high-order methods were not made for this. In tests, one should consider large, strongly-coupled systems such as those of molecular dynamics and discretizations of PDEs (see Section 11). In any event, iterating to roundoff is wasteful; would prefer to only solve the equations to within the local truncation error.

- Develop iterations for the standard implicit symplectic methods which are symplectic and preserve equivariances of the original method after any number of iterations.

6 Generating functions

On a linear symplectic space there are many different kinds of generating functions. In addition to the four traditional types, Feng [34, 35] has provided a general discussion and described several non-conventional examples. These different types of generating functions may have different applications. For example, to generate the identity map from a generating function of the first ("q, Q") kind requires a singular generating function [20].

- What are the advantages of the specific types of generating functions? In particular, what are the advantages or disadvantages of using a singular generating function?

Ge [42] has shown that generating functions of the first kind respect symmetries well for cotangent bundles of nonlinear configuration spaces. Namely, they can be used to generate integrators which preserves the momentum corresponding to *all* cotangent lift symmetries, not just linear ones as for generating functions of the third ("q, P") kind. This leads to the Ge-Marsden [46] generating function for Lie-Poisson systems. Indeed, Ge [43] has generalized this construction to show that there is a generating function theory and a Hamilton-Jacobi equation, which in theory can be used to construct symplectic integrators for arbitrary Poisson manifolds. This construction uses the symplectic groupoid which exists locally over every Poisson manifold [23]; the attendant groupoid structures give rise naturally to a generating function for the Poisson map.

For regular quadratic Lie algebras (those with a nondegenerate Ad-invariant quadratic form), there is a generating function which generates the identity transformation on the dual of the Lie algebra. However, it uses the coadjoint action, not of the identity group element, but of a family of group elements $g(\mu)$ such that each $g(\mu)$ is in the coadjoint isotropy group of μ. Namely, $Ad^*_{g(\mu)}\mu = \mu$. Reconstruction of this identity map to the cotangent bundle will give nontrivial group motions that we feel are unnecessary. Consequently, although numerical experiments shows that the method works well [21], we feel it desirable to construct a generating function for the identity map using $g(\mu) \equiv 1$.

- Is there a generating function for Lie-Poisson maps which generates the identity map via the coadjoint action of the identity group element?

A possible solution lies in the more general construction of Lagrangian submanifolds by Morse families [74]. Bear in mind that such a technique will not be of much use if it entails a substantial increase in computational complexity, as generating function methods already have inadequate efficiency [9].

For matrix Lie groups, the group can be thought of as a submanifold of a linear space which is acted on linearly by the group. One can integrate in the linear space with constraints to the submanifold, and one such algorithm [87, 107] has a generating function of the third kind. Indeed, there is a generating function that generates the identity map by the coadjoint action of the identity group element in this case. See Section 10 for further discussion of symplectic constraint algorithms.

7 Multistep methods

The failure of symplectic methods to use information from past time steps leads to their needing more function evaluations than nonsymplectic methods, particular at orders more than four. Unfortunately, the most useful definition of symplecticity for multistep methods has not yet been arrived at. Some experimentation with actual systems is called for.

Start with a Hamiltonian system on a symplectic manifold M. An n-step method on M may be written as a one-step method on M^n. The continuous system can also be extended to one on M^n with the product symplectic structure. Ge and Feng [41] have investigated the circumstances under which one can preserve the product symplectic structure on M^n. Eirola and Sanz-Serna [32] and Ge [44] have investigated more general symplectic structures on M^n, and Bochev and Scovel [11] have shown that the symplecticity of their methods follows from preservation of quadratic invariants.

But not every ODE on M^n arises from one on M in this way, so preserving the product symplectic structure cannot be expected to lead to good results. For example, the two-step method $x_{i+1} = x_{i-1} + 2tX(x_i)$ preserves the product symplectic structure on M^2 [35, 32], but has none of the nice properties one associates with symplecticity. Applied to integrable m-degree-of-freedom systems, the method does preserve invariant tori, but they have dimension $2m$ in the $4m$-dimensional space M^2, and do not project to subsets of M close to the original tori, even as the time step tends to zero. What is needed are some extra integrals in the new directions, so that the invariant tori have dimension m and do project nicely.

Feng [136] has proposed the completely different criterion that the induced one-step method (the map $m\colon M \to M$ satisfying the multistep method) should be symplectic in the usual sense. This sounds extremely appealing, but unfortunately Tang [136] has shown that no linear multistep methods can satisfy such a stringent criterion. Of course *some* general linear methods satisfy this criterion, namely, those derived from the composition of symplectic one-step methods; are there any others?

- Which multistep methods satisfy Feng's criterion?

One also needs to consider how multistep methods are functionally dependent on the initialization technique. Consider a two-step method for simplicity, in which x_3 is given as a function of x_1 and x_2 by combining the formulas for two identical one-step methods m. If the initial conditions are selected by $x_2 = m(x_1)$, the sequence of points is that determined by the symplectic map m. However, if the initial conditions are selected by another mapping $x_2 = f(x_1)$, this is not the case. We took the composition of two midpoint rules in the form

$$x_3 = x_1 + t\Big(X\big(\frac{x_1 + x_2}{2}\big) + X\big(\frac{x_2 + x_3}{2}\big)\Big),$$

which is symplectic in Feng's sense: its equivalent one-step method is the midpoint rule. Applied to the pendulum, we observed that as f is perturbed from the midpoint rule, each invariant curve bifurcates into two curves, with successive iterates alternating between the two. This suggests that the map corresponding to two time steps might be symplectic for all f.

MacKay [76] has proposed the criterion that if ξ^i and η^i are vectors such that $\omega(\xi^1, \eta^1) = \cdots = \omega(\xi^n, \eta^n) = c$, where ω is the symplectic two-form, then one should have $\omega(\xi^{n+1}, \eta^{n+1}) = c$ also.

- Find a criterion for symplecticity of multistep methods, and methods satisfying it, that leads to methods with nice properties for Hamiltonian systems.

8 Preservation of integrals

To preserve integrals one generally appeals to the theorem of Ge [42] that to do so up to a constant it is necessary and sufficient for the algorithm to be equivariant with respect to the symmetries generating the integrals. (See also [79], where conditions are given under which the constant can be removed.) But there may be other approaches. Suris [134] has shown how making the Hamiltonian depend parametrically on the time step can lead to completely integrable versions of the standard map (i.e., $x_{j+1} - 2x_j + x_{j-1} = t^2 f(x_j, t)$). Moser and Veselov [94] showed that a Lagrangian discretization of the free rigid body on $T^*SO(n)$ is not only completely integrable but has all the same integrals as the continuous system, except the usual momentum is not conserved. On the other hand, Lewis and Simo [71] have constructed an energy-momentum-preserving symplectic integrator. Curiously, viewing the free rigid body as a free particle constrained to lie in $TSO(n) \subset T\mathbb{R}^{n\times n}$ and applying the symplectic constraint algorithm of Reich [107] gives the same integrator as Lewis and Simo [87], and on the reduced space, the same integrator as Moser and Veselov. We do not know if this is more than a coincidence. Because the integrals do not depend on the time step, they are still preserved by higher-order maps formed by composition.

There are also the integrable discrete versions of integrable evolution equations such as the nonlinear Schrödinger equation and the Korteweg-de Vries equation. Some have integrable time discretizations too, although these are not usually spatially local [3]. Such equations are derived by discretizing the scattering problem of the continuous system, an approach which could be investigated from the point of view of symplectic integration. They usually preserve not the original symplectic structure, but a nearby one—an understandable price to pay for conserving energy exactly. The QR eigenvalue method can be viewed as a symplectic integrator conserving all integrals of the Toda lattice in Lie-Poisson form [135]. This, and the Moser-Veselov matrix factorization techniques [94], should be understood from the point of view of symplectic integration.

If a system were sensitive to errors in a known integral, it might be worth enforcing it implicitly. One would add the integral as a constraint, satisfied by a Lagrange multiplier, which would naturally be zero for the time-continuous problem. One needs an integrator which leaves invariant all the level sets of the integral, and is still symplectic in the usual sense—a slightly different criterion than that used in integration with constraints (see Section 10).

- In which ways can integrals be conserved? Is it worth weakening the symplectic condition by requiring only preservation of a perturbed symplectic structure, if it enables conservation of more integrals?

Many implementors of symplectic schemes have noticed that the energy, although not conserved exactly, does not drift secularly. A theorem of Ge [46] shows that for systems with no integrals apart from the energy, a map preserving both the energy and the symplectic structure would have iterates lying on the true orbit. Such a map is unlikely to be computable. The nice energy behavior of symplectic schemes has been explained by Neishtadt [96], who proves (among other things) that if the perturbation due to the integrator is of size ε, then the energy error will be bounded by $\mathcal{O}(\varepsilon)$ for time $\mathcal{O}(e^{c/\varepsilon})$.

As ε increases, this time decreases, and can eventually limit the feasible time-span of the simulation. Often a critical step size is observed beyond which the energy behavior suddenly worsens. (Related phenomena are seen in Arnold diffusion.) The energy can even blow up in finite time [130]. To avoid this, it seems that the step size must be small compared to the shortest period of the system, a severe restriction if the system is a discretization of a PDE [130, 131].

- Quantify this observed energy breakdown, and find ways to avoid it.

An alternative route is to abandon symplecticity, and require instead strict energy conservation. This approach has been explored by (among others) Simo and coworkers [79, 129, 130, 131]. It is possible to simultaneously conserve momentum, and such "energy-momentum" integrators are very promising. At first sight, however, the resulting map on the energy-momentum level set is not constrained in any way. It might not preserve a volume form, for example, and it might possess attractors, so the implications for long-time simulations are not clear. They are very effective, though, in providing unconditionally stable methods for conservative PDEs [129]: the usual methods for improving stability introduce large dissipation.

9 Variable time steps

Standard methods can sometimes beat symplectic methods for pointwise accuracy, even at low order, if the system contains widely varying time scales in different parts of phase space. The reason is that the standard methods can adaptively change the time step. If one tries to, say, halve the time-step in a symplectic integration when $\|H''\|$ exceeds a certain threshold, the resulting map is not even continuous at the threshold hypersurface, and hence not symplectic. KAM tori are broken and good long-time behavior is lost [15].

Just as the time-t map of a canonical Hamiltonian system is symplectic, so its time-$t(q,p)$ is symplectic with respect to a non-canonical symplectic form. This suggests that this latter form should be preserved by a numerical method, and that $t(q,p)$ should be depend locally and smoothly on q and p. This approach has not been developed to our knowledge. Another, similar, approach would be to integrate $JdH/\|JdH\|$, which also preserves a non-canonical symplectic form. It has been suggested that the ultimate criterion is that the numerical method should be the time-$t(q,p)$ flow of a Hamiltonian system close to the original one. For this and further discussions, see [121, 127].

In the integration of Hamiltonian systems with a nonstandard symplectic structure, such as the integrable discrete nonlinear Schrödinger equation [2], Darboux charts are useful to bring the system to the standard canonical form. Our preliminary investigations show that often the mapping to symplectic variables is explicit but its inverse is not; the inverse is required to specify the Hamiltonian in the new variables. This presents algorithmic difficulties that require further investigation.

To date the only successful approach is that of Skeel and Biesiadecki [128] and MacEvoy and Scovel [75]. They partition the Hamiltonian so that it is a sum of fast and slow terms, the fast terms being zero outside some region of phase space. The slow terms are integrated with a large time step unless that step puts that orbit in the fast region, in which case it must be retaken with a smaller time step. MacEvoy and Scovel [75] join the two regions smoothly by a partition of unity. To date these algorithms have not been developed to the point where one can make a fair comparison with standard variable time step schemes. Simó [125] has suggested taking the Kepler problem (a particle in $\mathbb{R}^2$ under a central inverse-square force) with eccentricity 0.999999 as a good test of a variable time step method.

- Develop variable time step symplectic integrators so that they are competitive for pointwise accuracy with standard methods for the Kepler problem, while retaining the good long-time behavior of constant time step symplectic methods.

10 Constrained systems

10.1 Dirac constraints Suppose that we are given a Hamiltonian on a symplectic manifold M and a submanifold $N \subset M$ to which we would like to constrain the dynamics. Suppose that we have a found a set of constraint functions $\phi : M \to V$ (where V is a linear space) such that $N = \{x \in M : \phi(x) = 0\}$ and the matrix $\{\phi_i, \phi_j\}$ is nonsingular. The problem is to construct a map from N to N approximating the constrained dynamics and preserving the induced symplectic structure on N. Frequently the system in invariant under a symmetry group, in which case the integrator should preserve the corresponding momenta.

This has been addressed and solved only in the case of *holonomic* constraints [63, 69, 87, 107, 108], and when the symmetry group acts by cotangent lifts of *linear* transformations. That is, when $M = T^*Q$ is a cotangent bundle, Q is a linear space, and the constraints are divided into two sets, the primary constraints $R = \{q \in Q : \phi_1(q) = 0\}$, and the secondary constraints $\phi_2(q,p) = d\phi(q)\partial_p H(q,p) = 0$. The constraint methods induce a symplectic map on T^*R.

The problem for the more general class of non-holonomic constraints is still open, as is the question of the correct analogue of symplectic integration for non-holonomically constrained Lagrangian systems.

- Develop momentum-preserving constrained symplectic integrators for general constraints and for general symmetry groups.

For Hamiltonian ODEs, Dirac constraints are equivalent to the intrinsic procedure of constraining a Hamiltonian vector field by merely restricting both the Hamiltonian and the symplectic form to the desired submanifold (call this vector field X_N). Is there an analogous procedure for maps? (call the constrained map φ_N). (Reich [110] has shown that for holonomic constraints, an arbitrary symplectic integrator for the unconstrained problem can be projected to a symplectic map on the constraint space.) The operation should at least commute with the previous procedure under the operation of taking flows, that is, $\exp(tX_N) = \exp(tX)_N$.

At present, the general theorems concerning momentum preservation require the constraint functions to be group invariant. However, in the case of the Lagrange top, McLachlan and Scovel [87] found only degenerate invariant constraint functions. Since nondegeneracy is essential for the constraint algorithm, they dropped one of the invariances; but it turned out that the integrator preserved its momentum anyway.

- What special properties are required of the constraint function in order that a momentum-preserving symplectic integrator may be determined?

10.2 Fewer implicit steps The currently recommended algorithms for holonomically constrained separable Hamiltonian systems [4, 63, 69, 87, 107] involve solving a set of implicit equations to enforce the primary constraints each time q is updated. For a higher-order method formed by composition, this means many implicit steps. (The secondary constraints can be enforced with an explicit step.) Is it possible to do better by only enforcing the constraints at the end of each complete time-step?

- Are there constrained symplectic integrators with fewer implicit steps than those currently in use?

10.3 Generating functions It is easy to determine the exact Lagrange multipliers for the continuous system, but at the cost of more derivatives and matrix inversions. In return, the number of iterations on the implicit step is reduced. At present we are using Lagrange multipliers extrapolated from previous time steps, combined with simple or Newton-chord iteration to roundoff error.

- Can one modify the Hamiltonian or predict the value of the Lagrange multipliers to make the implicit step more efficient? Are there methods which are symplectic after a finite number of iterations on the multipliers?

Indeed, the generating function picture is far from complete.

- Understand the symplectic geometry of the generating function method for holonomic constraints.

It is hoped that a full understanding of this problem will lead to a method for nonholonomic constraints.

10.4 Automatic preservation of quadratic constraints In many physical systems, such as rigid body and molecular dynamics, the constraints are quadratic functions, being based on preservation of Euclidean lengths. Because symplectic Runge-Kutta methods, such as the midpoint rule, preserve quadratic integrals [22, 26, 120], these seem to be good candidates to integrate such quadratically-constrained systems.

Specifically, the constraints on the configuration variables are quadratic and the secondary constraints $\phi(q,p)$ have quadratic extensions to the full space. In some examples [87], we have successfully modified the Hamiltonian away from the constraint surface so that the flow of the Hamiltonian leaves the level sets of the primary constraints invariant and the zero level set of the secondary constraint invariant. However, we were unable to to determine a modification such that *all* the level sets of the secondary constraints are invariant. Therefore, the quadratic invariant preserving methods will not automatically preserve the secondary constraints.

- For which quadratically-constrained systems can the Hamiltonian and the constraint functions be extended to a neighborhood of the constraint surface such that the constraints are true quadratic integrals of the extended Hamiltonian system?

Another application involves Lie-Poisson systems. Many Lie groups can be written as submanifolds of linear spaces defined by quadratic constraints. For example, $SO(n) = \{q \in \mathbb{R}^{n\times n} \colon q^t q = 1\}$. In this case symplectic Runge-Kutta methods can be used to approximate the exponential map and compute group elements. For $\mathfrak{so}(n)$, the midpoint rule is the Cayley transform between $\mathfrak{so}(n)$ and $SO(n)$: for $A \in \mathfrak{so}(n)$, $(I-A)^{-1}(I+A) \in SO(n)$. It is tempting to use this fact in conjunction with the coadjoint action to construct Lie-Poisson integrators for matrix Lie groups determined by quadratic constraints. For example, consider the integrator $z \mapsto Z$ where

$$Z = Ad^*_{R(dH((z+Z)/2))} z,$$

and $R(x)$ is the Runge-Kutta integrator applied to the Lie algebra element x. The range of R is in the group, so that it is natural to ask if this integrator is Lie-Poisson. Our preliminary calculations indicate that it is not, but that does not mean that there is not some alternative Lie-Poisson form. There are many other integrators based on this idea; one should test them numerically first.

- Use quadratic-preserving Runge-Kutta methods to generate Lie-Poisson integrators.

11 Approximation of infinite-dimensional Poisson manifolds

For Hamiltonian field theories, the phase space is an infinite-dimensional Poisson manifold. For an introduction to this subject, see Olver [98]; for applications to fluid mechanics, see Salmon [119] and Arnold and Khesin [8]. To numerically simulate these equations a finite dimensional truncation is necessary.

- Should a finite dimensional truncation of an infinite dimensional Hamiltonian system be Hamiltonian?

Specifically, we want to be able to let $\triangle t = f(\triangle x)$ (f being determined by stability criteria) and study

$$\text{“} \lim_{\triangle x \to 0} \lim_{t \to \infty} \text{”} \tag{11.1}$$

Usually in numerical analysis these two limits are reversed, or t is held fixed. But most real computations use (11.1), because global errors can be kept small only for impractically short times. It is implicitly hoped that the qualitative and statistical behavior of the PDE is mimicked by the finite-dimensional model, Hamiltonian or not.

In the hope that there are positive answers the the above question, we now discuss the approximation of Poisson manifolds. After the truncation of the phase space, there is still the question of choosing the best Hamiltonian.

If the Poisson bracket is constant, i.e., independent of the field variables, truncation is trivial, because any antisymmetric matrix is Poisson. As in finite dimensions, this is the most common case. Thus we immediately get Hamiltonian truncations for nonlinear wave equations, the nonlinear Schrödinger equation, the Davey-Stewartson and Boussinesq equations, and so on. In fact, with periodic boundary conditions, all of these have a two-term splitting, so the design of the Hamiltonian method is finished. Some questions of numerical analysis are considered in [83]. If the resulting Courant-type stability restriction (roughly two time-steps per period of the fastest wave) are too severe, the fast modes should be removed in the Hamiltonian, not by the integrator.

For equations of the form $\dot{u} = \partial_x(\mathcal{L}u + f(u))$, where $\mathcal{L}$ is a linear operator, it is annoying that being able to solve $\dot{u} = \partial_x f(u)$ by characteristics is no help: we must solve the discretized equation $\dot{\mathbf{u}} = D\mathbf{f}(u)$. If D has bandwidth b above the diagonal, then we can do a $b+1$-piece splitting (e.g., odd-even splitting for $b = 1$), and $b/(b+1)$ of the unknowns evolve under a given piece. During a (first-order) time step, each unknown is stepped b times, which in unappealing—particularly if D is full, as for spectral methods. We have the paradoxical situation that an implicit method using FFTs would be asymptotically faster than an explicit method.

The simplest truncation for Hamiltonian systems on T^*Q, where Q is the configuration space, is to determine a Hamiltonian system on T^*N where N is a finite-dimensional submanifold of Q. This is done in the Lagrangian description by restriction to the tangent space $TN \subset TQ$. Then the Legendre transform gives a Hamiltonian system on T^*N. This is the technique proposed by Holm et al.[53].

However, it can destroy symmetries of the system such as the particle relabeling symmetries of fluid mechanics.

Many other evolution equations of physics—Eulerian inviscid fluids, the Vlasov-Poisson equation, the Korteweg-de Vries equation—can be written as Lie-Poisson systems. Their phase space is the dual of some infinite-dimensional Lie algebra. To spatially truncate such systems to a set of Lie-Poisson ODEs leads naturally to the question of finding finite-dimensional approximations of these Lie algebras. The approximation should converge to the infinite-dimensional algebra in some useful sense (perhaps its structure coefficients converge pointwise) and inherit as much structure as possible: all the discrete Casimirs (such as helicity in the 3D Euler equations), and an increasing number of the continuous ones (such as the generalized enstrophies in the 2D Euler equations) as the approximation improves. It is hoped that in this way statistical properties of the discrete system will approximate those of the continuous system, although this is little more than a hunch at present [10, 62, 145]. For example, how does symplectic integration of Zeitlin's truncation of the Euler equation converge to Joyce-Montgomery [60] states? Is there a Joyce-Montgomery equation for negative temperature states on $\mathfrak{su}(n)$? What is the effect of different Hamiltonians, such as the one presented in Dowker and Wolski [28]? How important is it to preserve the higher Casimirs in the numerical simulation?

Truncation of Poisson manifolds is a very difficult and fundamental problem, and little progress has been made. There is no general theory. Simple finite differences and spectral methods do not work, even elegant ones such as Arakawa's energy- and enstrophy-preserving schemes [5, 6]. We can only summarize the presently known approaches.

11.1 The sine bracket For the algebra of Hamiltonian vector fields on the torus $\mathbb{T}^2$, the Lie bracket in Fourier space takes the form

$$[\omega_{\mathbf{m}}, \omega_{\mathbf{n}}] = (\mathbf{m} \times \mathbf{n})\omega_{\mathbf{m}+\mathbf{n}} \tag{11.2}$$

where $\mathbf{m} = (m_1, m_2)$ is a wavenumber and $\mathbf{m}\times\mathbf{n} = m_1 n_2 - m_2 n_1$. The sine-bracket truncation retains the modes with $-M \le m_1, m_2 \le M$ and perturbs the bracket to

$$[\omega_{\mathbf{m}}, \omega_{\mathbf{n}}]_{sin} = \varepsilon^{-1} \sin(\varepsilon \mathbf{m} \times \mathbf{n})\omega_{mod(\mathbf{m}+\mathbf{n})} \tag{11.3}$$

where $N = 2M + 1$, $\varepsilon = 2\pi/N$, and $mod(\mathbf{m})$ reduces $\mathbf{m}$ modulo N to the lattice $-M \le m_1, m_2 \le M$ [33, 54, 55]. Notice how the sine bracket converges to the continuous bracket as $N \to \infty$ for fixed $\mathbf{m}$, $\mathbf{n}$. (11.3) turns out to define the Lie bracket of $\mathfrak{su}(N)$, so this algebra has $N-1$ Casimirs, the nth one of which tends to the nth Casimir of (11.2) as $N \to \infty$ for fixed n. Clearly the precise sense in which $SU(N)$ approximates the group of symplectic maps of $\mathbb{T}^2$ remains to be established!

This algebra provides a Lie-Poisson truncation of, among other systems, the 2D Euler equations with periodic boundary conditions [10, 28, 89, 145]. It seems to be the only such truncation, and is related to the Moyal bracket of quantum mechanics [38]. It allows truncations of the shallow water equations in Eulerian [47] and Lagrangian [113] formulations, although these have not been tested. Perhaps one should first study the sine-bracket independently of its application to fluid mechanics, for instance, by choosing Hamiltonians which give smoother flows.

- Is the sine bracket useful in numerical methods?

11.2 Moment truncations For Vlasov-Poisson-type systems whose phase space consists of densities f on a base symplectic manifold P, one can define the moments $f^\alpha = \int_P x^\alpha f\, dx$ and the induced Poisson bracket $\{f^\alpha, f^\beta\} = f^\gamma$, where $x^\gamma = \{x^\alpha, x^\beta\}_P$. (Here α is a multiindex.) There are two trivial truncations of this infinite-dimensional Poisson space which may be useful. Notice that $|\gamma| = \max(0, |\alpha| + |\beta| - 2)$ so that the algebra has two subalgebras, one consisting of moments of order 0, 1, and 2, and one of moments of orders 2 and higher. In the first case no truncation is required, and Rouhi [114] has used this to approximate elliptical vortex patches; one can also get the "next correction" to point vortices in this way. The second case has an elementary finite-dimensional truncation to moments of order 2 through n [21], discussed below. Such a truncation is useful for the localized distributions that arise in accelerator dynamics [17]. But if one wishes to retain moments of order 0 through n, perhaps with multiple base points, a more sophisticated approach leading to mixed Eulerian-Lagrangian schemes, outlined in Section 11.4, is needed.

More applications of these truncations should be explored, if only because truncations are rare.

- Apply moment truncations to fluid mechanics and to general Hamiltonian PDEs.

11.3 Other approaches Unfortunately, the sine bracket exists only for the algebra of Hamiltonian vector fields on the sphere, the $2n$-torus, and possibly on other quotients of coadjoint orbits [54]. If one changes variables by the discrete Fourier transform, the sine bracket transforms to a bracket on functions on a lattice. However, this bracket is nonlocal.

- Is there a local Lie bracket on functions on a discrete representation of the plane approximating the usual Poisson bracket?

It is possible that there is none, due to the rigidity of Lie algebras. One cannot get rid of this problem by going to a nonlinear Poisson bracket since the Poisson structure transverse to the symplectic leaves is linearizable to a Lie-Poisson structure [139], so there must be Lie algebras present. However, if we could find a local Lie bracket, the next question would be how to deal with boundary conditions. For example, to do simulations of the Euler equations on the sphere it would be nice to have a Lie algebra which approximates the Poisson algebra of Hamiltonian functions which vanish on continents. This would also be useful for the boundary conditions imposed by the walls in an accelerator.

There is also the work of Dubrovin [30] on discrete Poisson brackets. For a certain bracket on the field variable $u : \mathbb{R} \to \mathfrak{g}$, where $\mathfrak{g}$ is a Lie algebra, he was able to replace the domain $\mathbb{R}$ by a discrete lattice only by deforming the range $\mathfrak{g}$ to a Poisson-Lie group. Workers in quantum groups have pointed out the similarity between the processes of quantization and discretization [101].

There is no known approximation for the algebra of divergence-free vector fields on $\mathbb{T}^3$, which would be needed to study the 3D Euler equations. To make the formulas easier one might study the algebra of vector fields on the circle, for which the Poisson operator is $u\partial_\theta + \partial_\theta u$ [84, 101].

However, one can leave the category of strict Poisson truncation. For the algebra of vector fields $u(\theta)$ on the circle, the change of variables $u = v^{1/2}$ makes the Poisson operator constant [83]. Similar changes of variables exist in other cases [91] but due to their hodograph nature may not be of practical use. One could search for canonical variables on particular symplectic leaves, or find symplectic (Clebsch) variables on the whole space [80, 144]. This latter approach is unappealing as the truncation will generally destroy the gauge symmetry of the symplectic variables, leading to an excessively large phase space. Lagrangian formulations of fluids can be canonical, so one should explore as many such formulations as one can find. The "velicity" method [14, 99] is interesting in this regard. See also Virasoro [137]. Lagrangian vorticity and velocity-vorticity formulations should be studied. The three-dimensional vortex methods currently in use do not appear to be canonical [140]; extending them to be so would be an important development. Symplecticity is important in stellar dynamics too [90], so Hamiltonian truncations of its Vlasov-Poisson formulation should be useful.

The method of Dirac constraints has proved useful in deriving simplified evolution equations [118]; is it possible to get a sensible Hamiltonian truncation by constraining the fluid motion to a finite-dimensional submanifold?

11.4 Augmentation and decoupling of semi-direct products Although there is no approximation of the group D of volume-preserving diffeomorphisms at present, there is a useful approximation of the dual of its Lie algebra as a Poisson manifold. Following Scovel and Weinstein [124] we augment to a slightly larger space and use an isomorphism to direct products of Poisson manifolds, each which have Poisson truncations.

Consider the subgroup D_0 consisting of those diffeomorphisms of the 3-torus $\mathbb{T}^3$ which preserve the origin. Let N_k denote the normal subgroup of D_0 whose elements d are such that the 1st through kth derivatives of $d(x) - x$ vanish at the origin. Then, $D_k = D_0/N_k$ can be thought of as an approximation to D_0. It is the finite-dimensional group consisting of the kth-order jets of volume-preserving diffeomorphisms which preserve the origin. Likewise, the dual of the Lie algebra $\mathcal{D}_k$ is a space of vector moments up to kth order.

However, what we wish to approximate is $\mathcal{D}^*$. We do this by pulling back the Hamiltonian to the semidirect Poisson manifold $\mathcal{D}^*Ⓢ\mathbb{T}^3$ by the Poisson map consisting of projection onto the first factor. Then pull back by a Poisson isomorphism $T^*\mathbb{T}^3 \oplus \mathcal{D}_0^* \to \mathcal{D}^*Ⓢ\mathbb{T}^3$ which decouples the semidirect product Poisson manifold into the direct product of Poisson manifolds. This determines a Hamiltonian system on $T^*\mathbb{T}^3 \oplus \mathcal{D}_0^*$ for which the first factor is finite dimensional and the second is truncatable. The new phase space has the physical interpretation of the replacement of the momentum density by a single particle plus the 0th through n moments of the distribution about that particle [124]. The determination of a Hamiltonian on the truncated space would follow by pulling back by the Poisson map $T^*\mathbb{T}^3 \oplus \mathcal{D}_k^* \to T^*\mathbb{T}^3 \oplus \mathcal{D}_0^*$ which is the adjoint to the one induced by the projection $\mathcal{D}_0 \to \mathcal{D}_k$. However, the Hamiltonian is not usually well defined on this image since an image point consists of sums of delta functions and their derivatives, which we call a multipole. Consequently, we pull back by another map $\phi : T^*\mathbb{T}^3 \oplus \mathcal{D}_k^* \to T^*\mathbb{T}^3 \oplus \mathcal{D}_0^*$ such that when it is composed with a natural projection $T^*\mathbb{T}^3 \oplus \mathcal{D}_0^* \to T^*\mathbb{T}^3 \oplus \mathcal{D}_k^*$ should give the identity. Probably ϕ should be the identity in the first factor.

- What are the optimal ϕ's for a specific physical system?

This technique appears to work well when the distribution in $\mathcal{D}^*$ is concentrated about a point and can be well represented by its first few moments. One can incorporate many of these multipoles by first pulling back to the direct sum $\sum \mathcal{D}^*$ by the map $\sum \mathcal{D}^* \to \mathcal{D}^*$ that simply sums up the distributions. Since the Lie-Poisson structure on $\mathcal{D}^*$ is linear, this summation is a Poisson map. Consequently, we have a progression of Hamiltonian truncations, from one multipole with a large number of moments to a large number of multipoles with a small number of moments.

- For a specific problem, what is the optimal balance between the number of multipoles and the number of moments retained for each?

In the truncation of the shallow water equation, Ge and Scovel [47] use a Poisson isomorphism

$$T^*F^* \oplus \mathcal{D}^*_\mu \to \mathcal{D}^* \circledS F^*$$

where F^* is the space of positive mass densities, $\mathcal{D}^*$ is the space of momentum densities, and $\mathcal{D}^*_\mu$ the space of vorticities. The direct product space can be truncated by truncating F^* by a finite element method and approximating $\mathcal{D}^*_\mu$ by $\mathfrak{su}(n)$ for n large [33]. The difficulty here is that the isomorphism involves a map determined to exist by a theorem of Moser [95] on volume elements. Liao et al.[72] have been developing a numerical approximation of this map based on a deformation technique, and Cullen et al.[24] have successfully used the Monge-Ampere equation. The latter method appears to be more amenable to differentiation.

- Construct a numerical approximation to a map corresponding to Moser's [95] theorem on volume elements that makes the Ge-Scovel [47] shallow water approximation implementable.

12 Theoretical performance

12.1 The KAM theorem One of the strongest, and at the same time the most rigorous, explanations for the good performance of symplectic integrators is the KAM theorem. When it applies, one can be sure that for some orbits at least—the quasi-periodic ones—there are nearby numerical trajectories for which the orbital error is bounded for all time.

Consider a smooth autonomous Hamiltonian of the form

$$H_0 + \varepsilon H_1 + (\Delta t)^p H_2 \tag{12.1}$$

where H_0 is integrable, $H_0 + \varepsilon H_1$ is not, and H_2 is the first term in the perturbation due to a symplectic integrator. If $\varepsilon = 0$, KAM theory implies that most of the invariant tori of the system H_0 are preserved by the integrator, and that the fraction preserved tends to 1 as $\Delta t \to 0$. What can we do when $\varepsilon \neq 0$? A direct application of KAM theory only shows that most of the tori of $H_0 + \varepsilon H_1$ are still preserved when $\Delta t \neq 0$, by comparing back to H_0. It seems that one should be able to get more; see for example the informal discussion in [84].

- Rigorously investigate Hamiltonians of the form (12.1) as $\Delta t \to 0$ for ε fixed but arbitrarily small.

12.2 Extensions of Hamiltonian stability theorems We discussed in Section 8 how symplectic integrators preserve energy well for long times. However, this result on the adiabatic invariance of the energy appears to have some validity outside the realm of Hamiltonian mechanics. We also suspect that the KAM and Nekhoroshev theorems have counterparts in larger categories than Hamiltonian systems. Let us be more precise.

Some proposed integrators are not actually symplectic but appear to have the same stability characteristics as symplectic methods. For example, consider the integrators of Channell and Scovel [20] based on generating functions determined from the Hamilton-Jacobi theory. They are not symplectic. This can be seen as follows. Pick an initial point z_0. Then, determine the generating function about z_0 to generate the symplectic map from z_0 to the next point. Now consider another point z_1 near z_0. In [20], its generating function is determined by a Taylor series about z_1, not z_0, so the generating functions do not match in general. Therefore, z_1 is advanced by a different symplectic map than z_0. Consequently, the method does not necessarily determine even a local symplectic map. However, their numerical experiments show that this method has the typical superior performance of symplectic integrators.

The domain where this integrator exists is probably something like the action groupoid $G \times Z$ where G is the group of symplectic maps and Z is the phase space, instead of the symplectic group relevant for the usual symplectic integrator. Actually, G is more likely to be the local symplectic mappings, but we prefer not to confuse the issue any more than necessary here. For a review of groupoids and algebroids see MacKenzie [77]. The partially-defined multiplication in $G \times Z$ is $(g_2, z_2)(g_1, z_1) = (g_2 g_1, z_1)$ if $z_2 = g_1(z_1)$. This is natural, since if one starts at z_1 and determines a flow map g_1, then on the next step one begins at $z_2 = g_1(z_1)$. This structure seems to be relevant for numerical analysis in general, not just for Hamiltonian mechanics. The analogue of the Lie algebra of Hamiltonians $\mathfrak{g}$ of G for the groupoid is the Lie algebroid $\mathfrak{g} \times Z$ which is a vector bundle with fibers $\mathfrak{g}$ and base Z. It is not a Lie algebra, but there is a Lie bracket on sections of this vector bundle. Now consider a Hamiltonian $H \in \mathfrak{g}$ and the constant section $Z \to \mathfrak{g} \times Z : z \mapsto (H, z)$.

- If the KAM or Nekhoroshev theorems are applicable to the Hamiltonian system determined by H, are there analogues of these stability theorems for sections of the action algebroid $\mathfrak{g} \times Z$ which are small perturbations of the constant section $z \mapsto (H, z)$? Is there an adiabatic invariance theorem for such an extension?

As mentioned above, there is numerical evidence to support the existence of these theorems. Further, since the evolution of H is by quasi-periodic motion, substitution of this motion into the z dependency of the perturbation gives a quasi-periodic time-dependent perturbation of H to use as a base motion.

12.3 Reversible adiabatic invariance theorem It is known that there is a KAM-type theorem for R-reversible systems (see also Section 14.2). However, in [75] and [83] it is observed that integrating an R-reversible Hamiltonian system by an R-reversible nonsymplectic integrator preserves the energy well over long simulations. See [112] for a review of the interactions between reversible and Hamiltonian systems.

- Is there an adiabatic invariance theorem for reversible perturbations of reversible Hamiltonian systems?

When the original Hamiltonian system is integrable, one can ask whether all the integrals of motion are adiabatic invariants for R-reversible perturbations.

- Is there a Nekhoroshev theorem for reversible perturbations of reversible integrable Hamiltonian systems?

Reversibility extends naturally to multistep methods, and Saha and Tremaine [115] have developed and applied such methods usefully to the solar system. Presumably one only obtains good behavior near the symmetry plane. More generally we propose:

- Study R-reversible perturbations of R-reversible Hamiltonian systems.

12.4 Chaos It is impossible to track orbits whose Lyapunov exponents are of order one for long times. Typically, the chaotic and quasiperiodic orbits are intermingled in Hamiltonian systems; indeed, there are no known completely hyperbolic polynomial Hamiltonians. So, at least the boundaries of the chaotic regions will be better preserved in a symplectic than in a nonsymplectic integrator. But what about the flow in the chaotic region itself?

Because individual orbits cannot be tracked accurately, the chaotic case should be symplectic integrators' forte. But so little is known analytically that it is hard to know what kind of result one should aim for. Instead, there is a result which argues *against* the necessity of symplectic methods: for Anosov systems, *all* nearby flows are conjugate, not just the symplectic ones. Some numerical work might resolve the apparent contradiction.

- Measure statistical properties of chaotic orbits, and their dependence on symplectic and nonsymplectic perturbations.

13 Dissipative systems

- In the numerical simulation of slightly dissipative perturbations of Hamiltonian systems, does one simply give up the Hamiltonian structure?

There is evidence to the contrary. Unpublished work of the second author with D. Umbarger used the fact that the Duffing oscillator has a time-dependent coordinate transformation into a time-dependent Hamiltonian system. They compared a fourth-order symplectic integrator in this moving frame with a Bulirsch-Stoer variable time step method: the two appeared to be of comparable quality.

Lewis and Marsden [70] describe a Hamiltonian-dissipative normal form decomposition of a vector field and discuss how the Hamiltonian part provides an organizing center of the system. Kirchgraber et al.[67] observe that in a simple case that if the integrator is symplectic when the dissipation is turned off, the numerical center manifold is stable under the dissipative perturbation. Mielke [88] describes the Hamiltonian structure inherited on center manifolds and Sleeman [132] shows that time-periodic solutions of reaction-diffusion equations can form Hamiltonian systems in the single spatial variable. Since the center manifold is a component of the system with the exponential decay factored out we ask the following questions.

- When is the flow on the center manifold of a dissipative system Hamiltonian? When is the flow on an inertial manifold Hamiltonian?

Such a situation is surely not true in general, but it would be interesting to know if there are any important special cases.

Grmela [48] and Morrison [92, 93] provide bracket formulations for Hamiltonian plus gradient systems and Vujanovic and Jones [138] have described a variational form for nonconservative systems. What is missing is an invariant structure whose preservation provides superior numerical simulations.

For a Hamiltonian system on M, the graph of the vector field is a Lagrangian submanifold of TM with respect to the tangent symplectic structure. This is the infinitesimal version of the fact that the graph of the flow map of a Hamiltonian system is Lagrangian in the symplectic manifold $M \times M_{-}$, where M_{-} is M with the negative of its symplectic form. Cantrijn [16] has shown that the graph of a gradient perturbation of a Hamiltonian system in TM is Lagrangian with respect to a symplectic form on TM which is close to the tangent one. Unfortunately, this symplectic formulation does not seem to produce invariants at the map level, so it is unclear what its implications are for numerical simulations.

- Which dissipative systems have flow invariants whose preservation produces superior numerical simulations?

For any dynamical system, one can double the dimension of the space and extend the ODE by cotangent lift to obtain a Hamiltonian system. This does no good numerically as invariant tori are typically half the dimension of the symplectic manifold, i.e., equal to the dimension of the original system. However, Nosé [100] shows that molecular dynamics formulations with constant temperature may be presented in Hamiltonian form by the addition of two more variables.

- What is the structure of a dynamical system which allows a Hamiltonian formulation by the addition of a small number of variables?

14 Assorted issues

The program of studying geometric structure in numerical methods has scarcely begun. We select just a few such structures for further comment.

14.1 The discrete Lagrangian method Discretization of the Lagrangian description appears to be very old. Maeda [78] and MacKay [76] outline formalisms for discrete Lagrangian methods and describe the corresponding symplectic method. Wu [143] illustrates the corresponding discrete variational problems corresponding to a large class of symplectic methods and applies shooting techniques to 2-point boundary value problems. Moser and Veselov [94] discretize the Lagrangian description of the free rigid body and determine a completely integrable approximation preserving the rigid body integrals.

It is possible that in many cases either the Lagrangian formulation or the the Hamiltonian formulation might yield a superior numerical simulation.

- Investigate the quality of Lagrangian integrators, and their equivalent Hamiltonian forms.

14.2 Reversible systems Many authors have pointed out how easy it is to construct *reversible* integrators [83, 84, 107, 122], and how important it is *a priori*. Most applications are in Hamiltonian systems. However, one should also

- Perform numerical experiments to investigate the importance of preserving reversibility with respect to one or multiple involutions.

This is important since one may not know the involutions explicitly, or even if they exist. Scovel [122] projects an arbitrary map onto R-reversible maps; this doubles the amount of work, so one would like to know whether it is worth it. See [112] for a review of reversibility and Hamiltonian systems.

14.3 Volume preservation Volume-preserving flows arise in particle-tracking in incompressible fluids and in the standard spectral truncation of the 2D Euler equations. (Interestingly, this latter model also preserves energy and enstrophy—how well are they preserved by a volume-preserving scheme?) The space of divergence-free vector fields seems to be too large: no standard numerical methods, such as Runge-Kuttas, are volume-preserving for all such vector fields 102. There is a generating function for volume-preserving maps 103, 122, but it involves evaluating definite integrals of the vector field and does not preserve fixed points in general. In many systems such as the ABC flows a splitting technique works well [51, 52, 102, 103, 122].

- Are there general volume-preserving integrators which use only the vector field?

- Develop efficient, general volume-preserving methods.

14.4 Symplectic lattice maps Often roundoff accumulation can become a problem even with symplectic methods. Symplectic lattice maps have been studied [106, 66, 65] and are a possible resolution of this problem. Earn and Tremaine [31] show that the elimination of roundoff error in this way can effect the statistical behavior of the simulation. Scovel [123] shows that often a symplectic map can be altered to a nearby symplectic map which preserves a lattice. The lattice maps should get more accurate as the lattice is refined. However, as the lattice gets finer, the preservation of the lattice causes a high-frequency perturbation which may cause diffusion problems in the zero-lattice-spacing limit.

- For specific Hamiltonian systems, is there an optimal lattice spacing for a symplectic lattice map approximation?

14.5 Standard methods For some dynamical systems, very high order Cowell methods are so accurate and so inexpensive that the local truncation errors can be made comparable to roundoff errors. Then (to within roundoff error) it is a symplectic multistep scheme in Feng's sense. See Hyman et al. [58] for a discussion of this matter.

- For a specific Hamiltonian system, and a fixed degree of accuracy of some property of the system, develop a criterion to determine when it is worth the extra effort to use a symplectic integrator versus a traditional method.

14.6 Shooting methods One-dimensional quantum mechanics problems often involve computing the eigenvalues of a Schrödinger operator. Leaving the eigenvalue as a parameter to be determined one can use a shooting method from the left-hand boundary to the right. The differential equations are Hamiltonian and the shooting technique has been studied by Wu [143].

- What are the benefits of using a symplectic integrator in a shooting method?

14.7 Elliptic equations Can one devise numerical approximations of the *Div*, *Grad*, and *Curl* operators that satisfy the vector identities $Div \circ Curl = 0$, $Curl \circ Grad = 0$ *exactly*? The answer is yes [27, 57]. Triangulate the basic manifold by a chain complex. Then the coboundary operator δ is the discrete analogue of the exterior derivative d and the discrete vector identities $\delta \circ \delta = 0$ are consistent with the fact that the de Rham map from the de Rham complex to the cochain complex is a chain map. In a closely related strategy, Hyman et al. [59] develop and test finite volume approximations to differential operators on nonuniform grids based on Stokes's theorem.

If we impose a Riemannian structure, we can take adjoints and compute a Laplacian. In fact, Hyman and Scovel [57] show that the Hodge theorem still applies here in the sense that there is a discrete Hodge decomposition and the dimension of the kernel of the discrete version of the Laplacian on k-forms is the dimension of the kth cohomology group of the manifold. However, at present these Laplacians are not local since the induced Hodge star operator is not local.

- Is there a local discrete Hodge star operator which has a local inverse?

- Are there chain map approximations to the exterior derivative that are more than first-order accurate?

- Can one quantify the benefits of using these techniques?

14.8 Linear algebra Although we are not qualified to comment extensively, we would like to direct attention to the active field of structure-preserving algorithms in numerical linear algebra [13]. Such algorithms for finding the eigenvalues for Hamiltonian or symplectic matrices, for example, are both faster and more accurate (because the resulting eigenvalues have the right symmetries) than standard methods. Another application is to solving algebraic Riccati equations [104]. This work may also feed back to differential equations, as in work on differential Riccati equations [25] and on computing Lyapunov exponents [49, 73].

15 Acknowledgements

We would like to thank the referees for their helpful comments and for providing several references, and our colleagues at the workshop who provided many corrections and references.

References

[1] Abia, L. and Sanz-Serna, J.M. [1993], *Partitioned Runge-Kutta methods for separable Hamiltonian problems*, Math. Comp., **60**, 617–634.

[2] Ablowitz, M. J. and Herbst, B.M. [1990], *On homoclinic structure and numerically induced chaos for the nonlinear Schrödinger equation*, SIAM J. Appl. Math., **50**, 339–351.

[3] Ablowitz, M. and Segur, H. [1980], *Solitons and the Inverse Scattering Transform*, SIAM.

[4] Andersen, H. C. [1983], *Rattle: a 'velocity' version of the shake algorithm for molecular dynamics calculations*, J. Comp. Phys., **52**, 24–34.

[5] Arakawa, A. [1966], *Computational design for long-term numerical integration of the equations of fluid motion: Two dimensional incompressible flow.* J. Comp. Phys., **1**, 119–143.

[6] Arakawa, A. [1981], *A potential enstrophy and energy conserving scheme for the shallow water equations*, Monthly Weather Review, **109**, 18–36.

[7] Arnold, V. I. [1964], *Instability of dynamical systems with many degrees of freedom*, Sov. Math. Dokl., **5**, 581–584.

[8] Arnold, V. I. and Khesin, B.A. [1992], *Topological methods in hydrodynamics*, Ann. Rev. Fluid Mech., **24**, 145–166.

[9] Benzel, S., Ge, Z., and Scovel, C. [1993], *Elementary construction of higher order Lie-Poisson integrators*, Phys. Lett. A, **174**, 229–232.

[10] Benzel, S., McLachlan, R. I., and Scovel, C., *Symplectic integration of the Euler fluid equations*, in preparation.

[11] Bochev, P. and Scovel, C. [1994], *On quadratic invariants and symplectic structure*, BIT, **34**, 337-345.

[12] Bourbaki [1989], *Lie Groups and Lie Algebras*, Chapters 1 – 3, Springer-Verlag, New York.

[13] Bunse-Gerstner, A., Byers, P., and Mehrmann, V. [1992], *A chart of numerical methods for structured eigenvalue problems*, SIAM J. Matrix Anal. Appl., **13**, 419–453.

[14] Buttke, T. F. [1992], *Lagrangian numerical methods which preserve the Hamiltonian structure of incompressible fluid flow*, Vortex Flows and Related Numerical Methods (Beale, J. T., Cottet, G.-H., and Huberson, S. eds.) Kluwer, Dordrecht, 1993.

[15] Calvo, M. P. and Sanz-Serna, J. M. [1993], *The development of variable-step symplectic integrators, with application to the two-body problem*, SIAM J. Sci. Comp., **14**, 936–952.

[16] Cantrijn, F. [1984], *Symplectic approach to nonconservative mechanics*, J. Math. Phys., **25**, 271–276.

[17] Channell, P. J. [1983], *The moment approach to charged particle beam dynamics*, IEEE Trans. Nucl. Sci., **NS-30**, 2607–2609.

[18] Channell, P. J. [1986], *Explicit integration of kick Hamiltonians in three degrees of freedom*, Los Alamos National Laboratory internal report AT-6: ATN-86-6.

[19] Channell, P. J. [1996], *A brief introduction to symplectic integrators and recent results*, these proceedings.

[20] Channell, P. J., and Scovel, J. C. [1990], *Symplectic integration of Hamiltonian systems*, Nonlinearity, **3**, 231–259.

[21] Channell, P. J., and Scovel, J. C. [1990], *Integrators for Lie-Poisson dynamical systems*, Physica D, **50**, 80–88.

[22] Cooper, G. J. [1987], *Stability of Runge-Kutta Methods for Trajectory Problems*, IMA J. Num. Anal., **7**, 1–13.

[23] Coste, A., Dazord, P., and Weinstein, A. [1987], *Groupoïdes symplectiques*, Publications du Département de Mathématiques, Université Claude Bernard-Lyon I, **2A**, 1–62.

[24] Cullen, M. J. P., Norbury, J., and Purser, R. J. [1991], *Generalized Lagrangian solutions for atmospheric and oceanic flows*, SIAM J. Appl. Math., **51**, 20–31.

[25] Dieci, L. and Eirola, T., *Positive definiteness in the numerical solution of Riccati differential equations*, Numer. Math., **67**, 303.

[26] Dieci, L., Russell, R. D., and Van Vleck, E. S. [1994], *Unitary integrators and applications to orthonormalization techniques*, SIAM J. Numer. Anal., **31**, 261–281.

[27] Dodzuik, J., *Finite difference approach to Hodge theory of harmonic forms*, Am. J. Math., **98**, 79–104.

[28] Dowker, J. S., and Wolski, A. [1992], *Finite model of two-dimensional ideal hydrodynamics*, Phys. Rev. A, **46**, 6417–6430.

[29] Dubrovin, B. A., *Theory of operators and real algebraic geometry*, Lect. Notes Math. 1334, 42–59.

[30] Dubrovin, B. A. [1989], *Differential-geometric Poisson brackets on a lattice*, Funct. Anal. Appl., **23**(2).

[31] Earn, D. J. D. and Tremaine, S. [1992], *Exact numerical studies of Hamiltonian maps: iterating without roundoff error*, Physica D, **56**, 1.

[32] Eirola, T. and Sanz-Serna, J. M. [1992], *Conservation of integrals and symplectic structure in the integration of differential equations by multistep methods*, Numer. Math., **61**, 281–290.

[33] Fairlie, Fletcher, P. and Zachos, C. K. [1990], *Infinite-dimensional algebras and a trigonometric basis for the classical Lie algebras*, J. Math. Phys., **31**(5), 1088–1094.

[34] Feng, K. [1986], *Difference scheme for Hamiltonian formalism and symplectic geometry*, J. Comput. Math., **4**, 279–289.

[35] Feng, K. and Qin, M.-Z. [1987], *The symplectic methods for the computation of Hamiltonian equations*, in Numerical Methods for partial differential equations, Lect. Notes Math. 1297, Springer, Berlin New York, 1–31.

[36] Feng, K. and Wang, D.-L. [1994], *Dynamical systems and geometric construction of algorithms*, in Contemporary Mathematics (Shi, Z.-C. and Yang, C.-C., eds.), **163**.

[37] Fischer, H.-C. [1992], *Automatic differentiation and applications*, Math. Sci. Eng., **189**, 105.

[38] Fletcher, P. [1990], *The Moyal bracket*, in Quantum Groups (1990), 143–157, World Scientific, Teaneck NJ, 1991.

[39] Fomenko, A. T. and Trofimov, V.V. [1988], *Integrable Systems on Lie Algebras and Symmetric Spaces*, Gordon and Breach, New York.

[40] Forest, É., Bengtsson, J., and Reusch, M.F. [1991], *Application of the Yoshida-Ruth techniques to implicit integration and multi-map explicit integration*, Phys. Lett. A, **158**, 99–101.

[41] Ge, Z. and Feng, K. [1988], *On the approximation of linear Hamiltonian systems*, J. Comp. Math., **6**, 88-97.

[42] Ge, Z. [1991], *Equivariant symplectic difference schemes and generating functions*, Physica D, **49**, 376–386.

[43] Ge, Z. [1990], *Generating functions, Hamilton-Jacobi equations and symplectic groupoids on Poisson manifolds*, Indiana Univ. Math. Journal, **39:3**, 859–876.

[44] Ge, Z. [1993], *On multiple-level symplectic integrators for Hamiltonian systems*, preprint.

[45] Ge, Z., Kruse, H.-P., Marsden, J., and Scovel, C., *The Shallow Water Approximation and Deformation of Poisson Brackets*, in preparation.

[46] Ge, Z. and Marsden, J. [1988], *Lie-Poisson Hamilton-Jacobi theory and Lie-Poisson integrators*, Phys. Lett. A, **133**, 135–139.

[47] Ge, Z. and Scovel, C. [1994], *Hamiltonian truncation of the shallow water equation*, Lett. Math. Phys., **31**, 1–13.

[48] Grmela, M. [1986], *Bracket formulation of diffusion-convection equations*, Physica D, **21**, 179–212.

[49] Habib, S. and Ryne, R.D., *Symplectic calculation of Lyapunov exponents*, Los Alamos report LA-UR-94-2169.

[50] Hairer, E., Murua, A. and Sanz-Serna, J. M. [1993], *The non-existence of symplectic multi-derivative Runge-Kutta methods*, preprint.

[51] Holm, D. D. and Kimura, Y. [1991], *Zero-helicity Lagrangian kinematics of three dimensional advection*, Phys. Fluids A, **3**, 1033–1038.

[52] Holm, D. D., Kimura, Y., and Scovel, C. [1990], *Lagrangian particle kinematics in 3-D convection*, in Nonlinear Structures in Physical Systems (Lam, L. and Morris, H.C. eds.), Springer, New York.

[53] Holm, D. D., Kuperschmidt, B. A., and Levermore, C. D. [1985], *Hamiltonian differencing of fluid dynamics*, Adv. Appl. Math., **6**, 52–84.

[54] Hoppe, J. [1989], *Diffeomorphism groups, quantization, and $SU(\infty)$*, Int. J. Mod. Phys. A, **4**(19), 5235–5248.

[55] Hoppe, J. [1990], *Infinite dimensional algebras and $2+1$ dimensional field theory. Yet another view of $gl(\infty)$. Some new algebras*, Rev. Math. Phys., **2**(2), 193–200.

[56] Hori, G. [1966], *Theory of general perturbations with unspecified canonical variables*, Publ. Astron. Soc. Japan, **18**, 287–296.

[57] Hyman, J. M., and Scovel, J.C. [1988], *Deriving mimetic difference approximations to differential operators using algebraic topology*, Los Alamos National Laboratory Report.

[58] Hyman, J. M., Kaula, W. M., Newman, W. I., and Scovel, C., *Integration schemes for n-body simulations*, these conference proceedings.

[59] Hyman, J. M., Knapp, R. J., and Scovel, J.C. [1992], *High order finite volume approximations of differential operators on nonuniform grids*, Physica D, **60**, 112–138.

[60] Kraichnan, R. H. and Montgomery, D. [1980], *Two-dimensional turbulence*, Rep. Prog. Phys., **43**, 547–619.

[61] Iserles, A. [1991], Bull. Greek Math. Soc., **32**, 3–20.

[62] Isichenko, M. B. [1993], *Can computer simulation predict the real behavior of turbulence?*, Institute for Fusion Studies report 600.

[63] Jay, L., *Symplectic partitioned Runge-Kutta methods for constrained Hamiltonian systems*, SIAM J. Numer. Anal., to appear.

[64] Kahaner, D. K. [1992], *Automatic differentiation comments from Dr. Andreas Griewank*, Sci. Inf. Bull., **17**(4), 79.

[65] Kaneko, K. [1988], *Symplectic Cellular Automata*, Phys. Lett. A, **129**, 9–16.

[66] Karney, C. F. F [1986], *Numerical Techniques for the Study of Long Time Correlations*, Particle Accelerators, **19**, 213-221.

[67] Kirchgraber, U., Lasagni, F., Nipp, K., and Stoffer, D. [1991], *On the application of invariant manifold theory, in particular to numerical analysis*, Int. Ser. Num. Math., **97**, 189–197.

[68] Koseleff, P.-V. [1993], *Relations among Lie formal series and construction of symplectic integrators*, in Applied Algebra, Algebraic Algorithms and Error-correcting Codes (Cohen, G. Mora, T., and Moreno, O. eds.), AAECC–10 (Puerto Rico, 1993), Springer Berlin New York.

[69] Leimkuhler, B. J. and Skeel, R.D. [1994], *Symplectic numerical integrators in constrained Hamiltonian systems*, J. Comp. Phys., **112**, 117–125.

[70] Lewis, D. and Marsden, J. [1989], *A Hamiltonian-dissipative decomposition of normal forms of vector fields*, in Proc. Int. Conf. Bifurcation Theory and its Numerical Analysis (Li, K., Marsden, J., Golubitsky, M., and Iooss, G., eds.), Xian Jiaotong University Press.

[71] Lewis, D., and Simo, J. C. [1994], *Conserving algorithms for the dynamics of Hamiltonian systems on Lie groups*, J. Nonlinear Sci., **4** , 253–299.

[72] Liao, G., Trong-Whay, P., and Su, J. [1992], *A numerical grid generator based on Moser's deformation method*, preprint.

[73] Liao, X.-H., Zhao, C.-Y., and Liu, L. [1993], *The use of symplectic algorithm in the computation of Lyapunov characteristic numbers*, Chinese Astronomy and Astrophysics, **17**, 449.

[74] Libermann, P. and Marle, C.-M. [1987], *Symplectic Geometry and Analytical Mechanics*, D. Reidel, Dordrecht.

[75] MacEvoy, W. and Scovel, C. [1993], *Integrators with variable stepsize for Hamiltonian systems*, submitted to J. Comp. Appl. Math..

[76] MacKay, R. [1992], *Some aspects of the dynamics and numerics of Hamiltonian systems*, in The Dynamics of Numerics and the Numerics of Dynamics (Broomhead, D. S. and Iserles, A. eds.), Oxford.

[77] Mackenzie, K. [1987], *Lie Groupoids and Lie Algebroids in Differential Geometry*, LMS Lecture Notes Series 124, Cambridge University Press.

[78] Maeda, S. [1982], *Lagrangian formulation of discrete systems and concept of difference space*, Math. Japonica, **37**, 345–356.

[79] Marsden, J. E., O'Reilly, O. M., Wicklin, F. J., and Zombro, B. W. [1991], *Symmetry, stability, geometric phases, and mechanical integrators*, Nonlinear Science Today, **1**, 4–21.

[80] Marsden, J. and Weinstein, A. [1983], *Coadjoint orbits, vortices and Clebsch variables for incompressible fluids*, Physica D, **7**, 305–323.

[81] Marsden, J. and Ratiu, T. [1992], *An Introduction to Mechanics and Symmetry*, preprint.

[82] McLachlan, R. I. [1993], *Explicit Lie-Poisson integration and the Euler equations*, Phys. Rev. Lett., **71**, 3043–3046.

[83] McLachlan, R. I. [1993], *Symplectic integration of Hamiltonian wave equations*, Numer. Math., **66** (1994), 465–492.

[84] McLachlan, R. I. [1995], *On the numerical integration of ordinary differential equations by symmetric composition methods (1993)*, SIAM J. Sci. Comp., **16**, 151–168.

[85] McLachlan, R. I., *Composition methods in the presence of small parameters*, BIT, to appear.

[86] McLachlan, R. I. and Atela, P. [1991], *The accuracy of symplectic integrators*, Nonlinearity, **5**, 541–562.

[87] McLachlan, R. I. and Scovel, C. [1995], *Equivariant constrained symplectic integration*, J. Nonlinear Sci., **5**, 233–256.

[88] Mielke, A. [1991], *Hamiltonian and Lagrangian Flows on Center Manifolds*, Lect. Notes in Math. 1489, Springer Verlag, Berlin.

[89] Miller, J., Weichman, P. B., and Cross, M. C. [1992], *Statistical mechanics, Euler's equations, and Jupiter's red spot*, Phys. Rev. A, **45**(4), 2328–2359.

[90] Miller, R. H. [1993], *Stellar dynamics*, preprint.

[91] Mokhov, O. I. [1989], *Vorticity equation of two-dimensional hydrodynamics of an incompressible fluid as a canonical Hamiltonian system*, Phys. Lett. A, **139**, 363–368.

[92] Morrison. P. J. [1984], *Bracket formulation for irreversible classical fields*, Phys. Lett. A, **100**, 423–427.

[93] Morrison. P. J. [1986], *A paradigm for joined Hamiltonian and dissipative systems*, Physica D, **18**, 410-419.

[94] Moser, J. and Veselov, A. P. [1991], *Discrete versions of some classical integrable systems and factorization of matrix polynomials*, Commun. Math. Phys., **139**, 217–243.

[95] Moser, J. [1965], *On the volume elements on a manifold*, Trans. Am. Math. Soc., **120**, 286–294.

[96] Neishtadt, A. I. [1984], *The separation of motions in systems with rapidly rotating phase*, J. Appl. Math. Mech., **48**, 133–139.

[97] Okunbor, D. I. [1992], *Canonical integration methods for Hamiltonian dynamical systems*, Thesis, Comp. Sci., U. Illinois Urbana-Champaign.

[98] Olver, P. J. [1993], *Applications of Lie groups to Differential Equations*, 2nd ed., Springer-Verlag, New York.

[99] Osedelets, V. I. [1989] *On a new way of writing the Navier-Stokes equation*, The Hamiltonian formalism. Comm. Moscow Math. Soc.

[100] Nosé, S. [1984], *A unified formulation of the constant temperature molecular dynamics methods*, J. Chem. Phys., **81**, 511–519.

[101] Polychronakos, A. P. [1991], *Aspects of q-Virasoro algebra*, in Quantum Groups (1990), World Scientific, Teaneck NJ.

[102] Qin, M.-Z. and Zhu, W.-J. [1993], *Volume-preserving schemes and numerical experiments*, Comp. Math. Appl., **26**, 33–42.

[103] Quispel, G. R. W., *Volume-preserving integrators*, preprint.

[104] Raines, A. C. and Watkins, D. S. [1994], *A class of Hamiltonian-symplectic methods for solving the algebraic Riccati equation*, Lin. Alg. Appl., **205**, 1045.

[105] Ramani, A., Grammaticos, B., and Bountis, T. [1989], Phys. Rep., **180**(3), 159–245.

[106] Rannou, F. [1974], *Numerical study of discrete plane area-preserving mappings*, Astron. Astrophys., **31**, 289–301.

[107] Reich, S. [1993], *Symplectic integration of constrained Hamiltonian systems by Runge-Kutta methods*, Tech. report 93–13, Univ. British Columbia.

[108] Reich, S. [1994], *Momentum preserving symplectic integrators*, Physica D, **76**, 375–383.

[109] Reich, S. [1993], *Numerical integration of the generalized Euler equations*, Tech. report 93–20, Univ. British Columbia.

[110] Reich, S. [1994], *Symplectic integration of constrained Hamiltonian systems*, preprint.

[111] Reutenauer, C. [1993], *Free Lie Algebras*, Clarendon Press, Oxford.

[112] Roberts, J. A. G. and Quispel, G. R. W. [1991], *Chaos and time-reversal symmetry: order and chaos in reversible dynamical systems*, Instituut voor Theoretische Fysica Report 91–32, Universiteit van Amsterdam, Valckenierstraat 65, 1018 XE Amsterdam, The Netherlands.

[113] Rouhi, A. and Abarbanel, H. D. I. [1993], *Symmetric truncations of the Shallow water equations*, Phys. Rev. E, **48**, 3643–3655.

[114] Saffman, P. G. [1993], *Vortex Motion*, Cambridge University Press, Cambridge.

[115] Saha, P. and Tremaine, S. [1992], *Symplectic integrators for solar system dynamics*, Astron. J., **104**, 1633–1640.

[116] Saha, P. and S. Tremaine, *Long-time planetary integration with individual time steps*, preprint.

[117] Saito, S., Sugiura, H., and Mitsui, T. [1992], *Family of symplectic implicit Runge-Kutta formulae*, BIT, **32**, 539–543.

[118] Salmon, R. [1988], *Semigeostrophic theory as a Dirac-bracket projection*, J. Fluid Mech., **196**, 345–358.

[119] Salmon, R. [1988], *Hamiltonian fluid mechanics*, Ann. Rev. Fluid. Mech., **20**, 225–256.

[120] Sanz-Serna, J. M. [1988], *Runge-Kutta schemes for Hamiltonian systems*, BIT, **28**, 877–883.

[121] Sanz-Serna, J. M. [1992], *Symplectic integrators for Hamiltonian problems: An overview*, in Acta Numerica 1992, 243–286, Cambridge University Press, Cambridge, NY.

[122] Scovel, J. C. [1991], *Symplectic numerical integration of Hamiltonian systems*, in The Geometry of Hamiltonian Systems (Ratiu, T. ed.), 463–496, MSRI Publ. 22, Springer-Verlag, New York.

[123] Scovel, C. [1991], *On symplectic lattice maps*, Phys. Lett. A, **159**, 396–400.

[124] Scovel, C. and Weinstein, A. [1994], *Finite dimensional Lie-Poisson approximations to Vlasov-Poisson equations*, Comm. Pure. Appl. Math., **47:5**, 683–709.

[125] Simó, C. [1992], personal communication.

[126] Skeel, R. D. [1993], personal communication.

[127] Skeel, R. D. and Gear, C. W. [1992], *Does variable stepsize ruin a symplectic integrator?*, Physica D, **60**, 311–313.

[128] Skeel, R. D. and Biesiadecki, J. J. [1994], *Symplectic integration with variable stepsize*, Ann. Numer. Math., **1**, 191–198.

[129] Simo, J. C. and Armero, F. [1994], *Unconditional stability and long-term behavior of transient algorithms for the incompressible Navier-Stokes and Euler equations*, Comp. Meth. Appl. Mech. Eng., **111**, 111–154.

[130] Simo, J. C. and Tarnow, N. [1992], *The discrete energy-momentum method. Conserving algorithms for nonlinear elastodynamics*, ZAMP, **43**, 757–792.

[131] Simo, J. C., Tarnow, N., and Wong, K. K. [1992], *Exact energy-momentum conserving algorithms and symplectic schemes for nonlinear dynamics*, Comp. Meth. Appl. Mech. Eng., **100**, 63–116.

[132] Sleeman, B. D. [1989], *Complexity in biological systems and Hamiltonian dynamics*, Proc. R. Soc. Lond. A, **425**, 17–47.

[133] Stofer, D. M. [1988], *Some geometric and numerical methods for perturbed integrable systems*, Thesis, ETH-Zürich.

[134] Suris, Yu. B. [1989], *Integrable mappings of the standard type*, Funct. Anal. Appl., **23**(1), 74–75.

[135] Symes, W. W. [1982], *The QR algorithm and scattering for the finite nonperiodic Toda lattice*, Physica D, **4**, 272–280.

[136] Tang, Yi-Fa [1993], *The symplecticity of multi-step methods*, Comp. Math. Appl., **25**, 83–90.

[137] Virasoro, M. A. [1981], *Variational principle for two-dimensional incompressible hydrodynamics and quasigeostrophic flows*, Phys. Rev. Lett., **47**, 1181–1183.

[138] Vujanovic, B. D. and Jones, S. E. [1989], *Variational Methods in Nonconservative Phenomena* Academic Press, Boston.

[139] Weinstein, A. [1983], *The local structure of Poisson manifolds*, J. Diff. Geom., **18**, 523–557.

[140] Winckelmans, G. S. and Leonard, A. [1993], *Contributions to vortex particle methods for the computation of three-dimensional incompressible unsteady flows*, J. Comp. Phys., **109**, 247–273.

[141] Wisdom, J. and Holman, M. [1991], *Symplectic maps for the N-body problem*, Astron. J., **102**(4) 1528–1538; II: stability analysis, Astron. J., **104**(5), 2022–2029.

[142] Wisdom, J., Holman, M., and Touma, J. [1996], *Symplectic correctors*, these proceedings.

[143] Wu, Y. [1990], *The discrete variational approach to the Euler-Lagrange equation*, Computers Math. Applic., **20**, 61–75.

[144] Zakharov, V. E. [1990], *The algebra of integrals of motion of two-dimensional hydrodynamics in Clebsch variables*, Funct. Anal. Appl., 189–196.

[145] Zeitlin, V. [1991], *Finite-mode analogues of 2D ideal hydrodynamics: Coadjoint orbits and local canonical structure*, Physica D, **49**, 353–362.

Fields Institute Communications
Volume 10, 1996

Symplectic Integrators for Systems of Rigid Bodies

Sebastian Reich
Konrad-Zuse Zentrum
Heilbronner Straße 10
D–10711 Berlin, Germany

Abstract. Recent work reported in the literature suggest that for the long-term integration of Hamiltonian dynamical systems one should use methods that preserve the symplectic structure of the flow. This has especially been shown for the numerical treatment of conservative many-particle systems as they arise, e.g., in molecular dynamics and astronomy. In this paper we use recent results by McLachlan [10] and Reich [13] on the symplectic integration of Lie-Poisson systems to derive an explicit symplectic integrator for rigid bodies moving under the influence of external forces. In case that rigid bodies are interconnected by joints, the resulting holonomic constraints can be treated in the same way as described by Reich [12] for general Hamiltonian systems with holonomic constraints.

1 Introduction

Much recent research has gone into developing numerical discretization schemes that inherit the symplectic structure of a Hamiltonian vector field. It has been observed that symplectic methods with fixed step-size possess better long-term stability properties than non-symplectic methods or symplectic methods with varying step-size. It has also been shown that these methods often preserve the Hamiltonian to high accuracy over long periods of time.

So far most applications of symplectic schemes are related to many-particle systems; i.e., systems of bodies moving under the influence of conservative forces such that the bodies can be modeled as point particles. Such problems arise, e.g., in astronomy and molecular dynamics.

Only a few results are available for conservative rigid body systems. In [5], Ge and Marsden derived integrators that preserve the natural Lie-Poisson structure on the duals of Lie algebras. Based on this, an implicit algorithm for the Euler equation of rigid body dynamics was given. More recently, McLachlan and Scovel [11] and Reich [14] showed that Lie-Poisson integrators can also be obtained from the

1991 *Mathematics Subject Classification.* 65L05, 70-08, 70F.
Key words and phrases. Hamiltonian systems, symplectic discretization, multibody systems.

symplectic integration of Hamiltonian systems with holonomic constraints. Using this result, symplectic integrators were obtained, e.g., for the free rigid body and the heavy top. An altogether different approach has been taken by Barth and Leimkuhler in [3]. They replace a given rigid body by a system of point particles whose mutual distances are kept constant. This so called particulation of a rigid body reduces the symplectic integration of a multi-body system to the symplectic integration of a constrained many-particle system which, e.g., can be done by means of the RATTLE variant of the Verlet scheme [1]. This approach allows very easily to include various types of joints and is not restricted to special classes of rigid bodies. It has the disadvantage that additional constraints and variables have to be introduced. In addition to this, the particulation of a rigid body is not unique and an optimal choice seems to depend on the particular problem to be solved. Still another approach to the symplectic integration of rigid bodies is provided by the use of Euler angles [9]. This approach, however, has the disadvantage that the resulting Hamiltonian function is not separable and that, therefore, the symplectic integration of the corresponding equations of motion requires implicit methods such as the Gauss Runge-Kutta methods [6]. It is also not obvious how to include joints such that the schemes remain symplectic and preserve the resulting constraints.

In this paper we follow the idea of Barth and Leimkuhler to derive a symplectic integrator for general multi-body systems. However our starting point is a different one: It is well-known that the motion of a rigid body is characterized by the superposition of a translation and a rotation and that the differential equations describing the translation of the center of mass and the rotation about the center of mass are Hamiltonian. Based on this Hamiltonian formulation, we derive an explicit symplectic scheme for rigid bodies moving under the influence of external conservative forces. The interconnection of several rigid bodies by joints can then be handled as described by Barth and Leimkuhler [3]; i.e., we model joints by linear or quadratic holonomic constraints and solve the corresponding constrained equations of motion by a proper modification of the unconstrained scheme as described by Reich [12] for general Hamiltonian systems with holonomic constraints.

2 Equations of Motion for a Single Rigid Body

In this section we give the Hamiltonian formulation for the equations of motion for a single rigid body moving in R^3 under the influence of an external force. This result naturally extends to unconstrained systems of rigid bodies moving in an external force field.

Consider a rigid body free to move in R^3. A reference configuration $\mathfrak{B}$ of the body is the closure of an open set in R^3. Points in $\mathfrak{B}$, denoted by $X \in \mathfrak{B}$, are called material points. A configuration of $\mathfrak{B}$ is a mapping $\phi : \mathfrak{B} \to R^3$ which is smooth, orientation preserving, and invertible on its image. The points of the target space R^3 of ϕ are called spatial points and denoted by $x \in R^3$. A motion of $\mathfrak{B}$ is a time dependent family of configurations, written $x(t) = \phi(X,t) = \phi_t(X)$. Rigidity of the body means that the distance between points of the body are fixed as the body moves. Let us assume that the center of mass of $\mathfrak{B}$ is at the origin and let us denote the motion of the center of mass by $q(t) = \phi_t(0)$. Then any motion of $\mathfrak{B}$ is the superposition of $q(t)$ and a rotation about the center of mass; i.e.

$$x(t) \quad = \quad Q(t)X + q(t) \tag{1}$$

where $Q(t) \in SO(3)$.

Let $m \in R^{+}$ be the total mass of the rigid body and $M \in R^{3\times 3}$ the inertial tensor which, for simplicity, we assume to be diagonal. Furthermore, let a conservative force

$$F(x_o(t)) = \nabla V(x_o(t)) \tag{2}$$

$V : R^3 \to R$, act on the rigid body where

$$x_o(t) = Q(t)X_o + q(t)$$

and $X_o \in \mathfrak{B}$ is a fixed point in $\mathfrak{B}$. Upon introducing canonical momenta $p = m\dot{q}$ and $P = \dot{Q}J$ where the diagonal matrix $J \in R^{3\times 3}$ is defined by

$$M_{ii}^{-1} = \frac{1}{2}\sum_{i\neq j} J_{jj}^{-1},$$

the motion of the rigid body in R^3 is Hamiltonian with the Hamiltonian function

$$H_c(q,p,Q,P) = \tfrac{1}{2}(p^t m^{-1} p) + \tfrac{1}{2}\mathrm{tr}(PJ^{-1}P^t) + V(q,Q) + \tfrac{1}{2}\mathrm{tr}\left(\{Q^tQ - I\}\Lambda\right) \tag{3}$$

where tr denotes the trace operator and $\Lambda \in R^{3\times 3}$ is a symmetric matrix implicitly determined by the holonomic constraint $Q \in SO(3)$, i.e.

$$0 = Q^tQ - I\,. \tag{4}$$

[9],[11]. Thus the motion for the center of mass is simply given by

$$\begin{aligned} \dot{q} &= m^{-1}p \\ \dot{p} &= -\nabla_q V(q,Q) \end{aligned} \tag{5}$$

where $V(q,Q)$ is defined by

$$V(q,Q) = V(QX_o + q)\,.$$

and the corresponding equations of motion for the variable Q are

$$\begin{aligned} \dot{Q} &= PJ^{-1} \\ \dot{P} &= -\nabla_Q V(q,Q) - Q\Lambda \\ 0 &= Q^tQ - I \end{aligned} \tag{6}$$

which define a Hamiltonian vector field on the phase space

$$\mathcal{M} = \{(Q,P) : Q^tQ - I = 0,\ Q^tPJ^{-1} + J^{-1}P^tQ = 0\}$$

Note that $\nabla_q V(q,Q) = F(q,Q)$ and $\nabla_Q V(q,Q) = F(q,Q)X_o^t$.

Now $\mathcal{M} \neq T^*SO(3)$ in general. However, as shown by McLachlan and Scovel [11], the motion on $T^*SO(3)$ can be obtained from the motion on $\mathcal{M}$ by putting $\tilde{Q} = Q$ and $\tilde{P} = \mathrm{Pr}(Q)P$ where $\mathrm{Pr}(Q)$ is the orthogonal projector onto $T_Q^*SO(3) \subset R^{3\times 3}$.

To make the equations (6) more transparent, let us rewrite them on $SO(3) \times \mathfrak{so}(3)^*$. Here $\mathfrak{so}(3)^*$ denotes the dual of the Lie algebra of skew-symmetric matrices [16]. This transformation can be achieved by introducing the body angular momentum $\Pi = Q^tP \in \mathfrak{so}(3)^*$ [9]. Using the standard isomorphism between R^3 and $\mathfrak{so}(3)^*$ denoted by skew : $R^3 \to \mathfrak{so}(3)^*$, we identify $\Pi \in \mathfrak{so}(3)^*$ with $\pi \in R^3$,

$\{Q^tF(q,Q)X_o^t - X_oF(q,Q)^tQ\}/2$ with $Q^tF(q,Q)\times X_o$, and $\{J^{-1}\Pi+\Pi J^{-1}\}/2$ with $M^{-1}\pi$ where M is the inertial tensor. Then we obtain the equations

$$\dot{\pi} \;=\; \pi \times M^{-1}\pi - \{Q^tF(q,Q)\}\times X_o \tag{7}$$

and

$$\dot{Q} \;=\; Q\,\mathrm{skew}\,(M^{-1}\pi) \tag{8}$$

Note that for $F=0$, (7) becomes the standard Euler equation for the free rigid body [9].

In the following section we will discuss numerical methods for the symplectic integration of the constrained Hamiltonian system corresponding to (3)-(4).

3 Symplectic Integration of a Single Rigid Body

A Hamiltonian system on $R^n \times R^n$ with holonomic constraints of the form $g(q)=0$, where $g: R^n \to R^m$ is a smooth map with full rank Jacobian in the neighborhood of its zero-set $g^{-1}(0)$, is characterized by the equations

$$\begin{aligned} \dot{q} &= \nabla_p H(q,p) \\ \dot{p} &= -\nabla_q H(q,p) - G(q)^t\lambda \\ 0 &= g(q) \end{aligned} \tag{9}$$

where $H: R^n\times R^n \to R$ is the Hamiltonian of the unconstrained system, $G(q) = g_q(q)$, and

$$H_c \;=\; H + \lambda^t g$$

the Hamiltonian of the constrained system. The solutions of (9) define a symplectic flow [2] on the constraint manifold

$$\mathcal{M} \;=\; \{(q,p) : g(q)=0,\, G(q)\nabla_p H(q,p) = 0\}$$

Recently it has been shown by Jay [7] and Reich [12] that for arbitrary Hamiltonian functions H there exist discretizations of (9) which preserve the constraint manifold $\mathcal{M}$ and are symplectic. Even more recently, McLachlan and Scovel [11] extended this result to the symplectic integration of the corresponding constrained system on the cotangent manifold

$$T^*\mathcal{N} \;=\; \{(q,p) : g(q)=0,\, G(q)p=0\}$$

In case that the Hamiltonian H is separable, i.e. of the form

$$H(q,p) \;=\; T(p) + V(q)\,,$$

we can, e.g., use the scheme RATTLE [1] which is a generalization of the Verlet scheme for unconstrained systems. It was shown to be symplectic and constraint-preserving by Leimkuhler and Skeel [8].

Since the Hamiltonian (3) is separable, the RATTLE discretization can be directly applied to the symplectic integration of the corresponding constrained Hamiltonian system. To see this, we have to use the correspondences $q\mapsto(q,Q)$ and $p\mapsto(p,P)$ and the constraint function g is given by (4). Thus we obtain the following

Algorithm 3.1

Step 1.

$$\begin{aligned} q_{n+1} &= q_n + hm^{-1}p_{n+1/2} \\ p_{n+1/2} &= p_n - (h/2)F(q_n, Q_n) \end{aligned}$$

Step 2.

$$\begin{aligned} Q_{n+1} &= Q_n + hP_{n+1/2}J^{-1} \\ P_{n+1/2} &= P_n - (h/2)\{F(q_n, Q_n)X_o^t + Q_n\Lambda_n\} \\ 0 &= Q_{n+1}^t Q_{n+1} - I \end{aligned}$$

Step 3.

$$\begin{aligned} P_{n+1} &= P_{n+1/2} - (h/2)\{F(q_{n+1}, Q_{n+1})X_o^t + Q_{n+1}\mu_{n+1}\} \\ 0 &= Q_{n+1}^t P_{n+1} + P_{n+1}^t Q_{n+1} \end{aligned}$$

Step 4.

$$p_{n+1} = p_{n+1/2} - (h/2)F(q_{n+1}, Q_{n+1})$$

where $F(q, Q) = \nabla V(q + QX_o)$.

Step 2 requires the solution of a nonlinear system of equations in 6 variables (since Λ is assumed to be symmetric). This can be avoided by applying recent results by McLachlan [10] and Reich [13] on the explicit symplectic integration of the Euler equations on the $\mathfrak{so}(3)^*$. Specifically:

Step 2 and 3 represent a second order symplectic integrator for the Hamiltonian

$$H_r(Q, P) = \frac{1}{2}\mathrm{tr}\,(PJ^{-1}P^t) + V(q, Q) + \frac{1}{2}\mathrm{tr}\,(\{Q^tQ - I\}\Lambda)$$

where q is treated as a parameter. Upon rewriting H_r as

$$H_r(Q, P) = H_e(Q, P, \Lambda) + V(q, Q) \tag{10}$$

with

$$H_e(Q, P, \Lambda) = \frac{1}{2}\mathrm{tr}\,(PJ^{-1}P^t) + \frac{1}{2}\mathrm{tr}\,(\{Q^tQ - I\}\Lambda)$$

we see that H_e is the Hamiltonian of a free rigid body fixed at the origin and the potential V corresponds to an external force acting on the body. Using the notation introduced in Section 2, the equations of motion of a free rigid body in terms of the body angular momentum are

$$\dot{\pi} = \pi \times M^{-1}\pi \tag{11}$$

where the Hamiltonian is now the kinetic energy of the body; i.e.

$$T(\pi) = \frac{1}{2}\pi^t M^{-1}\pi$$

As pointed out by McLachlan [10] and Reich [13], (11) can be integrated numerically by rewriting (11) as the sum of three Euler equations $\dot{\pi} = \pi \times \nabla T_i(\pi)$ where $T_i(\pi) = \frac{1}{2}\pi_i^2 M_{ii}^{-1}$. For each of these equations we have $\pi_i = const.$ and therefore each equation can be solved exactly. The solutions, denoted by $\exp(tD_{T_i})\,\pi$, are rotations about the π_i-axis with constant angular velocity $\pi_i M_{ii}^{-1}$. For example, for $i = 1$, we have to solve the system of linear differential equations

$$\begin{aligned} \dot{\pi}_1 &= 0 \\ \dot{\pi}_2 &= +\pi_1 M_{11}^{-1} \pi_3 \\ \dot{\pi}_3 &= -\pi_1 M_{11}^{-1} \pi_2 \end{aligned}$$

By applying the Baker-Campbell-Hausdorff formula [16], one can show that the scheme

$$\pi_{n+1} = A_h(\pi_n)\, \pi_n$$

with

$$A_h(\pi) = \exp(\frac{h}{2} D_{T_1}) \circ \exp(\frac{h}{2} D_{T_2}) \circ \exp(h D_{T_3}) \circ \exp(\frac{h}{2} D_{T_2}) \circ \exp(\frac{h}{2} D_{T_1})$$

is of second order in the step-size h and naturally preserves the Lie-Poisson structure of the Euler equation [9]. Let us now return to the symplectic integration of the corresponding Hamiltonian system on $T^*SO(3)$. We make use of the same splitting of the kinetic energy $T(\pi)$. But now we have to solve for each $T_i(\pi)$ the system of linear differential equations

$$\begin{aligned} \dot{\pi} &= \pi \times \nabla T_i(\pi) \\ \dot{Q} &= Q \,\mathrm{skew}\,(\nabla T_i(\pi)) \end{aligned}$$

Composing the single steps as done above for the Euler equation, we finally obtain the second-order symplectic scheme

$$\begin{aligned} Q_{n+1} &= Q_n\, B_h(\pi_n) \\ \pi_{n+1} &= A_h(\pi)\, \pi_n \end{aligned} \tag{12}$$

with A_h as above and B_h properly defined.

The motion due to the potential V in (10) is given by the differential equation

$$\begin{aligned} \dot{\pi} &= -\{Q^t F(q, Q)\} \times X_o \\ \dot{Q} &= 0 \end{aligned} \tag{13}$$

This differential equation can be solved exactly and Step 2 and 3 in Algorithm 3.1 can now be replaced by a proper composition of (12) and the exact time-h-flow of the differential equation (13). Thus we obtain the following explicit

Algorithm 3.2

Step 1.

$$\begin{aligned} q_{n+1} &= q_n + h m^{-1} p_{n+1/2} \\ p_{n+1/2} &= p_n - (h/2) F(q_n, Q_n) \end{aligned}$$

Step 2.

$$\bar{\pi}_n = \pi_n - (h/2)\{Q_n^t F(q_n, Q_n)\} \times X_o$$

$$\begin{aligned} Q_{n+1} &= Q_n B_h(\bar{\pi}_n) \\ \bar{\pi}_{n+1} &= A_h(\bar{\pi}_n)\,\bar{\pi}_n \end{aligned}$$

$$\pi_{n+1} = \bar{\pi}_{n+1} - (h/2)\{Q_{n+1}^t F(q_{n+1}, Q_{n+1})\} \times X_o$$

Step 3.

$$p_{n+1} = p_{n+1/2} - (h/2)F(q_{n+1}, Q_{n+1})$$

If necessary, P_{n+1} can be computed from Q_{n+1} and π_{n+1} by means of

$$P_{n+1} = Q_{n+1}\Pi_{n+1}$$

where, using the standard isomorphism between R^3 and $\mathfrak{so}(3)^*$, we transform $\pi_{n+1} \in R^3$ back into $\Pi_{n+1} \in \mathfrak{so}(3)^*$. Let us denote the resulting scheme by

$$\begin{pmatrix} q_{n+1} \\ Q_{n+1} \\ p_{n+1} \\ P_{n+1} \end{pmatrix} = \Psi_h \begin{pmatrix} q_n \\ Q_n \\ p_n \\ P_n \end{pmatrix} \tag{14}$$

Note that this scheme is now explicit in Q and P. The scheme is of second order. This can be seen from the time reversibility of the scheme. Furthermore, by construction, the scheme is symplectic and preserves the constraint $Q \in SO(3)$.

4 Equations of Motion for Constrained Systems of Rigid Bodies

So far we have considered the motion of unconstrained rigid bodies under the influence of external forces. In this and the following section we generalize this to the situation that rigid bodies are interconnected by joints.

There are two classes of joints; those that do not allow for a relative translation and those that do not allow for a relative rotation. The first type of joints is obtained whenever two bodies are joined at a single location. If no further constraints are imposed, the resulting joint is called a free or spherical joint. Other types of joints include the universal, revolve, prismatic etc. joint [3]. In this paper we only consider the simplest form of a joint; the spherical joint. As discussed by Barth and Leimkuhler in [3], any other, more complicated joint can be treated in a similar fashion.

Let us consider two rigid bodies with Hamiltonian H_i, $i = 1, 2$, with, for simplicity of exposition, no external forces acting on them; i.e., the Hamiltonian functions H_i are of the form

$$H_i(q_i, p_i, Q_i, P_i) = \frac{1}{2}\mathrm{tr}\,(P_i J_i^{-1} P_i^t) + \frac{1}{2} p_i^t m_i^{-1} p_i + \frac{1}{2}\mathrm{tr}\,(\{Q_i^t Q_i - I\}\Lambda_i)$$

together with the constraint

$$0 = Q_i^t Q_i - I$$

Let us assume that these two rigid bodies are linked by a spherical joint. The position of the joint relative to the reference configurations $\mathfrak{B}_i$ shall be given by the vectors $X_i \in \mathfrak{B}_i$. Thus we obtain the linear holonomic constraint

$$0 = Q_1 X_1 + q_1 - Q_2 X_2 - q_2 \tag{15}$$

and the corresponding constraint function is

$$g(q_1, q_2, Q_1, Q_2) := Q_1 X_1 + q_1 - Q_2 X_2 - q_2$$

As for constrained many-particle systems, the modified Hamiltonian, describing the motion of the two interconnected rigid bodies, is then given by

$$H_c = H_1 + H_2 + \lambda^t g$$

where $\lambda \in R^3$. Following (9) and applying the correspondences $q \mapsto (q_1, q_2, Q_1, Q_2)$, $p \mapsto (p_1, p_2, P_1, P_2)$, and $H \mapsto H_1 + H_2$, the resulting equations of motion are

$$\begin{aligned} \dot{q}_i &= m_i^{-1} p_i \\ \dot{p}_i &= (-1)^i \lambda \\ \dot{Q}_i &= P_i J_i^{-1} \qquad\qquad i = 1,2 \\ \dot{P}_i &= -Q_i \Lambda_i + (-1)^i \lambda X_i^t \\ 0 &= Q_i^t Q_i - I \end{aligned} \tag{16}$$

$$0 = Q_1 X_1 + q_1 - Q_2 X_2 - q_2$$

The Lagrange multiplier λ can be obtained by differentiating the constraint (15) twice with respect to time. This yields the equation

$$\Big(\sum_i X_i^t J_i^{-1} X_i + m_i^{-1}\Big)\, \lambda = -Q_1 \Lambda_1 J_1^{-1} X_1 + Q_2 \Lambda_2 J_2^{-1} X_2$$

which, since $(\sum_i X_i^t J_i^{-1} X_i + m_i^{-1}) > 0$, can be solved for $\lambda \in R^3$.

For more complicated joints we will, in general, obtain quadratic holonomic constraints $g = 0$ in the variables q_i and Q_i [3]. These can be treated in the same way as described above for the constraint (15).

5 Symplectic Integration of Constrained Systems of Rigid Bodies

Let Ψ_h be a first or second order symplectic integrator for the unconstrained system

$$\begin{aligned} \dot{q} &= \nabla_p H(q,p) \\ \dot{p} &= -\nabla_q H(q,p) \end{aligned}$$

In [12], Reich showed that this symplectic integrator can be modified such that the resulting scheme provides a symplectic integrator for the corresponding constrained Hamiltonian systems (9). Upon substituting for Ψ_h the symplectic integrator (14) and taking (15) as the holonomic constraint $g(q) = 0$, we can generalize this method to the symplectic integration of two rigid bodies interconnected by a spherical joint. Again we have to use the correspondences $q \mapsto (q_1, q_2, Q_1, Q_2)$ and $p \mapsto (p_1, p_2, P_1, P_2)$. This results in

Algorithm 5.1 *For* $i = 1, 2$*:*

Step 1.

$$\begin{aligned}\bar{p}_{i,n} &= p_{i,n} + (-1)^i (h/2)\lambda_n \\ \bar{P}_{i,n} &= P_{i,n} + (-1)^i (h/2)\lambda_n X_i^t\end{aligned}$$

Step 2.

$$\begin{pmatrix} q_{i,n+1} \\ Q_{i,n+1} \\ p_{i,n+1} \\ P_{i,n+1} \end{pmatrix} = \Psi_h \begin{pmatrix} q_{i,n} \\ Q_{i,n} \\ \bar{p}_{i,n} \\ \bar{P}_{i,n} \end{pmatrix}$$

with λ_n *chosen such that*

$$0 = Q_{1,n+1}X_1 + q_{1,n+1} - Q_{2,n+1}X_2 - q_{2,n+1}$$

The resulting nonlinear equation for the variable λ_n can be solved by a modified Newton method. For the approximate Jacobian we can simply use $A = \sum_i X_i^t J_i^{-1} X_i + m_i^{-1}$, $A \in R^+$.

It is straightforward to generalize this algorithm to multi-body systems interconnected by joints that can be modeled by holonomic constraints of the form

$$0 = g(q_l, Q_l, q_k, Q_k)$$

Here subscript l (k) denotes the lth (kth) body.

6 Symmetries and Conservation of Momentum Maps

For symplectic algorithms preservation of momentum follows from the equivariance of the algorithm with respect to the corresponding symmetry group [4]. Now the schemes considered in Section 3 are composition methods with each substep equivariant to a given symmetry of the Hamiltonian (3). Thus the schemes derived in Section 3 preserve the corresponding momentum map. The same is true for systems of rigid bodies interconnected by joints whenever the Hamiltonian (3) for each rigid body and the holonomic constraints due to joints are invariant under certain symmetries.

7 Heavy Top

Consider the motion of a rigid body fixed at a stationary point subject to the action of a uniform gravitational field. In terms of the reference configuration $\mathfrak{B}$, let the body be fixed at a point $x_o = QX_o + q$, $X_o \in \mathfrak{B}$. For simplicity we take $x_o = (0,0,0)$ and $X_o = (0,0,-z)$, $z > 0$. Since the point x_o can be considered as a spherical joint of the heavy top with the ground, we obtain the holonomic constraint

$$0 = g(q, Q) = QX_o + q \tag{17}$$

Furthermore, since the gravitational force acts only on the center of mass, the potential V in (3) is given by

$$V(q, Q) = c\, k^t q$$

where $k = (0,0,1)^t$ and c is the constant of gravitation. Thus the Hamiltonian of a heavy top is given by

$$\begin{aligned} H_c(q,Q,p,P) &= \frac{1}{2}\mathrm{tr}\,(PJ^{-1}P^t) + \frac{1}{2}p^t m^{-1} p + c\,k^t q + \\ &+ \frac{1}{2}\mathrm{tr}\,(\{Q^tQ - I\}\Lambda) + \lambda^t(QX_o + q) \end{aligned}$$

subject to the constraints $Q \in SO(3)$ and (17).

It is well-known [9] that, due to symmetry, these equations can be reduced to a Lie-Poisson system on $\mathfrak{se}(3)^*$. Thus, in principle, any Lie-Poisson integrator can be applied. Here we are interested in the direct symplectic integration of the unreduced equations.

Following the approach taken in this paper, we derive first an symplectic integrator for the 'free' heavy top; i.e., for the heavy top equations of motion without the constraint (17). This can be done as described in Section 3. Then the constraint is incorporated as described in Section 5 for spherical joints. Because (17) can be considered as a spherical joint of the heavy top with the ground at $x_o = (0,0,0)$, the index i takes only the value $i = 1$ which we drop for simplicity of notation. Thus the overall scheme in the variable (q, Q, p, π) becomes

$$\begin{aligned} \bar{p}_n &= p_n - (h/2)\lambda_n \\ \bar{\pi}_n &= \pi_n - (h/2)\{Q_n^t\lambda_n\} \times X_o \end{aligned}$$

$$\begin{aligned} q_{n+1} &= q_n + h\,m^{-1}p_{n+1/2} \\ p_{n+1/2} &= \bar{p}_n - (h/2)\,c\,k \\ p_{n+1} &= p_{n+1/2} - (h/2)\,c\,k \end{aligned}$$

$$\begin{aligned} Q_{n+1} &= Q_n\,B_h(\bar{\pi}_n) \\ \pi_{n+1} &= A_h(\bar{\pi}_n)\,\bar{\pi}_n \end{aligned}$$

subject to the constraint

$$0 = Q_{n+1}X_o + q_{n+1}$$

Note that the above scheme also preserves the third component of angular momentum corresponding to the $SO(2)$ symmetry of the heavy top about its z-axis.

Acknowledgments.

I thank B. Leimkuhler and Ch. Lubich for valuable discussions.

References

[1] Anderson, H. C. [1983], *Rattle: A 'Velocity' Version of the Shake algorithm for Molecular Dynamics Calculations*, J. Comp. Phys., **52**, 24–34.

[2] Arnold, V. I. [1975], *Mathematical Methods of Classical Mechanics.* Springer-Verlag, NY.

[3] Barth, E. and Leimkuhler, B. [1996], *Symplectic Methods for Conservative Multibody Systems*, this volume.

[4] Ge, Z. [1991], *Symplectic Difference Schemes and Generating Functions*, Physica D, **49**, 376–386.

[5] Ge, Z. and Marsden, J. E. [1988], *Lie-Poisson Hamilton-Jacobi Theory and Lie-Poisson Integrators*, *Physics Letters A*, **133(3)**, 134–139.

[6] Hairer, E., Nørsett, S. P., and Wanner, G. [1987], *Solving Ordinary Differential Equations, I.* Springer-Verlag.

[7] Jay, L. [1996], *Symplectic Partitioned Runge-Kutta Methods for Constrained Hamiltonian Systems*, *SIAM J. Num. Anal.*, **33**.

[8] Leimkuhler, B. and Skeel, R. D. [1994], *Symplectic Numerical Integrators in Constrained Hamiltonian Systems*, J. Comp. Phys., **112**, 117–125.

[9] Marsden, J. E. and Ratiu, T. [1994], *An Introduction to Mechanics and Symmetry.* Springer-Verlag.

[10] McLachlan, R. I. [1993], *Explicit Lie-Poisson Integration and the Euler Equations*, Phys. Rev. Lett., **71**, 3043–3046.

[11] McLachlan, R. I. and Scovel, C. [1995], *Equivariant Constrained Symplectic Integration.* J. Nonlinear Science, **5**, 233-256.

[12] Reich, S. [1996], *Symplectic Integration of Constrained Hamiltonian Systems by Composition Methods*, SIAM J. Numer. Anal. **33**, 475-491.

[13] Reich, S. [1993], *Numerical Integration of the Generalized Euler Equation*, Techn. Report, Univ. of British Columbia.

[14] Reich, S. [1994], *Momentum Conserving Symplectic Integrator*, Physica D, **76**, 375–383.

[15] Sanz-Serna, J. M. [1992], *Symplectic Integrators for Hamiltonian Problems: An Overview*, Acta Numerica, **1**, 243–286.

[16] Varadarajan, V. S. [1994], *Lie Groups, Lie Algebras, and Their Representation.* Prentice-Hall, Englewood Cliffs.

Fields Institute Communications
Volume **10**, 1996

Backward Error Analysis of Symplectic Integrators

J. M. Sanz-Serna
Departamento de Matemática Aplicada y Computación
Universidad de Valladolid, Valladolid, Spain

Abstract. In this expository paper we study how numerical solutions to differential equations can be interpreted as exact solutions of nearby differential equations. The emphasis is on the interpretation of symplectic numerical solutions of Hamiltonian systems as exact solutions of a perturbed Hamiltonian problem.

1 Introduction

In numerical analysis, given a problem $\mathcal{P}$ with true solution $\mathcal{S}$ and given an approximate solution $\widetilde{\mathcal{S}}$, *forward error analysis* consists of estimating the distance between $\widetilde{\mathcal{S}}$ and $\mathcal{S}$. *Backward error analysis* consists of showing that $\mathcal{S}$ is the true solution of a problem $\widetilde{\mathcal{P}}$ which is close to $\mathcal{P}$. While backward error analysis has played a role of paramount importance in areas like numerical linear algebra, error analyses of numerical methods for differential equations have essentially been of the forward variety (see nevertheless [Beyn [1991]], [Sanz-Serna [1992]], [Eirola [1993]]).

Backward error analysis would be specially helpful for *long-time* integrations. In the long-time scenario, the outcome of any forward error analysis is that errors (in the traditional *forward* sense) are huge, regardless of the numerical method being used. On the other hand, a successful backward error analysis could show that a numerical simulation of a given system provides the true evolution of a nearby, perturbed system.

In this paper we are concerned with backward error analyses of numerical methods for the initial value problem

$$\dot{u} = f(u), \qquad t \geq 0, \tag{1.1}$$

$$u(0) = \alpha \in \mathcal{R}^D, \tag{1.2}$$

where for simplicity it is assumed that the vector field f is $\mathcal{C}^\infty$ in the whole of $\mathcal{R}^D$. Of particular significance for us is the Hamiltonian case where $D = 2d$, $u = [p, q]$,

1991 *Mathematics Subject Classification.* Primary: 65L05; Secondary: 70H05.

$p, q \in \mathcal{R}^D$ and $f = f(p,q)$ has components

$$f_i = -\frac{\partial H}{\partial q_i}, \quad f_{d+i} = \frac{\partial H}{\partial p_i}, \qquad i = 1, \ldots, d, \tag{1.3}$$

for a suitable real-valued Hamiltonian function $H = H(p,q)$. We assume that (1.1)–(1.2) is integrated by a one-step method

$$u_{n+1} = \psi_{h,f}(u_n), \qquad n = 0, 1, \ldots, \tag{1.4}$$

$$u_0 = \alpha, \tag{1.5}$$

where h is the (constant) step size and u_n is the numerical approximation at time $t_n = nh$. For instance

$$\psi_{h,f}(u_n) = u_n + hf(u_n) \tag{1.6}$$

corresponds to the well-known Euler rule. The numerical method $\psi_{h,f}$ is consistent of order $r \geq 1$ if, for all u in $\mathcal{R}^D$,

$$\psi_{h,f}(u) - \phi_{h,f}(u) = O(h^{r+1}), \qquad h \to 0, \tag{1.7}$$

where $\phi_{h,f}$ denotes the flow of (1.1), so that $\phi_{h,f}(u)$ is the value at time $t = h$ of the solution of (1.1) with initial value u at time $t = 0$. From the local error estimate (1.7), it follows ([Butcher [1987]], [Hairer et al. [1987]]), that, as $h \to 0$, the global errors $u_n - u(t_n)$ are $O(h^r)$ uniformly in bounded intervals $0 \leq t \leq t_{max}$ contained in the interval of existence of the true solution of (1.1)–(1.2).

The initial-value problem (1.1)–(1.2) has both the vector field f and the initial condition α as data. Therefore for the backward error analysis of (1.4)–(1.5) we may try (at least) two approaches. In one approach we look for a perturbed initial condition $\widetilde{\alpha}$ and compare the numerical points u_n, $n = 0, 1, \ldots,$ with the values $\phi_{t_n,f}(\widetilde{\alpha})$ of the solution of (1.1) with initial condition $\widetilde{\alpha}$. This is the approach that leads to the idea of shadowing (see [Sanz-Serna and Larsson [1993]] and its references) and will not be considered further in this paper. In a second, alternative approach, we keep α as an initial condition and look for a perturbed vector field $\widetilde{f}$ so that $u_n = \phi_{t_n,\widetilde{f}}(\alpha)$. This procedure is very much related to the method of modified equations [Griffiths and Sanz-Serna [1986]], [Warming and Hyett [1974]], a tool sometimes used in the analysis of numerical methods for evolutionary partial differential equations.

Let us illustrate the method of modified equation with the linear scalar equation

$$\dot{u} = \lambda u \tag{1.8}$$

integrated by Euler's rule (1.6). Clearly

$$\psi_{h,f}(u) - \phi_{h,f}(u) = (u + h\lambda u) - \exp(h\lambda)(u) = -\frac{h^2\lambda^2}{2}u - \frac{h^3\lambda^3}{6}u - \ldots,$$

so that, in (1.7), $r = 1$ (first order of consistency). Is it possible to find a perturbed vector field $\widetilde{f}_2$ so that Euler's rule is consistent of the *second order* with the equation $\dot{u} = \widetilde{f}_2(u)$? In symbols

$$\psi_{h,f}(u) - \phi_{h,\widetilde{f}_2}(u) = O(h^3), \qquad h \to 0. \tag{1.9}$$

On using the ansatz

$$\dot{u} = \widetilde{f}_2(u) = \widetilde{f}_2^h(u) = \lambda u + hF_2(u),$$

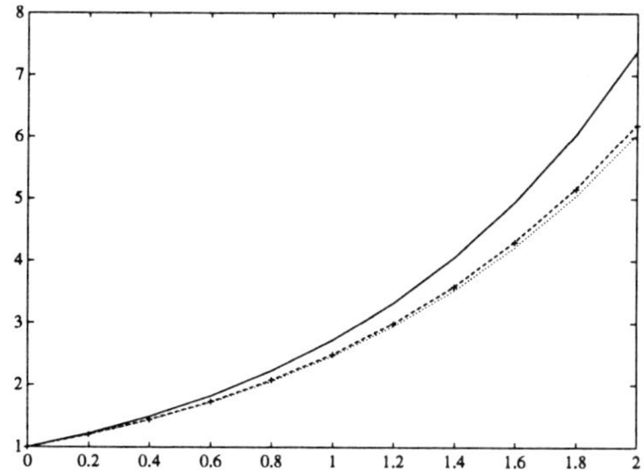

Figure 1 The initial value problem $du/dt = u$, $u(0) = 2$. Solid line: true solution.

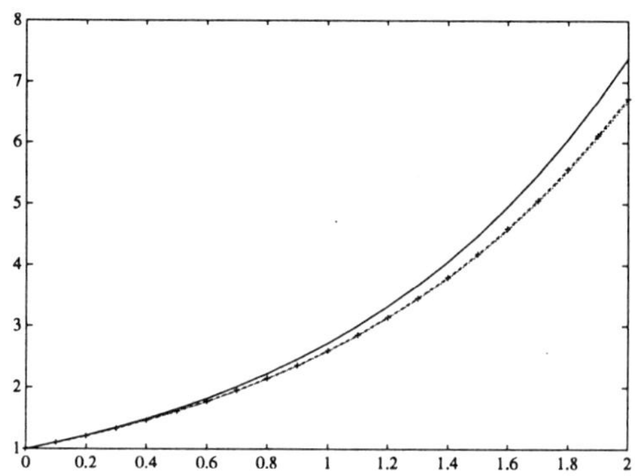

Figure 2 The case $h = 0.1$.

which implies

$$\ddot{u} = (\lambda + hF_2'(u))\dot{u} = (\lambda + hF_2'(u))\,(\lambda u + hF_2(u)),$$

we have

$$\begin{aligned}\psi_{h,f}(u) - \phi_{h,f}(u) &= (u + h\lambda u) \\ &\quad - \left(u + h[\lambda u + hF_2(u)] + \frac{h^2}{2}[\lambda^2 u + O(h)] + O(h^3)\right) \\ &= -\frac{h^2}{2}[2F_2(u) + \lambda^2 u] + O(h^3), \qquad h \to 0,\end{aligned}$$

and (1.9) leads to $F_2(u) = -\lambda^2 u/2$. By going from local errors to global errors as before, we now conclude that the Euler solution for (1.8) with $u(0) = \alpha$ is $O(h^2)$ away from the solution of the modified problem

$$\dot{u} = \left(\lambda - \frac{h\lambda^2}{2}\right) u, \quad u(0) = \alpha. \tag{1.10}$$

Figure 1 corresponds to $\lambda = 1$, $\alpha = 1$, $t_{max} = 2$ and $h = 0.2$. The solid line gives the exact solution $u(t) = \exp(t)$ and the crosses are the Euler solution. The dotted line provides the solution of (1.10). Figure 2 is identical to Figure 1, except that now $h = 0.1$. We see that the computed points, meant to approximate (1.8) provide a better approximation to the solution of the modified problem (1.10).

Better modified equations exist for our example. We may consider

$$\dot{u} = \widetilde{f}_3^h(u) = \left(\lambda - \frac{h\lambda^2}{2}\right) u + h^2 F_2(u) \tag{1.11}$$

and determine $F_3(u)$ to ensure

$$\psi_{h,f}(u) - \phi_{h,\widetilde{f}_3^h}(u) = O(h^4), \qquad h \to 0,$$

(consistency of the third order). Straightforward algebra leads to $F_3(u) = (\lambda^3/3)u$. The solution of (1.11) with $u(0) = 1$ is shown in Figs. 1–2 by means of a dashed line. We see that, within plotting accuracy, the Euler simulation of (1.8) gives the solution of the problem

$$\dot{u} = \lambda^h u, \quad \lambda^h = \lambda - \frac{h\lambda^2}{2} + \frac{h^2\lambda^3}{3}, \quad u(0) = \alpha$$

and a backward error interpretation is possible. If we imagine a modelling situation where the value of λ in (1.8) is not known exactly but rather arises from some measurement, then the backward error interpretation is useful because it is telling us that the effect of the numerical integration is to change the value of λ into a nearby value λ^h. If $|\lambda - \lambda^h|$ is of the order of the uncertainty in the measurement of λ, then the Euler's solution is as accurate as one may wish in this setting.

For (1.6) and (1.8), the modified equation ($N \geq 1$)

$$\dot{u} = \widetilde{f}_N^h(u) = \left(\lambda - \frac{h\lambda^2}{2} + \frac{h^2\lambda^3}{3} - \cdots \pm \frac{h^{N-1}\lambda^N}{N}\right) u,$$

is of order N, i.e.,

$$\psi_{h,f}(u) - \phi_{h,\widetilde{f}_N^h}(u) = O(h^{N+1}), \qquad h \to 0. \tag{1.12}$$

In this example, as $N \to \infty$, the vector fields $\widetilde{f}_N^h(u)$ converge to $(\log(1 + h\lambda)/h)u$. Then

$$\dot{u} = \widetilde{f}_\infty^h(u) = \frac{\log(1 + h\lambda)}{h} u$$

is an exact modified equation with $\psi_{h,f} \equiv \phi_{h,\widetilde{f}_\infty^h}$ [Beyn [1991]].

For general f it is still possible to find, for each $N \geq 1$, a modified equation $\dot{u} = \widetilde{f}_N^h(u)$ so that (1.12) holds. However, in general, the fields $\widetilde{f}_N^h$ do not converge as $N \to \infty$ and $\widetilde{f}_\infty^h$ cannot be defined. This is consistent with the fact that, for nonlinear f, $\psi_{h,f}$ is likely to present features (such as transversal crossings of separatrices) that cannot appear in any flow ϕ (see Section 4.10 of [Sanz-Serna [1991]]). Nevertheless, under suitable analyticity conditions, Neishtadt [Neishtadt [1984]] has proved that by letting the order N increase like $O(h^{-1})$, it is possible to have flows $\phi_{h,\widetilde{f}_{N(h)}^h}$ that approximate $\psi_{h,f}$ with exponentially small errors $O(\exp(-k/h))$, $k > 0$, as $h \to \infty$.

An outline of the remainder of the paper is as follows. In Section 2 we show how to systematically construct modified vector fields $\widetilde{f}_N^h$. The results presented are due to Hairer [Hairer [1994]], but the simple methodology used in the derivation of the formulae is taken from the Ph.D. thesis of A. Murua [Murua [1994]]. Section 3 looks at the particular case of Hamiltonian problems integrated by symplectic (canonical) methods [Sanz-Serna and Calvo [1994]]. The main result is that a method is symplectic if and only if the modified fields $\widetilde{f}_N^h$ corresponding to Hamilonian vector fields f are also Hamiltonian. Thus symplectic simulations change the Hamiltonian function, as distinct from nonsymplectic simulations, whose effect is to perturbe the Hamiltonian equation so as to render it non-Hamiltonian. Section 4 contains an application of the modified equation method to the study of error growth in the numerical solution of nonlinear oscillations. In the final Section 5 we consider an alternative approach where the numerical method is shown to exactly solve a nonautonomous system $\dot{u} = \widehat{f}(u,t)$, where, as $t \to 0$, the function $\widehat{f}$ approaches the true autonomous f.

2 Constructing modified equations

It is well known [Butcher [1987]], [Hairer et al. [1987]] that (rooted) trees are an important tool in the analysis of one-step methods. The trees with four of fewer nodes are depicted in Fig. 3. The symbol τ_1 denotes the only tree with one node. It is common to denote by $[\tau^1, \tau^2, \ldots, \tau^m]$ the tree that consists of the root and m leaving edges to which the trees τ^1, τ^2, ... , τ^m are attached. Thus in Fig. 3, $\tau_2 = [\tau_1]$, $\tau_{31} = [\tau_1, \tau_1]$, $\tau_{32} = [\tau_2]$, etc. For each tree τ, the integers $\rho(\tau)$ and $\alpha(\tau)$ respectively denote its order (number of nodes) and number of monotonic labellings. These functions can be computed recursively by the formulae $\rho(\tau_1) = \alpha(\tau_1) = 1$ and, for $\tau = [\tau^1, \ldots, \tau^m]$,

$$\begin{aligned} \rho(\tau) &= 1 + \rho(\tau^1) + \cdots + \rho(\tau^m), \\ \alpha(\tau) &= \frac{(\rho(\tau)-1)!}{\rho(\tau^1)! \cdots \rho(\tau^m)!} \alpha(\tau^1) \ldots \alpha(\tau^m) \frac{1}{\mu_1! \mu_2! \cdots}. \end{aligned}$$

The integers μ_i count the number of equal trees among τ^1, ... , τ^m. Finally, in connection with the system (1.1), an $\mathcal{R}^D$-valued function $F(\tau)(u)$ (elementary differential) is associated with each tree τ. The recursive definition of the $F(\tau)(u)$'s is $F(\tau_1)(u) = f(u)$ and for $\tau = [\tau^1, \ldots, \tau^m]$

$$F(\tau)(u) = f^{(m)}(u)(F(\tau^1)(u), \ldots, F(\tau^m)(u)),$$

where $f^{(m)}(u)$ represents the m-th Fréchet derivative of f evaluated at u.

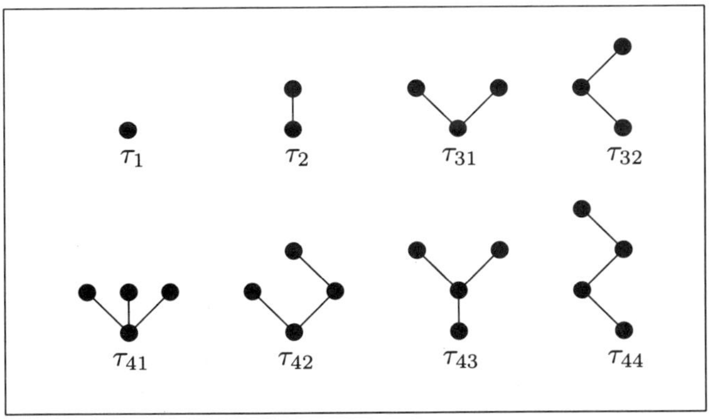

Figure 3 Trees of order $\leq$ 4.

With these notations, the formal Taylor expansion of the flow $\phi_{h,f}$ in powers of h is given by

$$\phi_{h,f}(u) = u + \sum_{\tau \in T} \frac{h^{\rho(\tau)}}{\rho(\tau)!} \alpha(\tau) F(\tau)(u),$$

where T denotes the set of all trees.

Numerical methods can be Taylor expanded in a similar way. For instance, for a Runge-Kutta method

$$\begin{aligned} U_i &= u_n + h \sum_{j=1}^{s} a_{ij} f(U_j), \\ u_{n+1} &= u_n + h \sum_{i=1}^{s} b_i f(U_i), \end{aligned}$$

the Taylor expansion is

$$\psi_{h,f}(u) = u + \sum_{\tau \in T} \frac{h^{\rho(\tau)}}{\rho(\tau)!} \alpha(\tau) \left(\gamma(\tau) \sum_{i=1}^{s} b_i \Phi_i(\tau) \right) F(\tau)(u),$$

where the recursive definitions of γ and Φ_i are $\gamma(\tau_1) = 1$, $\Phi_i(\tau_1) = 1$ and

$$\begin{aligned} \gamma(t) &= \rho(\tau)\gamma(\tau^1)\ldots\gamma(\tau^m), \\ \Phi_i(\tau) &= \sum_{j_1,\ldots,j_m} a_{ij_1}\Phi_{j_1}(\tau^1)\ldots a_{ij_m}\Phi_{j_m}(\tau^m). \end{aligned}$$

In view of the Taylor expansions above, [Hairer and Wanner [1974]] introduced the notion of a B-series. Given a real-valued mapping a defined in the union of T and the set $\{\emptyset\}$, a B-series $B(a,u)$ is a formal power series

$$a(\emptyset)u + \sum_{\tau \in T} \frac{h^{\rho(\tau)}}{\rho(\tau)!} \alpha(\tau) a(\tau) F(\tau)(u).$$

(The B-series depends on h and f, but this dependence is not made explicit in the notation.) Thus, the results just quoted imply that the true flow $\phi_{h,f}$ corresponds to $a \equiv 1$, while for a Runge-Kutta method $a(\emptyset) = 1$ and, for τ in T,

$$a(\tau) = \gamma(\tau) \sum_{i=1}^{s} b_i \Phi_i(\tau).$$

The Taylor expansion of most one-step methods used in practice is also a B-series. In the remainder of the paper we assume that we are dealing with a method $\psi_{h,f}(u)$ corresponding to a suitable B-series $B(a,u)$,

$$\psi_{h,f}(u) = B(a,u), \tag{2.1}$$

without specifying the exact nature of the method. We suppose throughout that the method is consistent, i.e., at least of order 1:

$$a(\emptyset) = 1, \qquad a(\tau_1) = 1. \tag{2.2}$$

Our aim is to construct a formal power series $\widetilde{f}$

$$\sum_{\tau\in T}\frac{h^{\rho(\tau)-1}}{\rho(\tau)!}\alpha(\tau)b(\tau)F(\tau)(u) \tag{2.3}$$

so that for each integer $N \geq 1$

$$\frac{du}{dt}=\widetilde{f}_N^h(u)=\sum_{1\leq\rho(\tau)\leq N}\frac{h^{\rho(\tau)-1}}{\rho(\tau)!}\alpha(\tau)b(\tau)F(\tau)(u) \tag{2.4}$$

provides a modified equation of order N.

An essential tool for our purposes is the formula for composition of B-series, see Theorem 11.6 in [Hairer et al. [1987]]. If a and b are B-series coefficients with $a(\emptyset) = 1$ then the composition $B(b, B(a, y))$ is a again a B-series $B(ab, y)$ whose coefficients $ab(\tau)$ can be found in a systematic way from the a's and b's. The formulae for the first $ab(\tau)$ are

$$\begin{array}{lcl} ab(\emptyset) & = & b(\emptyset), \\ ab(\tau_1) & = & b(\emptyset)a(\tau_1)+b(\tau_1), \\ ab(\tau_2) & = & b(\emptyset)a(\tau_2)+2b(\tau_1)a(\tau_1)+b(\tau_2), \\ ab(\tau_{31}) & = & b(\emptyset)a(\tau_{31})+3b(\tau_1)a(\tau_1)^2+3b(\tau_2)a(\tau)+b(\tau_{31}), \\ ab(\tau_{32}) & = & b(\emptyset)a(\tau_{32})+3b(\tau_1)a(\tau_2)+3b(\tau_2)a(\tau)+b(\tau_{32}). \end{array} \tag{2.5}$$

We introduce a real parameter λ and write the flow of the vector field in (2.3) as a B-series

$$\phi_{\lambda h,\widetilde{f}}(u)=u+\sum_{\tau\in T}\frac{h^{\rho(\tau)}}{\rho(\tau)!}\alpha(\tau)a_\lambda(\tau)F(\tau)(u).$$

Next we substitute this series into the equation

$$\frac{d}{dt}\phi_{t,\widetilde{f}}=\widetilde{f}(\phi_{t,\widetilde{f}});$$

in doing so the B-series of the right hand side is computed by the formula for composing B-series. In this way we find that the $a_\lambda(\tau)$ satisfy, for each tree τ,

$$\frac{d}{d\lambda}a_\lambda(\tau)=(a_\lambda b)(\tau). \tag{2.6}$$

Furthermore at $\lambda = 0$, $\phi_{0,\widetilde{f}}(u) = u$ and hence, for each τ,

$$a_0(\tau)=0. \tag{2.7}$$

The relations (2.6)–(2.7) allow us the computation of the $a_\lambda(\tau)$'s in terms of the $b(\tau)$'s when the latter are known. In our setting, the b coefficients are determined to ensure that, for each τ, at $\lambda = 1$,

$$a_1(\tau)=a(\tau), \tag{2.8}$$

to impose that, as formal power series, $\phi_{h,\widetilde{f}}$ and $\psi_{h,f}$ coincide.

The relations (2.6)–(2.8) make it possible to recursively compute the b coefficients. Let us illustrate this. For τ_1 we obtain from (2.6) and (2.5), $(d/d\lambda)a_\lambda(\tau_1) = b(\tau_1)$, so that, according to (2.7), $a_\lambda(\tau_1) = \lambda b(\tau_1)$. If we now impose (2.8), we obtain the relation $b(\tau_1) = a(\tau_1)$. We conclude, from the consistency assumption (2.2), that $b(\tau_1) = 1$, and therefore, as expected, $\widetilde{f}$ differs from f in $O(h)$ terms.

If we now go through the same steps for the next tree τ_2, we obtain

$$a(\tau_2) = b(\tau_1)^2 + b(\tau_2).$$

Note that for a method of order ≥ 2, $a(\tau_2) = 1$, which, in tandem with $b(\tau_1) = 1$, leads to $b(\tau_2) = 0$ and $\widetilde{f}$ and f differ in $O(h^2)$ terms.

The equations for finding $b(\tau_{32})$ and $b(\tau_{31})$ turn out to be

$$\begin{aligned} a(\tau_{31}) &= b(\tau_1)^3 + \frac{3}{2} b(\tau_2) b(\tau_1) + b(\tau_{31}), \\ a(\tau_{32}) &= b(\tau_1)^3 + 3b(\tau_2) b(\tau_1) + b(\tau_{32}). \end{aligned}$$

From here $b(\tau_{31}) = b(\tau_{32}) = 0$ for methods of order ≥ 3.

We summarize our findings in the following theorem, due to Hairer [Hairer [1994]].

Theorem 2.1 *Assume that an order r, $r \geq 1$, one-step method (2.1) can be formally Taylor expanded into a B-series $B(a,u)$. There is a unique B-series (2.3), differing from $f(u)$ in $O(h^r)$ terms, such that, for each integer $N \geq 1$, (2.4) provides a modified system of order N. The coefficients b can be recursively found as functions of the coefficients a.*

3 The Hamiltonian case

We now consider the Hamiltonian case (1.3). There has been much recent interest [Sanz-Serna and Calvo [1994]] in simulating (1.1)–(1.3) by so-called symplectic or canonical integrators, i.e., by methods for which $\psi_{h,f}$ preserves the differential form $dp \wedge dq$ and thus reproduce the main feature of Hamiltonian flows.

Calvo and the present author [Calvo and Sanz-Serna [1994]] have shown that, for methods $\psi_{h,f}$ given by a B-series as in (2.1), it is possible to check canonicity by looking at the corresponding coefficients $a(\tau)$. We use Butcher's notation and set

$$\tau \circ \tau^* = [\tau^1, \tau^2, \ldots, \tau^m, \tau^*]$$

if τ^* and $\tau = [\tau^1, \tau^2, \ldots, \tau^m]$ are trees. Then a B-series is canonical [Calvo and Sanz-Serna [1994]], if, for any pair of trees τ^1, τ^2,

$$\frac{a(\tau^1 \circ \tau^2)}{\gamma(\tau^1 \circ \tau^2)} + \frac{a(\tau^2 \circ \tau^1)}{\gamma(\tau^2 \circ \tau^1)} = \frac{a(\tau^1)}{\gamma(\tau^1)} \frac{a(\tau^2)}{\gamma(\tau^2)}. \tag{3.1}$$

On the other hand, Hairer [Hairer [1994]] has proved that the B-series vector field (2.3) (with f given by (1.3)) is Hamiltonian if and only if, for any τ^1, τ^2,

$$\frac{b(\tau^1 \circ \tau^2)}{\gamma(\tau^1 \circ \tau^2)} + \frac{b(\tau^2 \circ \tau^1)}{\gamma(\tau^2 \circ \tau^1)} = 0. \tag{3.2}$$

When (3.2) holds, it is possible to explicitly find the formal Hamiltonian $\widetilde{H}$ whose vector field is (2.3).

The key point is that if $\{a(\tau)\}$ are the coefficients associated with a numerical method as in (2.1) and $\{b(\tau)\}$ are the coefficients of the corresponding modified vector field as in Theorem 1, then (3.1) and (3.2) are equivalent [Hairer [1994]]. Therefore canonical methods can be characterized as those methods that when applied to Hamiltonian problems possess modified equations $\dot{u} = \widetilde{f}_N^h(u)$ that are Hamiltonian for all $N = 1, 2, \ldots$. The use of a symplectic integrator changes the

Hamiltonian function of the system being integrated; the use of a nonsymplectic integrator perturbes the differential equation so as to turn it non-Hamiltonian.

4 An application

We now illustrate the use of modified equations. We consider the well-known pendulum system with Hamiltonian function (energy) $H = (1/2)p^2 + 1 - \cos q$. Let (p_0, q_0) be an initial condition with energy H_0, $0 < H_0 < 2$, leading to a periodic solution. In phase plane the trajectory corresponds to the level set $H = H_0$; the period T_0 of the solution is an increasing function of H_0. Furthermore, we respectively denote by f_0 and g_0 the vector field f evaluated at (p_0, q_0) and the energy gradient at (p_0, q_0). The vectors f_0 and g_0 are mutually orthogonal by conservation of energy.

This initial value problem is integrated by a one-step method of order r with step length h, that for simplicity we assume to be of the form $h = T_0/\nu$, with ν a positive integer. Let $e_M(h)$ be the global error $u_n - u(t_n)$ after $n = M\nu$ steps, i.e. after simulating M periods of the solution. Then it is not too difficult to show, see [Calvo and Sanz-Serna [1993]], that

$$e_M(h) = Me_1(h) + \frac{1}{2}(M^2 - M)\Big(g_0, e_1(h)\Big)\,\delta_0 f_0 + O(h^{2r}), \qquad h \to 0,$$

where $(\cdot,\cdot)$ means inner product and δ_0 denotes the derivative of the period T with respect to the energy H evaluated at the initial condition. Therefore, ignoring the $O(h^{2r})$ remainder, the error $e_M(h)$ grows quadratically with M. The leading M^2 growth is in the direction of f_0, i.e., tangent to the solution at the initial point, thus corresponding to a *phase error*. However linear error growth with M is possible: if

$$(g_0, e_1(h)) = O(h^{2r}) \tag{4.1}$$

(i.e., the error after one period is almost orthogonal to the energy gradient), then

$$e_M(h) = Me_1(h) + O(h^{2r}), \qquad h \to 0.$$

In a nutshell, the way global errors build up is determined by the *direction* of the error $e_1(h)$. This is not suprising: if after one period the error $e_1(h)$ has a significant component in the direction of g_0, then the numerical solution has jumped in phase plane to a neighbouring trajectory corresponding to a different (say larger) value of the energy. Thereafter, the method, when evaluating the vector field f, picks up wrong information as to the solution period and is lead to believe that the motion is slower than it really is. As the integration proceeds the numerical solution keeps jumping to higher and higher energy levels and getting unduly slowing down. This is the mechanism leading to quadratic growth in the *phase* error. On the other hand, if $e_1(h)$ is essentially in the direction of f_0, then there is no energy error: the method is basically describing the right trajectory with a slightly distorted average velocity and errors grow linearly. These considerations apply to all nonlinear oscillators with one degree of freedom [Calvo and Sanz-Serna [1993]], to Kepler's problem [Calvo and Sanz-Serna [1993]], and even to some partial differential equations [Frutos and Sanz-Serna [1994]]. See [Cano and Sanz-Serna [1995]] for a comprehensive treatment.

We now use the method of modified equations to show that for canonical methods (4.1) holds. Let us construct a modified problem of order $2r$. By the results in the preceding section, this is a Hamiltonian problem with Hamiltonian $\widetilde{H}$. The modified solution $\widetilde{u}$ conserves $\widetilde{H}$ exactly and hence, Taylor expanding,

$$\begin{aligned} 0 = \widetilde{H}(\widetilde{u}(T_0)) - \widetilde{H}(u_0) &= (\widetilde{g}_0, \widetilde{u}(T_0) - u_0) + O(|\widetilde{u}(T_0) - u_0|^2) \\ &= (\widetilde{g}_0, \widetilde{u}(T_0) - u_0) + O(h^{2r}). \end{aligned} \tag{4.2}$$

Here $\widetilde{g}_0$ is the gradient of $\widetilde{H}$ at the initial point u_0 and we have used that

$$\widetilde{u}(T_0) - u_0 = \widetilde{u}(T_0) - u(T_0) = O(h^r), \tag{4.3}$$

due to the periodicity of the true solution and to the fact that the true and modified vector fields differ in $O(h^r)$ terms. From (4.2)–(4.3), along with $\widetilde{g}_0 - g_0 = O(h^r)$, we obtain

$$(g_0, \widetilde{u}(T_0) - u_0) = O(h^{2r}),$$

and finally, since $\widetilde{u}$ and the numerical solution differ in $O(h^{2r})$ terms, (4.1) holds.

5 An alternative approach

In an alternative approach to modified equations, we may compare $\psi_{h,f}(\alpha)$ with the value $\Phi_{\widehat{f}}(h,0)\alpha$, at time $t = h$, of the solution of a nonautonomous system $\dot{u} = \widehat{f}(u,t)$ with initial condition $u = \alpha$ at $t = 0$. It is not difficult to see that $\widehat{f}$ can be chosen to ensure that for all h and α,

$$\psi_{h,f}(u) = \Phi_{\widehat{f}}(h,0)\alpha. \tag{5.1}$$

In fact, consider the family $\mathcal{F}$ of curves $t \mapsto \psi_{t,f}\alpha$ (α is the parameter in the family). Differentiation with respect to t and elimination of α lead to a differential equation $\dot{u} = \widehat{f}(u,t)$ satisfied by all curves in the family. Then (5.1) holds.

For Euler's rule (1.6) applied to (1.8), the family $\mathcal{F}$ is

$$u = \alpha + t\lambda\alpha;$$

differentiation leads to

$$\dot{u} = \lambda\alpha$$

and, eliminating α, we find

$$\dot{u} = \widehat{f}(u,t) = \frac{\lambda}{1+t\lambda}u. \tag{5.2}$$

For a general system (1.1) integrated by a B-series method (2.1) we look for an $\widehat{f}$ of the form (cf. (2.3))

$$\widehat{f}(u,t) = \sum_{\tau\in T} \frac{t^{\rho(\tau)-1}}{\rho(\tau)!}\alpha(\tau)\widehat{b}(\tau)F(\tau)(u) \tag{5.3}$$

Differentiation with respect to t of the B-series for $\psi_{t,f}$ and substitution in $\dot{u} = \widehat{f}(u,t)$ show that for each tree τ

$$\rho(\tau)a(\tau) = (a\widehat{b})(\tau). \tag{5.4}$$

The $\widehat{b}$'s can be recursively found from (5.4). The formulae for the first $\widehat{b}$'s are

$$\begin{aligned}
\widehat{b}(\tau_1) &= a(\tau_1),\\
2\widehat{b}(\tau_1)a(\tau_1)+\widehat{b}(\tau_2) &= 2a(\tau_2),\\
3\widehat{b}(\tau_1)a(\tau_1)^2+3\widehat{b}(\tau_2)a(\tau_1)+\widehat{b}(\tau_{31}) &= 3a(\tau_{31}),\\
3\widehat{b}(\tau_1)a(\tau_2)+3\widehat{b}(\tau_2)a(\tau_1)+\widehat{b}(\tau_{32}) &= 3a(\tau_{32}).
\end{aligned}$$

Note that $\widehat{b}(\tau_1) = 1$ because we assume $a(\tau_1) = 1$ (consistency). For a method of order ≥ 2, $a(\tau_2) = 1$ and then $b(\tau_2) = 0$. If the order is ≥ 3, then $\widehat{b}(\tau_{31}) = \widehat{b}(\tau_{32}) = 0$. In general for a method of order r, $\widehat{f}$ differs from f in terms $O(t^r)$ as $t \to 0$. Furthermore, it is easy to check that, for a method of order r, the $\widehat{b}$'s corresponding to trees of order $r+1$ are related to the b's of Theorem 1 through the relation

$$\widehat{b}(\tau) = (r+1)b(\tau).$$

Hence, at the leading $O(h^r)$ order, $\widehat{f}(u,h) - f(u)$ and $\widetilde{f}_N^h(u) - f(u)$, $N > r$, only differ in a factor $r+1$. We have proved the following result.

Theorem 5.1 *Assume that an order r, $r \geq 1$, one-step method $\psi_{t,f}$ can be formally Taylor expanded into a B-series $B(a,u)$ (2.1). There is a unique $\widehat{f}(u,t)$ (5.3), differing from $f(u)$ in $O(t^r)$ terms ($t \to 0$), such that, for all stepsizes h and all points α, $\psi_{h,f}\alpha$ is the value at time $t = h$ of the solution of $\dot{u} = \widehat{f}(u,t)$ with initial condition $u(0) = \alpha$. The coefficients $\widehat{b}$ can be recursively found as functions of the coefficients a.*

Either by using the ideas in Hairer [Hairer [1994]] or by a general argument presented in [Sanz-Serna and Calvo [1994]], it is possible to prove that, in the Hamiltonian case, a method is canonical if and only if $\widehat{f}(u,t)$ turns out to be Hamiltonian. McLachlan and Atela [Mclachlan and Atela [1991]] then use the discrepancy between the true Hamiltonian H and the Hamiltonian $\widehat{H}$ of $\widehat{f}(u,t)$ as a measure of the accuracy of the numerical method.

A possible drawback of the approach in this section is that advancing n steps with the numerical method is not the same as going from $t = 0$ to $t = nh$ with the solution of $\dot{u} = \widehat{f}(u,t)$, because, since $\widehat{f}$ is nonautonomous, for the solution operator Φ

$$\Phi_{\widehat{f}}(h,0)\circ\cdots\circ\Phi_{\widehat{f}}(h,0) \neq \Phi_{\widehat{f}}(nh,0).$$

There is a way around this problem: for $0 \leq t < h$ we keep the function $\widehat{f}(u,t)$ found before and for $h \leq t < 2h$, $2h \leq t < 3h$, dots we repeat it periodically to get an h-periodic *discontinous* function $\widehat{f^h}(u,t)$. Now n steps with the method $\psi_{h,f}$ are equivalent to advancing from $t = 0$ to $t = nh$ with the differential system $\dot{u} = \widehat{f^h}(u,t)$. For a symplectic method applied to a Hamiltonian problem, the corresponding nonautonomous Hamiltonian $\widehat{H}^h$ has delta functions at the points $t = nh$, n integer. This is similar to the situation in [Wisdom and Homan [1991]].

Let us finally point out a connection with the approach in Section 2: the autonomous vector field $\widetilde{f}_N^h$ considered there is the result of eliminating *by averaging* the terms t^k, $k \geq N$ of the periodic function $\widehat{f^h}$ (cf. [Neishtadt [1984]]). For instance, from (5.2),

$$\dot{u} = [\lambda - t\lambda^2 + O(t^2)]u,$$

and averaging the term $-t\lambda^2$ over the period $0 \leq t \leq h$ we obtain $-h\lambda^2/2$ as in (1.10).

Acknowledgements

This research has been supported by project DGICYT PB92-254.

References

Beyn, W.-J. [1991], *Numerical methods for dynamical systems*, In Advances in Numerical Analysis (Light, W., ed.), **I**, 175–236. Clarendon, Oxford.

Butcher, J. C. [1987], *The Numerical Analysis of Ordinary Differential Equations.* John Wiley, Chichester.

Calvo, M. P. and Sanz-Serna, J. M. [1993], *The development of variable-step symplectic integrators, with application to the two-body problem*, SIAM J. Sci. Comput., **14**, 936–952.

Calvo, M. P. and Sanz-Serna, J. M. [1994], *Canonical B-series*, Numer. Math, **67**, 161–175.

Cano, B. and Sanz-Serna, J. M. [1995], *Error growth in the numerical integration of periodic orbits, with application to Hamiltonian and reversible systems*, SIAM J. Numer. Anal., to appear.

Eirola, T. [1993], *Aspects of backward error analysis of numerical ODEs*, J. Comput. Appl. Math., **45**, 65–73.

de Frutos, J. and J. M. Sanz-Serna, J. M. [1994], *Erring and being conservative*, In Numerical Analysis (Griffiths, D. F. and Watson, G. A. eds.), Longman, London, 75–88.

Griffiths, D. F. and Sanz-Serna, J. M. [1986], *On the scope of the method of modified equations*, SIAM J. Sci. Comput., **7**, 994–1008.

Hairer, E. [1994], *Backward analysis of numerical integrators and symplectic schemes*, Annals Numer. Math., **1**, 107–132.

Hairer, E., Nørsett, S. P., and Wanner, G. [1987], *Solving Ordinary Differential Equations I, Nonstiff Problems.* Springer, Berlin.

Hairer, E. and G. Wanner, G. [1974], *On the Butcher group and general multivalue methods*, Computing, **13**, 1–15.

McLachlan, R. I. and Atela, P. [1991], *The accuracy of symplectic integrators*, Nonlinearity, **5**, 541–562.

Murua, A. [1994], *Métodos simplécticos desarrollables en P-series*, Tesis, Universidad de Valladolid.

Neishtadt, A. I. [1984], *The separation of motions in systems with rapidly rotating phase*, J. Appl. Math. Mech., **48**, 133-139.

Sanz-Serna, J. M. [1991], *Two topics on nonlinear stability*, In Advances in Numerical Analysis (Light, W. ed.), Clarendon, Oxford. **I**, 147–174.

Sanz-Serna, J. M. [1992], *Numerical ordinary differential equations vs. dynamical systems*, In The Dynamics of Numerics and the Numerics of Dynamics (Broomhead, D.S. and Iserles, A., eds.), Clarendon, Oxford, 81–106.

Sanz-Serna, J. M. and Calvo, M. P. [1994], *Numerical Hamiltonian Problems.* Chapman and Hall, London.

Sanz-Serna, J. M. and Larsson, S. [1993], *Shadows, chaos and saddles.* Appl. Numer. Math., **13**, 181–190.

Warming, R.F. and Hyett, B.J. [1974], *The modified equation approach to the stability and accuracy of finite difference methods*, J. Comput. Phys., **14**, 159–179.

Wisdom, J. and Homan, M. [1991], *Symplectic maps for the N-body problem*, Astron. J., **102**, 1528–1538.

Fields Institute Communications
Volume **10**, 1996

Numerical Determination Of Caustics and their Bifurcations

T. J. Stuchi
Laboratório Nacional de Computação Científica-CNPq
Instituto de Física
Universidade Federal do Rio de Janeiro
Rio de Janeiro
Brazil

R. Vieira-Martins
Observatório Nacional-CNPq
Rio de Janeiro
Brazil

Abstract. We present some numerical results on bifurcations of caustics for the Hénon-Heiles Hamiltonian [1]. Caustic evolution, from regular to chaotic regimes, is followed through a parameter attached to the coupling term of this Hamiltonian. We find that evolution of the caustics can forecast the onset of chaotic behavior, giving additional information on the surviving tori. This is shown by comparing with the surface of section method. This method is being developed to find evidence of integrability/chaos for systems with more than two degrees of freedom [10].

1 Introduction

It is well known that bound integrable Hamiltonian orbits trace tori in phase space and these tori are Lagrangian manifolds. Lagrangian singularities are caustics of projections of a Lagrangian torus onto configuration space in our case [2]. In the general case, we find [9] that for a sufficiently large time, the focal families associated to each orbit have focal points sufficiently near the Lagrangian singularities of its torus. Therefore, if the system is integrable a set of these focal points draws caustics which can be identified by their typical curves. This method has been successfully applied to the Hénon-Heiles Hamiltonian [1], and was proposed [9] as an alternative to the surface of section method.

Theoretically speaking, we have the advantage that caustics are $n-1$ dimensional surfaces on the n dimensional configuration space, for n degrees of freedom. In fact, we are now starting to apply this method for systems with more than two degrees of freedom. As far as we know, the visualization of Poincaré sections of

1991 *Mathematics Subject Classification.* Primary: 58F05; Secondary: 70-08.

such a systems was abandoned since the works of Martinet and Magnenat [5] and Froeschelé [4] more than a decade ago. Moreover, as it is well known the caustics, or more precisely singularities, of Langrangian projections have been classified up to $n \leq 5$ into a few generic types. Furthermore we can have projections into the configuration space, where the physical phenomena actually takes place.

A similar method, which exhibits caustics by making Lagrangian projections of Poincaré sections for systems with three degrees of freedom was developed by Stuchi [8]; the application of this second method to mappings is straightforward.

Moreover, we stress that we have the advantage of capturing details of tori evolution which can not be captured by either the Poincaré section or the Lyapunov exponent methods. Such details are important, for example, for the semi-classical limit of quantum mechanics [6][7]. While revising this paper we learned about the work of Agekyan et. al. [3]. They use a different algorithm but also observed the early bifurcations of the caustics.

In this report we present a study of the evolution of caustics, from regular to chaotic situations, using a parameter, which when made equal to 1 yields the classical Hénon-Heiles Hamiltonian. In order to make this paper self contained, as far as the main ideas of the method are concerned, we present in Section 2 a summary of the method described in [9]. In Section 3 we show the results of the evolution and compare with the surface of section method.

2 Caustics of n Degrees of Freedom Hamiltonians

Hamilton's equations of motion can be written as

$$\dot{z} = \mathbf{J}\ \mathrm{grad} H(\mathbf{z}) \tag{2.1}$$

where $\mathbf{z}^T = (q_1, ..., q_n, p_1, ..., p_n)$ is the transpose of $\mathbf{z}$ and $\mathbf{J}$ is the standard symplectic matrix. Assume $\mathbf{z}_t$ is a given solution of (2.1), the associated variational equations are

$$\dot{\delta_z} = \mathbf{J}\ D^2 H(\mathbf{z}_t)\ \delta_z \tag{2.2}$$

where $D^2 H(\mathbf{z}_t) = \left(\frac{\partial H(\mathbf{z})}{\partial z_i z_j}\right)_{\mathbf{z}=z_t}$.

We have proved in [9], under fairly general conditions, that given $\epsilon > 0$, there exists an $M > 0$ such that for all $t > M$, the ratio by t of an arbitrary solution of the variational equations (mod $= 1$ for $t = 0$) differs, in general, from a tangent vector to the invariant torus by an amount smaller than ϵ.

Using the above result we can establish a general procedure for determining caustics of invariant tori of integrable Hamiltonian systems or surviving tori of perturbations thereof. It consists of simultaneous numerical integration of equations (2.1) for any set of initial conditions and n sets (or $n-1$ since one of them can be taken as the Hamiltonian vector field itself) of arbitrary initial conditions for the associated variational equations (2.2). At each point of the integrated solution grid, we evaluate the determinant of the matrix constructed with n components of the n solutions of the variational equations according to the desired Lagrangian projection. We then find the set of points where the ratio of this determinant by t^n is zero within the required accuracy.

This procedure allows determining the caustic surface of an invariant torus using a single trajectory together with its linearization. We point out that: i) if determinant$/t^n \to 0$ as $t \to \infty$, the hypothesis of the above theorem are no longer satisfied and the system is locally degenerated; ii) if determinant$/t^n \to \infty$ as $t \to \infty$ can be used as a criterion for a chaotic trajectory.

3 Evolution of Caustics from Regular to Chaotic Tori

We have applied the above procedure to the Hénon-Heiles Hamiltonian [1]

$$H(x, y, p_x, p_y) = \frac{p_x^2 + p_y^2}{2} + \frac{(x^2 + y^2)}{2} + \eta x^2 y - \frac{y^3}{3} \tag{3.1}$$

using η as a parameter to examine the evolution of the caustics for the fixed values $x = 0$, $y = .39$, $p_y = 0$, $E = 1/12$, and η is varied in the range $0 \leq \eta \leq 1.35$, i.e., from the completely separated case up to the case where most of the trajectories are chaotic. The evolution of the caustics ((x, y) projections) of these trajectories are shown in Figures 1 and 2. In Figure 3 we show the $x = 0$ surface of section ((y, p_y)-plane) of a set of trajectories for which $.35 \leq y \leq .44$, and the remaining initial conditions are the same as Figures 1 and 2. To complete the view of the tori through its caustics and their bifurcations, Figures 4 and 5 show some (p_x, p_y) projections for the same initial conditions as Figures 1 and 2. We have run η from .001, a completely degenerated projection, to 1.35 where the caustics no longer exist but some "scars" remain. When the chaotic behavior is widespread the method foretells that by its failure.

In Figures 1 and 2, for $\eta = 0$, the folds do not auto-intersect, showing cuspides. In this case the system is integrable as well as separated. The folds meet at hyperbolic umbilics at the corners. This singularity is not generic for two degrees of freedom. In fact, as it can be seen in the figure this singularity unfolds as the parameter η increases [6]. The graph for $\eta = .55$ shows the cuspidal edges which start to appear, and this singularity is generic for two degrees of freedom. At $\eta = .557098$ the cusps meet at the origin, and at $\eta = .5572$ the caustics separate in two different curves. These curves are projections "in and out" of the projection plane, i.e., depending on whether the determinant goes to zero from positive or negative values. Note that, in Figure 3 at $\eta = .557$ and $\eta = .557125$, the sections of these tori seem to cross from circulation to libration.

In the other extreme of the parameter range, Figure 2 at $\eta = 1.35$, the set of points "into and out of" the projection plane gets mixed up again, no longer going out at the external curve and into the projection plane at the auto-intersecting folds meeting at cuspid singularities, as in $\eta = .5527$ to $\eta = 1.33$. In fact, the onset of this kind of behavior is already visible at $\eta = 1.335$, when cuspides seems to bifurcate quickly with respect to η. Note that in the surface of section, Figure 3 at $\eta = 1.335$, there is a set of minute islands of a very large period. The method in this case has not yet been fully investigated. We only know that the caustics are about to disappear. The classical Hénon-Heiles Hamiltonian, $\eta = 1.0$, is shown in Figure 2.

A significative topological change in the caustics occurs already at $\eta = .557098$ for the (x, y) projection (Figure 1), where the degeneracy of the torus projection is broken. This change is preceded by a series of bifurcations which can also be seen in the projection onto momentum space (Figures 4 and 5).

The surfaces of section (Figure 3) confirm the rich features of the tori shown in Figures 1, 2, 4 and 5. However, most of these tori cannot be guessed simply by looking at the sections.

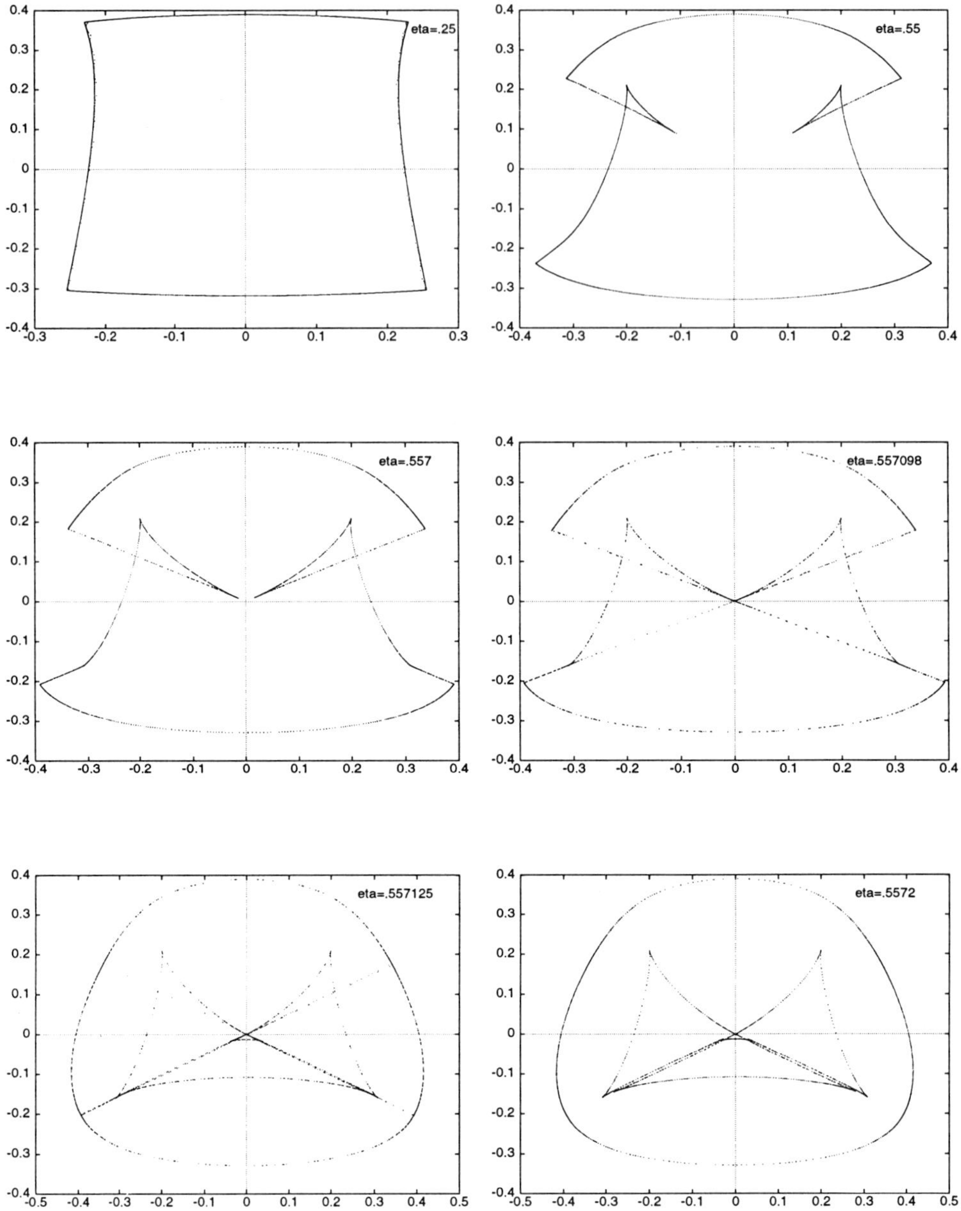

Figure 1 (x, y)-projections of the trajectory $x = 0$, $y = .39$, $p_y = 0$ for $E = 1/12$. η runs from .25, where we are still near the separated tori, to $\eta = .5572$. Figure 2 on the next page contiunes this for $\eta = .558$ to $\eta = 1.35$, where most of the trajectories are chaotic. 4000 caustic points are plotted in each graph. For $\eta \geq .5572$, the two sets of singular points (in and out of the (x, y)-plane) break up into two completely separated caustics. The case $\eta = 1.0$ is the classical Hénon-Heiles Hamiltonian.

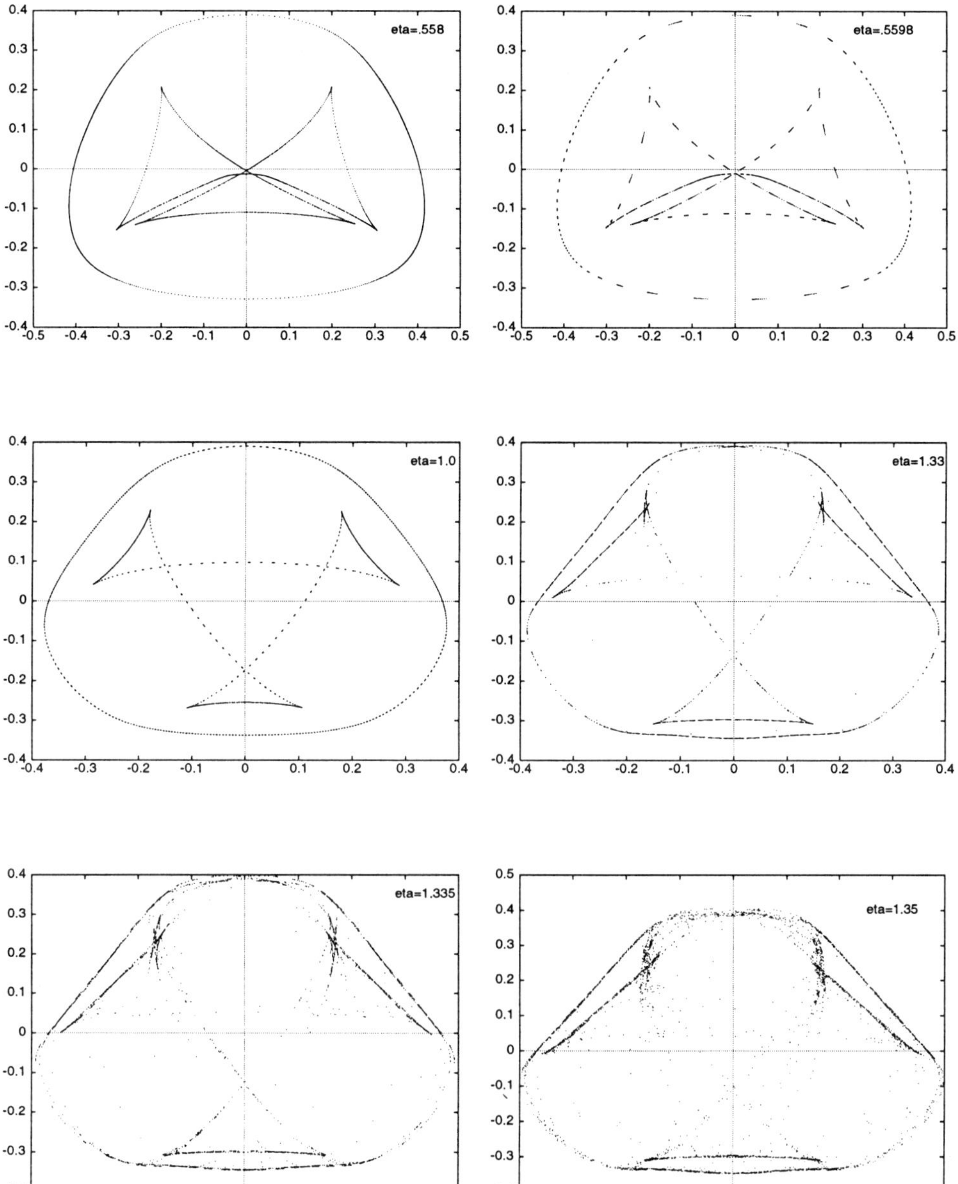

Figure 2 Continuation of Figure 1.

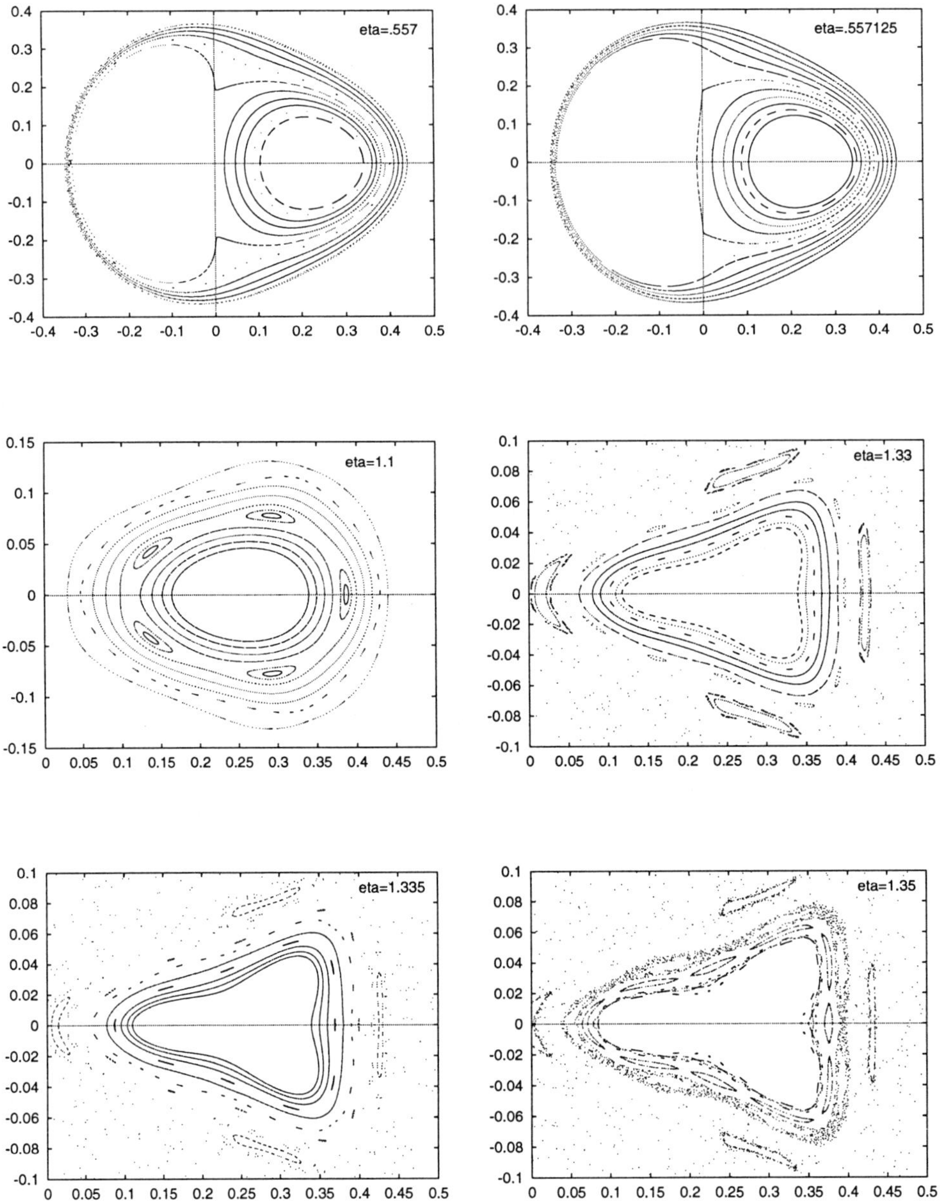

Figure 3 The $x = 0$ Poincaré section ((y, p_y)-plane) for a set of trajectories having y varying in the range $.34 \leq y \leq .44$, and the remaining initial conditions are as in Figure 1, to exhibit the neighborhood of the initial condition we are following. The tori shown in Figures 1 and 3 have initial conditions $y = .39$ and $p_y = 0$ in this section.

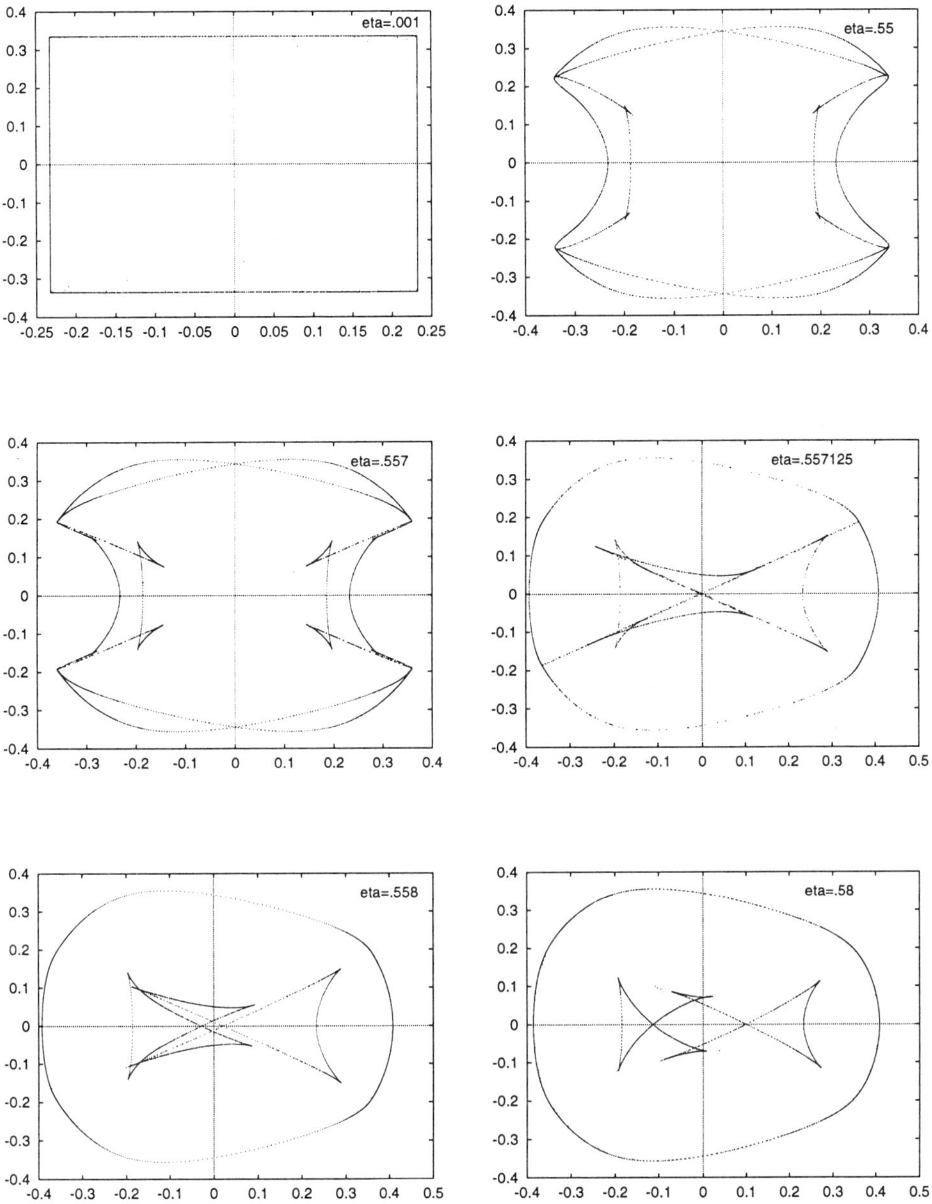

Figure 4 Evolution of the projections in the (p_x, p_y)-plane for some values of η and the same initial conditions of Figure 1. Continued in Figure 5. There is no degeneracy of the projections for most values of η. The two "in and out" curves just join at the major bifurcation points ($\eta = 0.0$ and around $\eta = .557125$) and of course at the onset of chaos. Note that the cuspid bifurcations start to appear again in the folds for $\eta = 1.2$

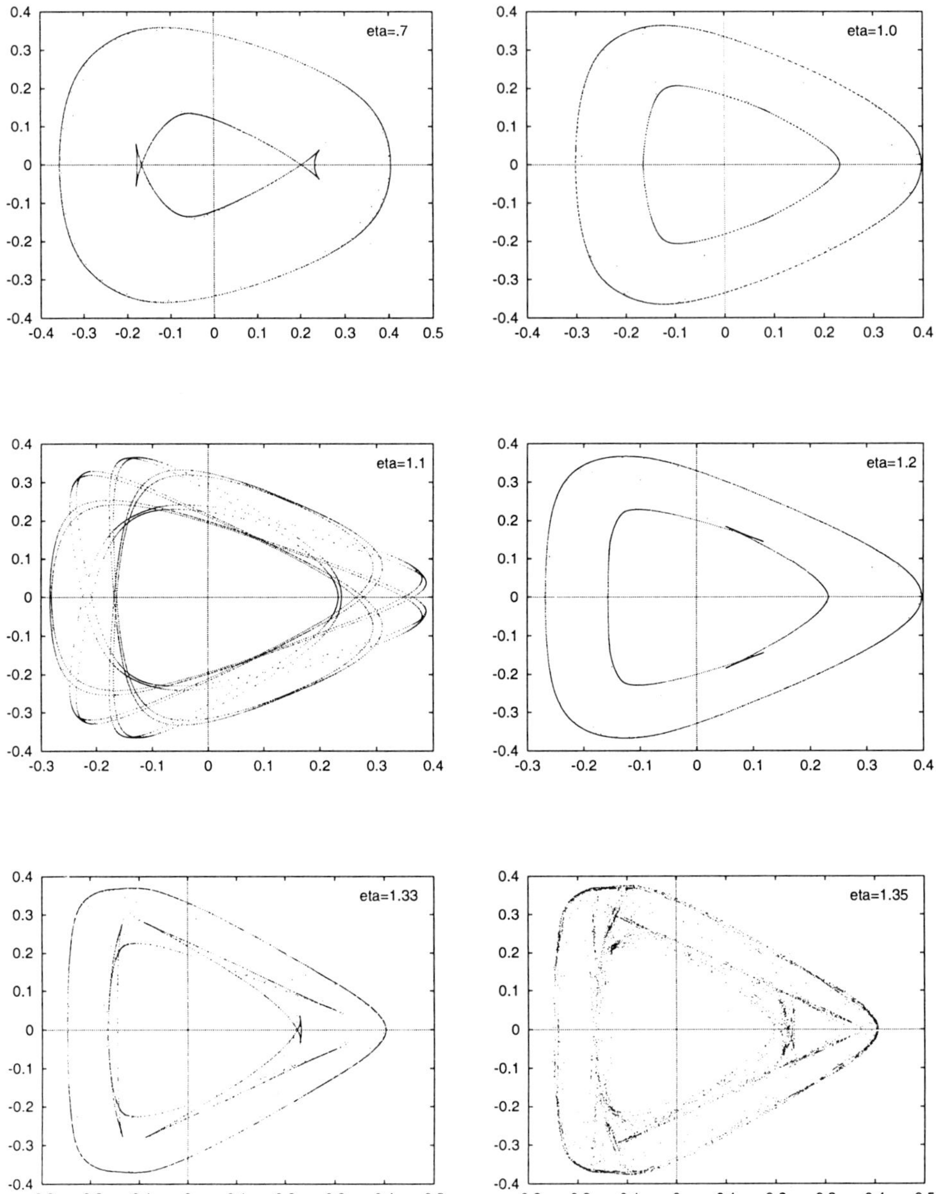

Figure 5 Continuation of Figure 4.

The degeneracy of the (p_x, p_y) projections breaks up for lower values of η. In fact, the caustics corresponding go "in and out" of the projection plane are mixed up only around $\eta = .557125$ (Figure 4) and at the onset of chaotic behavior. For this reason, the comparison of these two projections may be used as an aid to the visualization of the tori caustics. For example, in cases where the degeneracy makes it hard to decide whether the trajectory is regular or chaotic.

We point out that we are able to capture a thin long torus, Figure 5 at $\eta = .1.$, which appears as a set of islands in Figure 3. In Figure 5, at $\eta = 1.2$, the inner caustics show the A_2 singularities developing again and growing up to the onset of chaos, where they seem bifurcate quickly, as can already be seen in Figures 1 and 2 at the same parameter range. We are considering in our present research the possibility of a cascade of cusps anticipating chaotic behavior, through the passage by a periodic orbit of large period (Agekyan et al. [3]).

4 Final Comments

The method has been successfully applied to systems with two degrees of freedom, although its full potentiality has not yet been explored. We believe that it can be developed into a useful way of visualizing systems with more than two degrees of freedom.

The algorithm has been extended in order to catch the A_2 type singularities as we go along the trajectory. Its numerical implementation is in progress in order to examine systems with three degrees of freedom [10]. It is our belief that even higher order typical singularities for higher dimensions can be determined using the same approach we have been using so far.

Although, we can capture the A_2 singularities now, sections or projections of the caustics might be useful aids. For example the swallow-tail, which appears when $n = 3$, can be seen through sections of the projection which can be controlled by a parameter, in much the same fashion as described in this report. Moreover, we hope that the details which can be obtained with this method may be important not only for studies at the semi-classical limit of quantum mechanics, but also as a general indicator of integrability/chaos.

We thank Dr. Jair Koiller for many helpful discussions. T. Stuchi was partially supported by a grant from CNPq (Conselho Nacional de Pesquisas). We thank IBM and Dr. J. Marsden for our participation in this Workshop.

References

Hénon, M. and Heiles, C. [1964], *The Applicability of the third integral of motion: some numerical experiments*, Astron. J., **69**, 73–79.

Arnold, V. I. [1978], *Mathematical Methods of Classical Mechanics.* Springer, New York.

Agekyan, T. A., Myullayari, A. A., and Orlov, V. V. [1992], *Chaotic Motion and Structure of the Direction Field in the Hénon-Heiles Model*, Sov. Astron., **36(3)**, 236–240.

Froeschelé, C. [1970], *Numerical Study of Dynamical Systems with Three Degrees of Freedom*, Astron. Astrophys., **4**, 115–128.

Martinet, L. and Magnenat, P. [1981], *Invariant Surfaces and Orbital Behavior in Dynamical Systems with 3 Degrees of Freedom*, Astron. Astrophys., **96**, 68–77.

Ozorio de Almeida, A. M. and Hannay, J. H. [1982], *Geometry of Two Dimensional Tori in Phase Space: Projections, Sections and the Wigner Function*, Annals of Physics, **138**, 115-154.

Ramaswany, R. and Marcus R. A. [1981], *On the Onset of Chaotic Motion in Deterministic Systems*, J. Chem. Phys., **74(2)**, 1385.

Stuchi, T. J. [1991], Doctor Thesis, Instituto Tecnológico de Aeronáutica, São José dos Campos, Brazil.

Stuchi, T. J. and Vieira Martins, R. [1994]. *Caustics of Hamiltonian Systems: an Alternative to the Surface of Section Method*, Phys. Lett. A, **201**, 179–185.

Stuchi, T.J. and Vieira Martins, R. [1995], *Evolution of Caustics for systems with more than 2 degrees of freedom*, in preparation.

Fields Institute Communications
Volume **10**, 1996

Symplectic Correctors

J. Wisdom and M. Holman
Earth Atmospheric and Planetary Sciences
Massachusetts Institute of Technology
Cambridge MA USA 02139

J. Touma
Applied Mathematics
Massachusetts Institute of Technology
Cambridge MA USA 02139

Abstract. Symplectic integration algorithms typically yield trajectories that exhibit spurious oscillation in energy and state variables. In the delta function formulation of symplectic integration these oscillations have a clear origin, and canonical transformations can be made to remove them. The accuracy of symplectic integrators is substantially improved when combined with these symplectic correctors. The methods developed here are generally applicable to the integration of perturbed dynamical systems, but illustrated here by applications to the planetary n-body problem.

1 Introduction

And so I turn to the abyss
Of necromancy, try if art
Can voice or power of spirits start,
To do me service and reveal
The things of Nature's secret seal,
And save me from the weary dance,
Of holding forth in ignorance.
Then shall I see, with vision clear,
How secret elements cohere,
And what the universe engirds.
– Goethe's *Faust*

In the delta function formulation of symplectic integration (Chirikov [1979], Wisdom [1982] [1983], Wisdom [1988], Wisdom and Holman [1991] [1992]), the

1991 *Mathematics Subject Classification.* Primary: 65L05; Secondary: 58F05, 70-08, 70F15.

mapping approximation is derived by adding high frequency terms to the Hamiltonian in such a way that periodic Dirac delta functions are formed; the evolution of the resulting mapping Hamiltonian is a composition of the evolutions governed by the separately integrable pieces of the original Hamiltonian. The added high-frequency terms are argued to be unimportant to the long-term evolution by the averaging principle, provided stepsize resonances do not overlap and are avoided. The delta function formulation leads to the same algorithms as the Lie series composition formulation, but the motivation is different. In the delta function formalism the mapping is explicitly derived from a mapping Hamiltonian, which is motivated by the averaging principle. The Lie series composition formulation is motivated by the desire to have symplectic algorithms with a high order match of the Taylor series of the solutions. Recent reviews of the symplectic integration literature are given by Sanz-Serna [1992] and McLachlan [1995a].

Analysis of the mapping Hamiltonian with techniques from non-linear dynamics (most notably the resonance overlap criteria) allow stability criteria to be developed for the integration algorithms (Wisdom and Holman [1992]). It is not at all apparent how a similar discussion of the non-linear stability of the integration algorithms could be carried out in the pure Lie series composition formulation. Stepsize resonances are clearly exhibited by the algorithms, but they have no clear origin within the Lie series formulation. On the other hand, stepsize resonances have a clear origin in the delta function formalism, and furthermore the delta function formalism gives a quantitatively correct description of the stepsize resonances (Wisdom and Holman [1992]).

Trajectories computed with symplectic integration algorithms typically display spurious oscillations in energy and state variables. These oscillations cloud the issue of determining the accuracy of the computed trajectories. The averaging principle and the oscillatory nature of the errors suggests that the mapping method is more accurate than a naive view of the apparent errors suggests. The Lie series formulation gives no clear insight into the origin of the oscillations nor provides any means of removing them except through going to higher order and/or smaller stepsize. The mapping Hamiltonian in the delta function formalism gives a clear understanding of the origin of the spurious oscillations in energy and state variables. In this paper, we show that the evolution of the mapping and the real evolution are more closely related than is suggested by the agreement of the Taylor series. We show that it is a mistake to identify the mapping variables and the actual state variables. In fact, we show how to relate the mapping variables to the real variables and consequently remove the spurious oscillations, and in many cases dramatically improve the accuracy of the integrations without going to higher order or smaller stepsize. A preliminary sketch of some of these ideas was presented in Wisdom [1988], and they have already been used to some extent in Tittemore and Wisdom [1988] [1989] [1990]. The results presented here are deeper and the methods are generally applicable.

2 Perturbation Theory of Symplectic Maps

Consider a general Hamiltonian of the form

$$H = H_0(J_1, J_2) + \epsilon H_1(J_1, J_2, \theta_1, \theta_2), \tag{2.1}$$

where, for convenience, we have chosen canonical coordinates (J_i, θ_i) for which the unperturbed problem governed by H_0 is cyclic in the coordinates. We shall focus on problems with moderately small ϵ. Two degrees of freedom are enough to exhibit the problems and our solutions to those problems. In the end our formulation is not only independent of the number of degrees of freedom, but coordinate independent as well.

A simple mapping Hamiltonian is generated from H by adding high-frequency terms so that the Hamiltonian becomes

$$H_{\text{Map}} = H_0(J_1, J_2) + \epsilon H_1(J_1, J_2, \theta_1, \theta_2) 2\pi\delta_{2\pi}(\Omega t), \tag{2.2}$$

where

$$\delta_{2\pi}(t) = \sum_{n=-\infty}^{\infty} \delta(t - 2\pi n), \tag{2.3}$$

is a periodic sequence of Dirac delta functions with period 2π (Wisdom [1982]). Central to understanding the averaging motivation for the mapping method is the Fourier representation

$$\delta_{2\pi}(t) = \frac{1}{2\pi} \sum_{n=-\infty}^{\infty} \cos(nt). \tag{2.4}$$

Across each delta function the evolution is governed solely by H_1, and between the delta functions the evolution is governed solely by H_0. The mapping provides an approximation to the evolution of H by a composition of the evolutions governed by the pieces. If the component Hamiltonians are integrable and efficiently solvable, then the mapping provides an efficiently computable approximation to the evolution generated by the full Hamiltonian. As a conventional integration algorithm, the map presented above is only accurate to first order if the state of the system is recorded at integer multiples of the mapping period $t = n\Delta t = n2\pi/\Omega$. However if the state of the system is recorded halfway between the delta functions, the mapping is accurate to second order in the mapping stepsize. Construction of higher order mappings through the delta function formulation is discussed by Wisdom and Holman [1991].

To prepare for perturbation theory, we first express the perturbation as a multiply periodic Poisson series

$$H = H_0(J_1, J_2) + \epsilon \sum_{ij} a_{ij}(J_1, J_2) \cos(i\theta_1 + j\theta_2) \tag{2.5}$$

where, for convenience, we assume only cosine terms appear in the expansion of H_1. Using the Fourier representation of the periodic sequence of delta functions, the Poisson series for the mapping can be written

$$H = H_0(J_1, J_2) + \epsilon \sum_{ijn} a_{ij}(J_1, J_2) \cos(i\theta_1 + j\theta_2 + n\Omega t). \tag{2.6}$$

In this form it is apparent that the mapping Hamiltonian differs from the true Hamiltonian by the terms with $n \neq 0$. Now, we use ordinary perturbation theory to formally eliminate these terms and thus provide the connection between the mapping phase space variables and the real phase space variables.

We use Lie series in the extended phase space to facilitate the manipulations of canonical perturbation theory. Define the Lie derivative L_W by

$$L_W f = \{W, f\} = \sum_i \left\{ \frac{\partial W}{\partial q_i} \frac{\partial f}{\partial p_i} - \frac{\partial W}{\partial p_i} \frac{\partial f}{\partial q_i} \right\} \tag{2.7}$$

where, as indicated, the curly brackets denote the usual Poisson bracket. The operator $\mathrm{e}^{\epsilon L_W}$ generates canonical transformations. This is easy to see by replacing ϵ by t and W by $-H$, then the exponential generates the Taylor series in time of any dynamical variable, and time evolution is canonical. A function of the phase space coordinates evolves only as a result of the evolution of the coordinates; thus the Lie transform of a function is the function of the Lie transformed coordinates. Alternately stated, the inverse Lie transform of a function evaluated at the transformed coordinates gives the original function of the original coordinates. Thus if a canonical transformation is carried out by a Lie transform, the Hamiltonian governing the evolution of the transformed coordinates is the inverse Lie transform of the original Hamiltonian. Keep in mind that the manipulations are formal and the resulting series may not converge. To use Lie series with time dependent Hamiltonians it is convenient to go to the extended phase space, we promote the time t to canonical coordinate status, with associated canonical momentum T, and introduce a new time τ. The Hamiltonian in the extended phase space is obtained by adding T to H: $H' = H + T$. The equation of motion for t is $dt/d\tau = \partial H'/\partial T = 1$, confirming that the original dynamics is preserved. For an introduction to Lie series see Steinberg [1985].

We ask: What generator W can be used to eliminate the $n \neq 0$ terms from the mapping Hamiltonian? Applying $\mathrm{e}^{-\epsilon L_W}$ to H', we find, to first order in ϵ, that W must satisfy the equation

$$\{W, H_0 + T\} - \tilde{H}_1 = 0 \tag{2.8}$$

or

$$\omega_1 \frac{\partial W}{\partial \theta_1} + \omega_2 \frac{\partial W}{\partial \theta_2} + \frac{\partial W}{\partial t} - \sum_{ij,n\neq 0} a_{ij} \cos(i\theta_1 + j\theta_2 + n\Omega t) = 0, \tag{2.9}$$

where we have introduced the notation

$$\omega_i(J_1, J_2) = \frac{\partial H_0(J_1, J_2)}{\partial J_i}, \tag{2.10}$$

and represented the terms that must be added to H_1 to make the delta functions by $\tilde{H}_1 = H_1[2\pi\delta_{2\pi}(\Omega t) - 1]$. The solution of this equation is

$$W(J_1, J_2, \theta_1, \theta_2) = \sum_{ij,n\neq 0} \frac{a_{ij}(J_1, J_2) \sin(i\theta_1 + j\theta_2 + n\Omega t)}{i\omega_1 + j\omega_2 + n\Omega}. \tag{2.11}$$

Notice what has been accomplished. Formally, this transformation eliminates all terms that differ between the real Hamiltonian and the mapping Hamiltonian up to order ϵ^2. Contrast this with the Taylor series expansion for the generalized leap frog which contains error terms of order ϵ, the largest of which is generated by $\{H_0, \{H_0, H_1\}\}$. Even without further consideration of the perturbation theory, this is consistent with our suggestion that most of the error in the generalized leap frog is of an oscillatory nature and can be removed by a canonical transformation. However, we need a more practical representation of the generating function.

Note that n only appears in the sine function and in the denominator. This fact allows the sums over n to be carried out. By expanding the sine of the sum, we find

$$W=\sum_{ij,} a_{ij}\left[\sin(i\theta_1+j\theta_2)F(i\omega_1+j\omega_2,\Omega,t)+\cos(i\theta_1+j\theta_2)G(i\omega_1+j\omega_2,\Omega,t)\right] \quad (2.12)$$

where we have introduced the functions

$$F(\omega,\Omega,t)=\sum_{n\neq 0}\frac{\cos(n\Omega t)}{\omega+n\Omega}=\frac{2\omega}{\Omega^2}\sum_{n>0}\frac{\cos(n\Omega t)}{\left(\frac{\omega}{\Omega}\right)^2-n^2} \quad (2.13)$$

and

$$G(\omega,\Omega,t)=\sum_{n\neq 0}\frac{\sin(n\Omega t)}{\omega+n\Omega}=-\frac{2}{\Omega}\sum_{n>0}\frac{n\sin(n\Omega t)}{\left(\frac{\omega}{\Omega}\right)^2-n^2}. \quad (2.14)$$

These sums converge provided $\omega+n\Omega\neq 0$, for any n. The functions $F(\omega,\Omega,t)$ and $G(\omega,\Omega,t)$ are both periodic in time with period $2\pi/\Omega$, the mapping stepsize. Explicitly:

$$F(\omega,\Omega,t)=\frac{\pi}{\Omega}\frac{\cos(\omega t-\frac{\pi\omega}{\Omega})}{\sin\frac{\pi\omega}{\Omega}}-\frac{1}{\omega}, \quad (2.15)$$

for $0\leq\Omega t\leq 2\pi$, and

$$G(\omega,\Omega,t)=-\frac{\pi}{\Omega}\frac{\sin(\omega t-\frac{\pi\omega}{\Omega})}{\sin\frac{\pi\omega}{\Omega}}, \quad (2.16)$$

for $0<\Omega t<2\pi$, with $G=0$ for Ωt equal to integral multiples of 2π. The periodic extension of F is continuous but has discontinuous derivative for $\Omega t=2\pi n$. The periodic extension of G is discontinuous at $\Omega t=2\pi n$, and the sum converges to the midpoint of the discontinuity: $G(\omega,\Omega,2\pi n/\Omega)=0$. We have the limiting values $G(\omega,\Omega,0)=0$, $G(\omega,\Omega,0^+)=\pi/\Omega$, $G(\omega,\Omega,0^-)=-\pi/\Omega$, where 0^+ and 0^- indicate limits from above and below, respectively.

We note some further properties of F and G. First, both F and G have zero time average, as is evident from the original Fourier representation. We also note that

$$G(\omega,\Omega,t)=\frac{1}{\omega}\frac{d}{dt}F(\omega,\Omega,t). \quad (2.17)$$

This fact can be used to deduce the summed form of G from the form for F without explicitly performing the sum. The maximum value of F occurs at $\Omega t=\pi$:

$$F(\omega,\Omega,\pi/\Omega)=\frac{\pi}{\Omega}\frac{1}{\sin\frac{\pi\omega}{\Omega}}-\frac{1}{\omega}, \quad (2.18)$$

which, for small ω/Ω, is approximately

$$F(\omega,\Omega,\pi/\Omega)\approx\frac{\pi^2\omega}{6\Omega^2}. \quad (2.19)$$

The minimum value of F occurs at $\omega t = 0$; for small ω/Ω it has the approximate value:

$$F(\omega, \Omega, \pi/\Omega) \approx -\frac{\pi^2\omega}{2\Omega^2}. \tag{2.20}$$

Note that the magnitude of the function F is larger at $t = 0$ than at $t = \pi/\Omega$. The extreme values of G are given by the limiting values around $t = 0$. Recall that the stepsize of the mapping is $\Delta t = 2\pi/\Omega$, and note that the extreme values of F are proportional to Δt^2, whereas the extreme values of G are proportional to Δt. Both F and G have two zeros. $G(\omega, \Omega, 0) = G(\omega, \Omega, \pi/\Omega) = 0$. The zeros of F satisfy a transcendental equation.

The first lesson we haved learned by performing the sum over n is that, for small step size, the correction from map coordinates to dynamical coordinates is minimized at the midpoint of the interval between the delta functions, when the larger correction proportional to G is zero by virtue of the vanishing of G. Recording the state of the system midway between the delta functions minimizes the error of the mapping if the distinction between mapping variables and real state variables is ignored. Thus, from yet another point of view (see Wisdom and Holman [1991], for two others) we have discovered the generalized leap frog. The real lesson, though, is that the mapping variables and the actual variables are not the same, and that in fact they differ in an explicit calculable way. The leap frog is only the best solution if the distinction between these variables is ignored, and then the error is of order ϵ. Maintaining and accounting for the distinction between mapping variables and actual variables, the error of the mapping method is of order ϵ^2, and this is true at *any* time, not just at the time midway between the delta functions.

3 Illustration

A natural test application is to the pendulum. Consider the Hamiltonian

$$H = \frac{1}{2}J^2 + \epsilon \cos\theta. \tag{3.1}$$

We make a simple mapping approximation by introducing the delta functions:

$$H_{\text{Map}} = \frac{1}{2}J^2 + 2\pi\delta_{2\pi}(\Omega t)\epsilon\cos\theta. \tag{3.2}$$

The equations of motion are just Hamilton's equations. Crossing a delta function changes the momentum:

$$J' = J + \Delta t \epsilon \sin\theta, \tag{3.3}$$

where $\Delta t = 2\pi/\Omega$, the mapping stepsize, and between delta functions the angle rotates uniformly

$$\theta' = \theta + \Delta t J'. \tag{3.4}$$

At intermediate times between the delta functions the angle is rotated appropriately to that time. Of course, with a simple change of notation this is just the standard map (Chirikov [1979]), but without the usually assumed periodicity in the momentum variable. The usual standard map parameter is given by $K = \Delta t^2 \epsilon$.

In this case, since the perturbation has a trivial Fourier expansion, the generator for the corrector can be given in closed form:

$$W = [\sin(\theta) F(\omega, \Omega, t) + \cos(\theta) G(\omega, \Omega, t)], \tag{3.5}$$

with frequency $\omega(J) = \partial H_0 / \partial J = J$. The transformation from the mapping variables to the real variables is

$$J = \mathrm{e}^{\epsilon L_W} J_M \approx J_M + \epsilon \left. \frac{\partial W}{\partial \theta} \right|_M \tag{3.6}$$

and

$$\theta = \mathrm{e}^{\epsilon L_W} \theta_M \approx \theta_M - \epsilon \left. \frac{\partial W}{\partial J} \right|_M . \tag{3.7}$$

The subscript M indicates mapping variables. Using this Euler approximation to the evolution generated by W gives a corrector that at order ϵ is correct to all orders in Δt. The Euler formula is not symplectic, but this can hardly matter since the corrector is applied only to the output points and the corrected output points are not used further in the integration.

The maximum of the energy error for several methods is displayed in Figure 1. We compare the uncorrected first order map, the generalized leap frog, the corrected leap frog, and the fourth order method of Candy and Rozmus [1990], Yoshida [1990], and Forest and Ruth [1990]. Since the problem is artificial nothing guides the choice of parameters, so the parameters have been chosen to get the attention of the reader. Whether the idea of using correctors is useful will depend on application to problems of interest with appropriate parameters. On the principle that we should compare error of different methods for similar work, the stepsize used for the fourth order method is three times the stepsize for the leap frog, since three force evaluations are required for each fourth order step. Note the especially poor performance of the fourth order method. For this choice of parameters, the corrected map does substantially better than the other methods. Note also that the output time of the corrected map is not limited to the midpoint between the delta functions. That is, the corrected map is as accurate at any output phase.

4 Development of F and G

In general the perturbation will not have a simple Poisson series, so the practical use of the transformation to and from mapping coordinates will rest on the development of more practical expressions for the generating function. To this end, expand F and G as power series in the ratio ω/Ω (which we presume here to be less than 1). We find

$$F(\omega, \Omega, t) = \frac{2\omega}{\Omega^2} \sum_{n>0} \frac{\cos(n\Omega t)}{\left(\frac{\omega}{\Omega}\right)^2 - n^2} = -\frac{2\omega}{\Omega^2} \sum_{i=0}^{\infty} C_{2i+2}(\Omega t) \left(\frac{\omega}{\Omega}\right)^{2i} \tag{4.1}$$

and

$$G(\omega, \Omega, t) = -\frac{2}{\Omega} \sum_{n>0} \frac{n \sin(n\Omega t)}{\left(\frac{\omega}{\Omega}\right)^2 - n^2} = \frac{2}{\Omega} \sum_{i=0}^{\infty} S_{2i+1}(\Omega t) \left(\frac{\omega}{\Omega}\right)^{2i}, \tag{4.2}$$

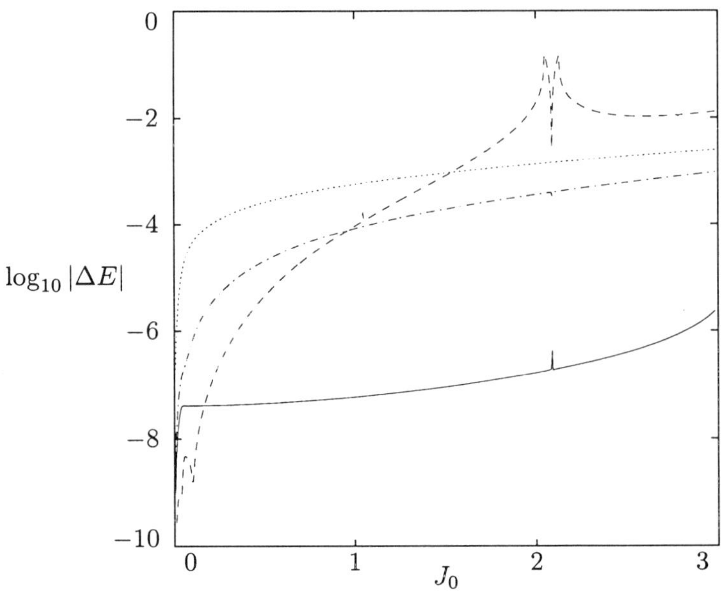

Figure 1 The maximum energy error for various methods is plotted as a function of the initial momentum. The dotted line is the uncorrected map. The dot-dashed line is the generalized leap-frog. The dashed line is the fourth order composition method. The solid line is corrected map. The performance of the corrected map is substantially better than that of the other methods. The parameters are: $\epsilon = 0.001$ with $\Delta t = 1$. The initial angle in all cases is $\theta = \pi$. The error plotted is the maximum over 10000 iterations.

where

$$C_i(\tau) = \sum_{n>0} \frac{\cos(n\tau)}{n^i} \tag{4.3}$$

and

$$S_i(\tau) = \sum_{n>0} \frac{\sin(n\tau)}{n^i}. \tag{4.4}$$

Both C_i and S_i are 2π periodic in τ, and they have zero average. The functions C_i and S_i have the important properties: $S_i(\tau) = -dC_{i+1}/d\tau$ and $C_i(\tau) = dS_{i+1}/d\tau$. Note that in F and G the index of C is always even and the index of S is always odd. We introduce the function E_i such that $E_{2i} = (-1)^i 2C_{2i}$ and $E_{2i+1} = (-1)^i 2S_{2i+1}$, then the E_i are a set of 2π periodic functions with zero average possessing the interesting property that $E_i(\tau) = dE_{i+1}/d\tau$. These properties determine all of the E_i given E_1. The sum for E_1 is recognized as the Fourier series for a line: $E_1(\tau) = \pi - \tau$ for $0 < \tau < 2\pi$, and $E_1(0) = 0$. Thus the E_n are in fact all polynomials. It turns out that the E_n can be written in terms of the Bernoulli polynomials

$$E_n(\tau) = -\frac{(2\pi)^n}{n!} B_n(\tau/2\pi). \tag{4.5}$$

The generating relation for the Bernoulli polynomials is

$$\frac{e^{xt}}{e^t - 1} = \sum_{n=0} B_n(x) \frac{t^{n-1}}{n!} \tag{4.6}$$

The first few are:

$$\begin{aligned} B_1(x) &= x - \tfrac{1}{2} \\ B_2(x) &= x^2 - x + \tfrac{1}{6} \\ B_3(x) &= x^3 - \tfrac{3}{2}x^2 + \tfrac{1}{2}x \\ B_4(x) &= x^4 - 2x^3 + x^2 - \tfrac{1}{30}. \end{aligned} \tag{4.7}$$

The representation of E_i in terms of Bernoulli polynomials is only valid in the interval $0 < t < \Delta t$, and is otherwise periodically extended. For consistency, we must add

$$E_0(\tau) = 2\pi\delta_{2\pi}(\tau) - 1. \tag{4.8}$$

We return now to the development of the generator W. Substituting for F and G their expansions in terms of the C and S polynomials, we find

$$\begin{aligned} W = \sum_{ij,k\geq 0} a_{ij} \Big\{ &\sin(i\theta_1 + j\theta_2) \left[-\frac{2\omega}{\Omega^2} C_{2k+2}(\Omega t) \left(\frac{\omega}{\Omega}\right)^{2k} \right] \\ &+ \cos(i\theta_1 + j\theta_2) \left[\frac{2}{\Omega} S_{2k+1}(\Omega t) \left(\frac{\omega}{\Omega}\right)^{2k} \right] \Big\}. \end{aligned} \tag{4.9}$$

Notice that each term can be written as a multiple Poisson bracket of H_0 with terms of H_1. Performing the sum over i and j first, the terms can be regrouped into H_1. We have then

$$W = \sum_{k>0} \frac{E_k(\Omega t)}{\Omega^k} L_{H_0}^{k-1} H_1 = -\Delta t \sum_{k>0} \frac{1}{k!} B_k \left(\frac{t}{\Delta t}\right) \Delta t^{k-1} L_{H_0}^{k-1} H_1 \tag{4.10}$$

This is a practical relation, as desired. The generator is written as a Taylor series in the stepsize the terms of which involve successively higher order Poisson brackets of H_0 with H_1 with rapidly decreasing coefficients. Notice that the expression is independent of the number of degrees of freedom, and coordinate free as well. The derivation is easily extended to n-degrees of freedom, with the same result.

5 Lie Series Correspondence

Lie series simplify the formulation of higher order composition methods (Forest and Ruth [1990], Yoshida [1990] and Wisdom and Holman [1991]). In this section we present the Lie series correspondence of the correctors introduced in the last section.

Let $A = \Delta t L_{H_0}$ and $B = \Delta t L_{H_1}$. Taylor series for the evolution of the actual system for one timestep are generated by the operator e^{A+B}. The generalized leap frog approximation to the evolution for one timestep is generated by $e^{A/2}e^{B}e^{A/2}$. The error is third order in the timestep; the generalized leap frog is, in this sense, a second order integrator. More formally,

$$e^{A/2}e^{B}e^{A/2} - e^{A+B} = -\frac{1}{24}[A,[A,B]] + \frac{1}{12}[B,[B,A]] + o(\Delta t^4). \tag{5.1}$$

A gazillion such higher order integrators are known (see, for example, Koseleff [1993]). Note that even though the error is of order Δt^3 it is first order in ϵ. In fact, there are error terms which are first order in ϵ at all orders in Δt.

In the previous sections we found, however, that the mapping method, including the generalized leap frog as a special case, is actually better than the error in the Taylor series suggests. That is, the evolution generated by the mapping has calculable and removable high frequency oscillations. It is a mistake to identify the mapping variables and the actual variables. More properly, we must not only derive the mapping approximation to the evolution, but we must also derive the relationship of the mapping variables to the actual variables.

In the Lie series formulation, the operator that corresponds to the mapping evolution to a time after a delta function of $t = \alpha\Delta t$, where $0 \leq \alpha \leq 1$ is

$$\mathrm{e}^{(1-\alpha)A}\mathrm{e}^{B}\mathrm{e}^{\alpha A}. \tag{5.2}$$

Of course, this approximates the true evolution only to first order in the stepsize, the local truncation error is proportional to $\epsilon\Delta t^2$, unless the output point is midway between the delta functions ($\alpha = 1/2$) and then the local truncation error is of order $\epsilon\Delta t^3$. We now realize though that the mapping variables do not correspond to the true variables and before evolving with the mapping we must transform to mapping variables and after the mapping step we must transform back to the real variables. The generator W of the Lie transformation that transforms to mapping variables has been derived in previous sections. Let $C(\alpha) = L_W(\alpha)$ be the Lie derivative corresponding to the generator W, then

$$\mathrm{e}^{-C(\alpha)}\mathrm{e}^{(1-\alpha)A}\mathrm{e}^{B}\mathrm{e}^{\alpha A}\mathrm{e}^{C(\alpha)}, \tag{5.3}$$

generates an evolution which is correct at order ϵ to all orders in Δt. The fact that the corrector depends on the phase of the output α illustrates the high-frequency nature of the corrections. If only a single output phase is considered, as, for example, in the leap frog, the high-frequency oscillatory character of the error is not apparent. It is "strobed" away. We introduce a notation for the commutator with respect to the operator A: $\mathcal{L}_A B = [A, B] = AB - BA$. To carry out the corresponding canonical transformation we need the operator L_W. The Lie derivative with respect to the Poisson bracket of two generating functions is the commutator of two Lie derivatives with respect to the individual generating functions. Thus the expression for L_W can be immediately written down in terms of A and B:

$$C(\alpha) = L_W(\alpha) = -\sum_{k>0} \frac{B_k(\alpha)}{k!}\mathcal{L}_A^{k-1}B \tag{5.4}$$

For the special case of $\alpha = 1/2$ this is

$$C = \frac{1}{24}\mathcal{L}_A B - \frac{7}{5760}\mathcal{L}_A^3 B + \cdots \frac{2^{2n-1}-1}{(2n)!2^{2n-1}} B_{2n}\mathcal{L}_A^{2n-1}B, \tag{5.5}$$

where B_n are the Bernoulli numbers. This corrector eliminates the error terms that are first order in ϵ to all orders in Δt. Use of the corrector does not affect substantially the efficiency of the mapping if output is uncommon, because successive steps can be combined: the effect of the e^{-C} is reversed by the following e^{C}, and the two factors of $\mathrm{e}^{\alpha A}$ and $\mathrm{e}^{(1-\alpha)A}$ may be combined into a single factor of e^{A}. Only when output is desired do we actually have to carry out the e^{C} step. McLachlan (personal communication [1994]) has pointed out that Butcher [1969] developed a similar idea to extend the order of explicit Runge-Kutta methods.

Though, no doubt, the complete expression for C could be derived entirely with the commutator algebra of A and B, the delta function formalism has provided the motivation for the approach as well as given us the complete expression to all orders in Δt by simple analysis. Independent of α the third order error term is $-(1/24)\mathcal{L}_B^2 A$, which is second order in ϵ. At this point, this is the real error of the method, the other errors were only apparent errors because the mapping coordinates were being mistaken for the actual coordinates.

6 Lie Series Implementation

The generator for the transformation from mapping coordinates to real coordinates has been written in terms of successive Poisson brackets of H_0 and H_1. Since this transformation need be done only at the output points, the efficiency of the implementation of the corrector is unimportant. The most direct implementation is to numerically integrate Hamilton's equations corresponding to the generator. A more convenient approach is to represent the corrector as a composition of the readily computable evolutions governed by H_0 and H_1. In this section, we present a representation of the corrector as such a composition. We optimize efficiency of derivation rather than attempting to find the most efficient composition. That can be done later.

We note a key formula for manipulating Lie series involving correctors, the effect of a corrector on a kernel:

$$e^C e^K e^{-C} = e^{K+\mathcal{L}_C K+\frac{1}{2}\mathcal{L}_C^2 K+\cdots} = e^{e^{\mathcal{L}_C} K}. \tag{6.1}$$

This formula is used repeatedly in the following development. With the knowledge that there is only one independent n^{th} order bracket involving a single factor of K, it is easy to prove this relation by matching the coefficients of the $C^n K$ terms in the expansion (see, for example, Belinfante and Kolman [1989]). The corrector formula provides a convenient base form from which to generate Lie series products because of the simplicity with which higher order terms are determined.

We here assume output of the mapping has been taken midway between the delta functions. Thus we shall derive a product corrector for the leap frog. For this case, we recall that the corrector has only brackets involving an even number of factors of H_0 and H_1. A convenient form that has this property is the product of two corrector forms. Let

$$e^{X(a_1,b_1)} = e^{a_1 A} e^{b_1 B} e^{-a_1 A}. \tag{6.2}$$

All terms in X are proportional to B. Now let

$$e^Z = e^{X(a_1,b_1)} e^{X(-a_1,-b_1)}. \tag{6.3}$$

Constructed in this way, Z has only terms with an even number of factors of A and B, at order ϵ. Composition of such forms maintains this property.

We write the corrector as a product of factors of Z:

$$e^C = e^{Z(a_1,b_1)} e^{Z(a_2,b_2)} \cdots e^{Z(a_n,b_n)}. \tag{6.4}$$

Let c_m be the coefficients of $\mathcal{L}_A^m B$ in the corrector. The constraint equations are easily written down using the corrector formula:

$$c_m = 2 \sum_{i=1}^{n} \frac{a_i^m b_i}{m!}. \tag{6.5}$$

As a set of linear equations for the b_i, the coefficients form a Vandermonde matrix, which has a simple explicit inverse. Given a set of a_i all of the $\mathcal{L}_A^m B$ constraint equations can be satisfied by choosing the b_i to be solutions of these linear equations. We do not want the corrector to introduce error terms of order ϵ^2, because this same corrector will be used subsequently with higher order methods. If we select $a_i = -a_{n+1-i}$ for $0 < i \leq n$, then $b_i = -b_{n+1-i}$, then the coefficient of the $\mathcal{L}_B^2 A$ term is automagically zero as required. The coefficient of the $\mathcal{L}_B \mathcal{L}_A^2 B$ term is also zero. This form then satisfies all of the constraints for arbitrary a_i. Calculation of the inverse Vandermonde matrix is carried out with rational arithmetic to avoid any possible numerical difficulties. A sample solution is:

$$a_1 = 1\alpha \quad ; \quad b_1 = \frac{45815578591785473}{24519298961757600}\beta$$

$$a_2 = 2\alpha \quad ; \quad b_2 = \frac{-104807478104929387}{80063017017984000}\beta$$

$$a_3 = 3\alpha \quad ; \quad b_3 = \frac{422297952838709}{648658702692000}\beta$$

$$a_4 = 4\alpha \quad ; \quad b_4 = \frac{-27170077124018711}{112088223825177600}\beta$$

$$a_5 = 5\alpha \quad ; \quad b_5 = \frac{102433989269}{1539673404192}\beta$$

$$a_6 = 6\alpha \quad ; \quad b_6 = \frac{-33737961615779}{2641809989145600}\beta$$

$$a_7 = 7\alpha \quad ; \quad b_7 = \frac{26880679644439}{17513784972684000}\beta$$

$$a_8 = 8\alpha \quad ; \quad b_8 = \frac{682938344463443}{7846175667762432000}\beta$$

with $a_i = -a_{17-i}$, and $b_i = -b_{17-i}$, for $8 < i \leq 16$, and $\alpha = \sqrt{7/40}$ and $\beta = 1/(48\alpha)$. The scaling of a_i and b_i by α and β was made to reduce the first two non-zero c_i (c_1 and c_3) to unit magnitude. Introduction of these scalings was convenient, but not necessary. This corrector has an error term of order $\mathcal{L}_A^{17} B$. Its performance for the pendulum for the parameters used above is nearly indistinguishable from that of the closed form expression for the corrector on the logarithmic scale. Correctors of lower order do not perform as well. We should keep in mind that different order correctors may be suitable for different problems. This implementation of the corrector is suitable for testing our ideas; optimal implementation of this and the other correctors derived below can be addressed in subsequent investigations.

7 Application to the N-Body Problem

The mapping method has been developed for the n-body problem by Wisdom and Holman [1991]. They tested the method with billion year integrations of the system of the outer planets (Jupiter to Pluto). Subsequently, Wisdom and Sussman [1992] used the mapping method to carry out 100 million year integrations of the whole solar system. These integrations confirmed the chaotic character of the evolution of the solar system (Laskar [1989]). Integrations of the whole solar system are extremely time-consuming and, practically speaking, would not have been possible without the speed of the mapping method. Unfortunately, the integrations were plagued by the high-frequency components. We now have a general prescription for removing them.

First consider the energy error. Most of the energy of the system is contributed by the massive planets, yet the small inner planets have the largest coordinate errors because for them the stepsize is a relatively larger fraction of the orbital period than for the outer planets. So the energy error is not a fair measure of the accuracy of an integration of the whole solar system. Nevertheless, it is interesting to examine it. Figure 2 shows the relative energy error (energy minus initial energy divided by initial energy) in the 100 million year integration of the whole solar system (Wisdom and Sussman [1992]). Early in the integration the energy error is of order a few parts in a billion, and there is no apparent increase in the magnitude of the error for the duration of the integration. Output was taken about every 20,000 years, and successive points are connected by lines for easy visibility. This presentation gives the false impression that the error varies on timescales of tens of thousands of years. The data are intact, and the corrector can be applied to the output even years after the integration was carried out. Figure 3 shows the energy error after the corrector has been applied to the output. Here the energy of the corrected output is compared to the energy of the corrected initial condition. Note that the scale has to be expanded by a factor of a hundred to see the variations in the energy after correction. The high frequency oscillation, which we have argued all along to be unimportant (Wisdom [1982] [1983], Wisdom [1988], Wisdom and Holman [1991] [1992], Sussman and Wisdom [1992]), has almost completely been removed. The remaining energy error shows a secular linear drift, and is probably real integration error. The slope of the relative energy drift is of order 2×10^{-11} per 100 million years, or about 4×10^{-21} per integration step. For comparison, the relative energy error in the Digital Orrery integration of the outer planets using a specially chosen "magic" stepsize was about 2×10^{-19} per integration step, and that was about a thousand times smaller than the error achieved in all previous long term integrations of the solar system (Sussman and Wisdom [1989]). The mapping achieves better energy conservation without the use of magic stepsizes. Both of these integrations use pseudo quadruple precision (compensated summation) in a small select subset of the calculation, but most operations were carried out in ordinary IEEE double precision with about 16 digits. On the average, in the mapping integration of the whole solar system it took about 40,000 integration steps to accumulate one bit of relative energy error! Of course, it is not completely fair to compare the errors achieved in the integration of different systems. The only other long-term direct integration of the whole solar system is the 3 million year integration of Quinn, Tremaine, and Duncan [1991] who used a high order symmetric integrator (Quinlan and Tremaine [1990]). These integrations also used select pseudo quadruple precision. The symmetric integrator is also subject to large energy oscillations, which may or may not be removable. On a 3 million year timescale, they estimate the energy error in their integration to already be 6×10^{-11}, which is larger than the energy error achieved by the mapping after 100 million years.

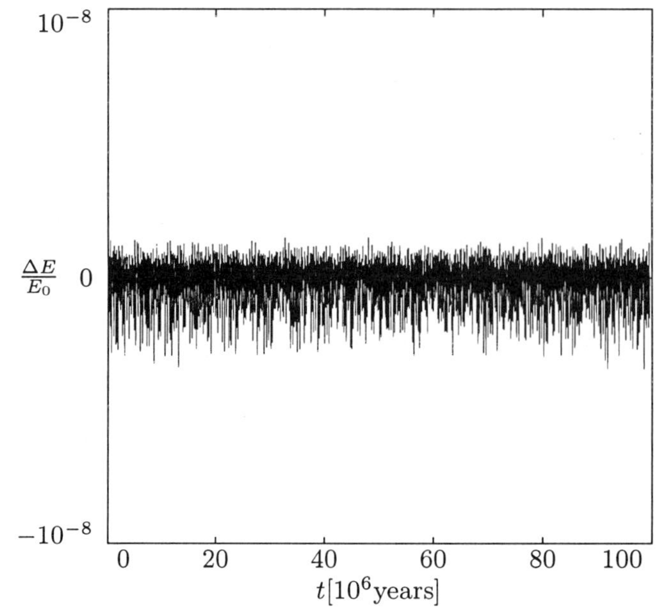

Figure 2 The energy error over 100 million years for the whole solar system. The energy error is large from the beginning, a behavior typical of symplectic integrators.

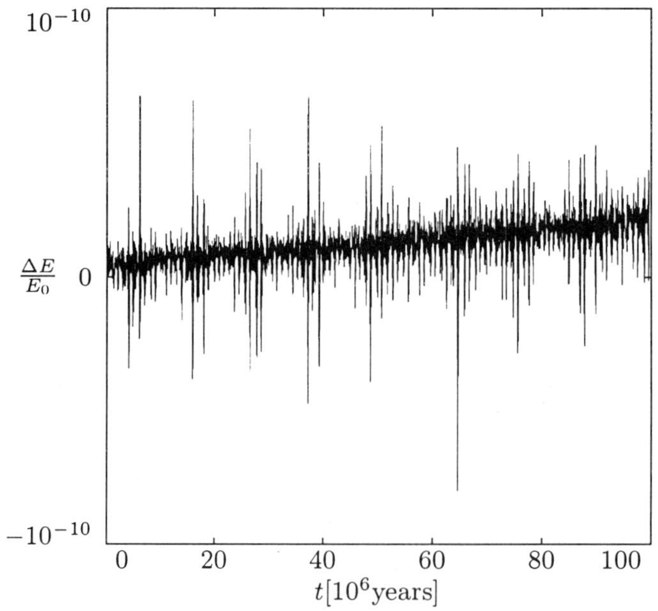

Figure 3 The energy error over 100 million years for the whole solar system after the corrector has been applied. The magnitude of the error is reduced by two orders of magnitude, and now a secular drift of the energy can be seen.

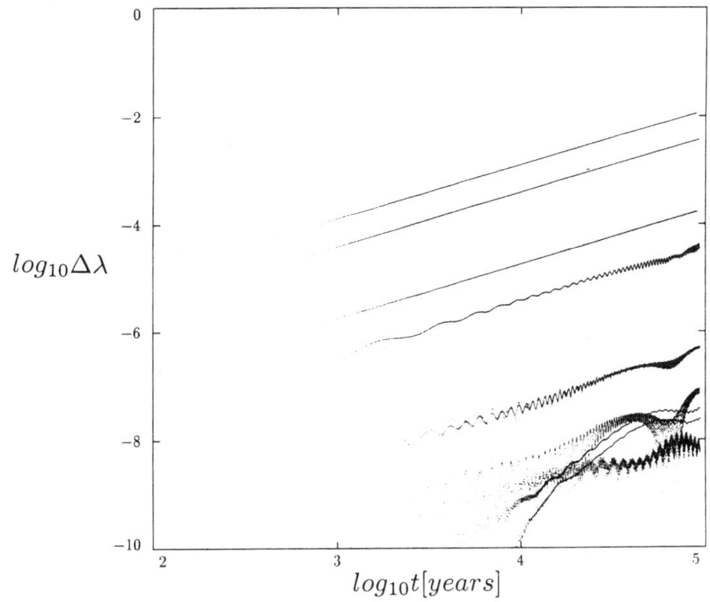

Figure 4 The longitude errors of the planets over 100,000 years using corrected initial conditions. The largest errors are in the inner planets, as expected. At the largest time displayed, the errors, from largest to smallest, are for the planets: Mercury, Earth, Venus, Mars, Saturn, Jupiter, Neptune, Pluto, Uranus.

Unfortunately, we now understand that the transformation from real coordinates to mapping coordinates should be carried out before the integration is begun. The adjustment of the initial conditions is small and does not substantially affect the conclusions of that study, since satisfactory agreement with other integrations was already observed, but the integration could have been better. The error in the initial conditions is most clearly manifest in the growth of the error of the longitudes of the planets, since small errors in the orbital frequencies produce secularly increasing longitude errors. Keep in mind that errors in longitude are not believed to be important to the long-term evolution of the solar system; indeed, Laskar removes the longitudes entirely through averaging. To examine the growth of longitude errors we compared a reference run with integrations with and without the corrector. In these runs we used a stepsize of 7.2 days, the same as the stepsize used in the 100 million year integrations, and no extended precision calculations were used. The reference run used the corrector, partial quadruple precision, and a stepsize reduced by a factor of 10 (from 7.2 days to 0.72 days). We found that the longitude errors grew linearly, both in the run with the corrector and in the run without the corrector. The errors in the run without the corrector are primarily due to using uncorrected initial conditions. The longitude errors in the corrected run are dominated by truncation error. The longitude errors in the corrected run are displayed in Figure 4. As expected, the largest longitude error is in the motion of Mercury. The longitude errors of the other planets are much smaller. The growth of the longitude error is linear on this timescale. Since the energy is experiencing a very small linear growth, we can expect that the growth of the longitude error is actually quadratic, but apparently the coefficient is too small to substantially affect the behavior. The effective relative error in the mean motion of Mercury using uncorrected initial conditions is 7×10^{-8}. With corrected initial conditions the effective mean motion error is 5×10^{-9}. For comparison, in the integration of Quinn, Tremaine, and Duncan, the longitude error of Mercury also grows linearly, with an effective mean motion error of about 4×10^{-9}. Keep in mind that the mapping integration took much larger steps: 7.2 day steps for the mapping, versus 0.75 day steps for the symmetric method. If the mapping step were reduced by this factor of 10 we could expect a factor of at least 100 improvement in the errors, assuming roundoff errors do not limit the improvement. The mapping calculations used for illustration here were carried out without extended precision, but the symmetric method used partial extended precision. Honestly, though this application of the corrector was very successful, we were surprised that the corrector did not reduce the longitude error of Mercury more than it did. Order arguments suggest there should have been much better improvement. We return to this issue in later sections.

We are speaking here of integration error. It should be noted that Sussman and Wisdom [1992] used a model for the general relativistic effects that represented the relativistic precession correctly, but made a small error in the mean motions. Though the model could have been improved (see the appendix in Saha and Tremaine [1992]), it was not considered important because the detailed longitudes are believed to be unimportant, and at the time it appeared that the mapping did not get the longitudes right anyway. Since the correctors have fixed the longitude errors, the modelling error is now larger than the integration error, though still probably unimportant physically.

The mapping method was derived as an efficient approximation to the evolution of the planets that might be only qualitatively correct because of the presence of the high-frequency terms. It turns out that the mapping is not only efficient, but, when used in conjunction with appropriate correctors, the mapping method surpasses the accuracy of competing integration schemes.

8 Extension to Higher Order

In this section we consider the problem of extending the method to higher order in ϵ.

Deriving mappings accurate to higher order in ϵ is facilitated by reinterpreting the transformation from mapping variables to real variables derived above. We shall call this the mapping transformation. Rather than asserting the Hamiltonian for the mapping and then deriving the correction that to some extent converts the mapping Hamiltonian back into the original Hamiltonian, we can use the inverse of the mapping transformation to go directly from the true Hamiltonian to the mapping Hamiltonian plus some error terms. We would like to keep track of the error terms and extend this process to higher order in ϵ.

We apply the transformation to mapping coordinates but now keep track of second order terms in ϵ. At this stage the mapping Hamiltonian is

$$\begin{aligned} H_{\mathrm{Map}} &= \mathrm{e}^{\epsilon L_W}(H_0 + T + \epsilon H_1) \\ &= H_0 + T + \epsilon\{W, H_0 + T\} + \tfrac{1}{2}\epsilon^2\{W, \{W, H_0 + T\}\} \\ &\quad + \epsilon H_1 + \epsilon^2\{W, H_1\} + o(\epsilon^3) \end{aligned}$$

The generating function W has been designed to insert high frequency terms so that perturbations become multiplied by delta functions to give local integrability. Let $\tilde{H}_1$ again denote the terms that must be added to H_1 to make $2\pi\delta_{2\pi}(\Omega t)H_1$. By construction we have $\{W, H_0 + T\} = \tilde{H}_1$ and

$$H_{\text{Map}} = H_0 + T + 2\pi\delta_{2\pi}(\Omega t)\epsilon H_1 + \epsilon^2\{W, H_1 + \frac{1}{2}\tilde{H}_1\}. \tag{8.1}$$

We shall denote the terms proportional to ϵ^2 by H_2; these terms need to be brought into a more useful form. To facilitate this, introduce the notation F_n and G_n for the Fourier coefficients of F and G:

$$F(\omega, \Omega, t) = \sum_{n>0} F_n(\omega, \Omega)\cos(n\Omega t) \tag{8.2}$$

$$G(\omega, \Omega, t) = \sum_{n>0} G_n(\omega, \Omega)\sin(n\Omega t) \tag{8.3}$$

We have:

$$\begin{aligned} H_2 &= \Big\{\textstyle\sum_{ij,n>0} a_{ij}[\sin(i\theta_1 + j\theta_2)F_n\cos(n\Omega t) \\ &+ \cos(i\theta_1 + j\theta_2)G_n\sin(n\Omega t)], H_1\left(1 + \frac{1}{2}\sum_{m\neq 0}\cos(m\Omega t)\right)\Big\}. \end{aligned}$$

Using the bilinearity of the Poisson bracket and the fact that T does not occur in this expression, we find

$$H_2 = \frac{1}{2}\{W, H_1\} + \frac{1}{2}2\pi\delta_{2\pi}(\Omega t)\{W(t=0), H_1\} \tag{8.4}$$

That is, the ϵ^2 terms are naturally written as a sum of oscillatory terms plus another set of terms multiplied by periodic delta functions.

The mapping Hamiltonian at this point is

$$\begin{aligned} H_{\text{Map}} = H_0 + T &+ 2\pi\delta_{2\pi}(\Omega t)\left[\epsilon H_1 + \tfrac{1}{2}\epsilon^2\{W(t=0), H_1\}\right] \\ &+ \tfrac{1}{2}\epsilon^2\{W, H_1\} + o(\epsilon^3) \end{aligned}$$

The oscillatory term can be pushed to higher order with a second corrector transformation. Let W_2 be the Lie series generator for this canonical transformation. Applying $e^{\epsilon^2 L_{W_2}}$ to H_{Map}, and requiring that the ϵ^2 oscillatory terms are killed, we find the determining condition for W_2:

$$\{W_2, H_0 + T\} + \frac{1}{2}\{W, H_1\} = 0. \tag{8.5}$$

This equation has the solution

$$W_2 = \frac{1}{2}\sum_{k>0}\frac{E_{k+2}(\Omega t)}{\Omega^{k+2}}\sum_{p=1}^{k} L_{H_0}^{k-p}L_{H_1}L_{H_0}^{p}H_1, \tag{8.6}$$

as may be verified by substitution. This corrector removes all oscillatory terms of order ϵ^2. Specializing to the case $\Omega t = \pi$ or $t = \Delta t/2$ the leading term in the second corrector is

$$W_2 = -\frac{7}{5760}\Delta t^3 L_{H_1}L_{H_0}^2 H_1, \tag{8.7}$$

where we have used the fact that $L_{H_0}L_{H_1}L_{H_0}H_1 = L_{H_1}L_{H_0}^2 H_1$.

After using the second corrector the mapping Hamiltonian is

$$H_{\mathrm{Map}} = H_0 + T + 2\pi\delta_{2\pi}(\Omega t)\left[\epsilon H_1 + \frac{1}{2}\epsilon^2\{W(t=0), H_1\}\right] + o(\epsilon^3) \tag{8.8}$$

In general the leading term in the energy error is of order ϵ^3 and is proportional to the fourth order bracket $L_{H_1}^2 L_{H_0} H_1$. If we specialize to $\Omega t = \pi$, then the leading error terms are still of order ϵ^3 but contain the two independent fifth order brackets $L_{H_0} L_{H_1}^2 L_{H_0} H_1$, and $L_{H_1} L_{H_0} L_{H_1} L_{H_0} H_1 = L_{H_1}^2 L_{H_0}^2 H_1$ that are of order ϵ^3.

McLachlan [1995b] introduces a useful notation to describe the order of integration methods for perturbed problems which distinguishes the order of the error at each order of ϵ. The order is described by the list of the powers of the highest matching term in the Taylor series, for successive powers of ϵ. If the order is described as $(n_1, n_2, \ldots)$ the terms $\epsilon^i \Delta t^{n_i}$ are correctly represented. Termination of the list indicates that all following terms are the same order as the last term. The methods developed in the previous section are, in this notation, $(\infty, 2)$, when the explicit closed form generator can be used. The particular Lie series composition formula we have presented for the corrector gives a composition integrator of order $(16, 2)$, but the method described can be used to achieve any desired order to first order in ϵ. When it is straightforward to extend the method to any desired order we shall continue to speak of the the order as ∞. In McLachlan's notation, the order of the mapping Hamiltonian developed in this section is generally $(\infty, \infty, 3)$, but with $\alpha = 1/2$ the order of the mapping Hamiltonian is extended to $(\infty, \infty, 4)$.

9 Lie Series Correspondence

In this section we translate the results of the last section into the Lie series language introduced earlier. For simplicity we consider only the case of $\alpha = t/\Delta t = 1/2$.

We have deduced

$$\mathrm{e}^{A+B} = \mathrm{e}^{-C}\mathrm{e}^{-C_2}\mathrm{e}^{K}\mathrm{e}^{C_2}\mathrm{e}^{C}, \tag{9.1}$$

where the kernel of the mapping is

$$\mathrm{e}^{K} = \mathrm{e}^{A/2}\mathrm{e}^{B'}\mathrm{e}^{A/2}, \tag{9.2}$$

where,

$$B' = B + \frac{1}{24}\mathcal{L}_B\mathcal{L}_A B - \frac{1}{1440}\mathcal{L}_B\mathcal{L}_A^3 B + \cdots \tag{9.3}$$

the corrector C is the same as before,

$$C = L_W = \frac{1}{24}\mathcal{L}_A B - \frac{7}{5760}\mathcal{L}_A^3 B + \cdots \tag{9.4}$$

and

$$C_2 = -\frac{7}{5760}\mathcal{L}_B\mathcal{L}_A^2 B + \cdots. \tag{9.5}$$

Recombining terms we can check that the integrator has the declared properties, and confirm that the error in the Hamiltonian is of order $\epsilon^3\Delta t^5$. We find that the leading error terms in representing $L_H = A + B$ are

$$\frac{1}{360}\mathcal{L}_B^2\mathcal{L}_A^2 B + \frac{11}{5760}\mathcal{L}_A\mathcal{L}_B^3 A + \frac{1}{720}\mathcal{L}_B^4 A. \tag{9.6}$$

In general it will not be possible to evaluate the kernel directly, and we are naturally led to develop approximations by composing Lie series. Whereas the efficiency of the corrector is not critical at all, it is particularly important to find expansions of the kernel with as few steps as possible, so that the mapping will be efficient. Here, however, we are interested in testing the ideas and leave the optimal expansion of the kernels for later investigation.

One key constraint on the expansion of the kernel is that B' must not have any terms of the form $\mathcal{L}_A^m B$ for any $m > 0$. One way to guarantee this (but perhaps not the only way) is to use the corrector formula again

$$\mathrm{e}^{B'} = \mathrm{e}^{Y} \mathrm{e}^{B} \mathrm{e}^{-Y}, \tag{9.7}$$

with all terms in Y containing at least one factor of B. The form of B' severely constrains Y. In fact, up to terms of order ϵ^3 we can write $Y = \beta B + L_{W(t=0)}/2$, where β is arbitrary and will not enter the final kernel. The task of representing the kernel is reduced to representing Y:

$$Y = \beta B + \sum_{k>0} \frac{B_k(0)}{k!} \mathcal{L}_A^k B. \tag{9.8}$$

Now, except for the term involving β, Y contains only odd powers of $\mathcal{L}_A$ multiplying B. One way to obtain this is to represent e^Y as a product of two exponentials with one exponential argument obtained from the other by reversing the sign of both A and B, thus preserving only the even terms in the product at order ϵ. The product will have ϵ^2 terms but these become ϵ^3 when wrapped around the central factor of e^B. One way for each of these factors to be represented as exponentials with all terms in the exponential proportional to B, is to again use the corrector formula. Using X as defined before

$$\mathrm{e}^{X(a_1,b_1)} = \mathrm{e}^{a_1 A} \mathrm{e}^{b_1 B} \mathrm{e}^{-a_1 A}. \tag{9.9}$$

All terms in X are proportional to B. In terms of X the simplest expression for Y satisfying the stated constraints is

$$\mathrm{e}^{Y} = \mathrm{e}^{X(a_1,b_1)} \mathrm{e}^{X(-a_1,-b_1)}. \tag{9.10}$$

Combining factors, we find that the coefficients a_1 and b_1 must satisfy $a_1 b_1 = -1/48$ to match the $k = 2$ term, and the $k = 1$ and $k = 3$ terms are zero as required, but the $k = 4$ term cannot simultaneously be matched. So with this Y the error in the kernel is of order $\epsilon^2 \Delta t^5$. A convenient choice is: $a_1 = 1/8$ and $b_1 = -1/6$. The full kernel is then:

$$\mathrm{e}^{K} = \mathrm{e}^{\frac{5}{8}A} \mathrm{e}^{-\frac{1}{6}B} \mathrm{e}^{-\frac{1}{4}A} \mathrm{e}^{\frac{1}{6}B} \mathrm{e}^{\frac{1}{8}A} \mathrm{e}^{B} \mathrm{e}^{-\frac{1}{8}A} \mathrm{e}^{-\frac{1}{6}B} \mathrm{e}^{\frac{1}{4}A} \mathrm{e}^{\frac{1}{6}B} \mathrm{e}^{\frac{3}{8}A} \tag{9.11}$$

We cannot say if the same order (including the absence of spurious $\mathcal{L}_A^m B$ terms) could have been obtained with fewer factors. We consider ours a "trial" solution, to test the ideas, and leave open the issue of whether there is a better product kernel at this order.

Consider next the problem of representing the correctors. We are interested in the problem of determining a corrector that is consistent in order with the kernel just derived. Thus, we require that the ϵ^2 terms up to fourth order in Δt be correctly represented, so that the error in the integrator remains of order $\epsilon^2 \Delta t^5$. The second corrector contributes terms of order $\epsilon^2 \Delta t^5$ to the integrator, so it may be ignored for the moment. The corrector we have already derived is thus of the required order.

Though the mapping Hamiltonian is correct to order $(\infty, \infty, 4)$, the practical implementation of this mapping in terms of the composition of Lie series that we have found is limited to order $(\infty, 4)$. It is a fourth order method with error proportional to ϵ^2. Implementation of the kernel to higher order as a composition is straightforward, but not pursued here. In this sense the method is still theoretically $(\infty, \infty, 4)$.

10 Special Cases

Some simplifications occur for Hamiltonians of the form $H_0 = T(p) + V_0(q)$ with quadratic T and $H_1 = V_1(q)$. The n-body problem has this form, where H_0 represents the Keplerian motion of each planet with respect to the Sun, and H_1 represents the planetary perturbations.

The Hamiltonian that governs the evolution across the delta functions is modified by a term proportional to $\{W(t = 0), H_1\}$. The leading term of this series is proportional to $\{H_1, \{H_1, H_0\}\}$. For Hamiltonians of the special form just described this bracket depends only on the coordinates and can thus be integrated along with H_1 which also depends only on the coordinates. Unfortunately, the remaining terms in the expansion of $\{W, H_1\}$ are not zero, and not obviously integrable. Failure to include these terms results in a fourth order integrator (truncation error proportional to $\epsilon^2 \Delta t^5$) with an energy error of order $\epsilon^2 \Delta t^4$. Nevertheless, taking the special form of the first term into account provides a very efficient, very accurate fourth order integrator. The truncation error is of order ϵ^2, and there is only a single modified force evaluation per step. By comparison, the standard fourth order Runge-Kutta method has four force evaluations per step, and the Forest-Ruth fourth order symplectic scheme has three force evaluations; both of those methods have local truncation error that is first order in ϵ.

Explicitly, let

$$H_0 = \sum_i \frac{p_i^2}{2m_i} + V_0(q_1, ..., q_n), \tag{10.1}$$

and

$$H_1 = V_1(q_1, ..., q_n). \tag{10.2}$$

The Hamiltonian governing the evolution across the delta functions has the two leading terms

$$H_\delta = H_1 + \Delta t^2 \frac{1}{24}\{H_1, \{H_1, H_0\}\} = V_1 + \Delta t^2 \frac{1}{24} \sum_i \frac{1}{m_i} \left(\frac{\partial V_1}{\partial q_i}\right)^2. \tag{10.3}$$

The modified "kick" is

$$\Delta p_j = -\Delta t \frac{\partial H_\delta}{\partial q_j} = -\Delta t \frac{\partial V_1}{\partial q_j} - \Delta t^3 \frac{1}{12} \sum_i \frac{1}{m_i} \frac{\partial V_1}{\partial q_i} \frac{\partial^2 V_1}{\partial q_i \partial q_j}. \tag{10.4}$$

For the n-body problem the extra terms are not expensive, since all the required square roots and inverses already have been computed to evaluate the original kick, so the modified kick involves no more than summing products of previously computed quantities.

Koseleff [1993] mentions that in the special case of $H = T(p) + V(q)$ there is a similar possibility of using a modified potential, but apparently, without using the correctors, order four can only be reached with two modified force evaluations. Here we reach fourth order with one modified force evaluation.

Wisdom and Holman [1991] raised the question of whether there were higher order methods that did not take backward steps. Their motivation was to maintain a Hamiltonian description of the algorithm. With negative steps, the representation as a delta function Hamiltonian is lost since the time order cannot be encoded in the scalar Hamiltonian. Yoshida [1993] subsequently answers the question negatively, by reference to a theorem by Suzuki [1991]. These theorems notwithstanding, we have here fulfilled both of the original goals we earlier set for ourselves. We have an accurate fourth order method that has a Hamiltonian representation, and furthermore involves only one force evaluation per step. It is interesting to note that the kernel involves no backward steps. In the Lie series implementation of the corrector, the backward steps required by the theorem are all hidden in the corrector. It is worth pointing out, though, that the corrector could also be implemented as a numerical integration of the generator, and then there would be no backward steps at all. Of course, in this case the theorems do not apply.

Next, we present an interesting alternate form of the modified kick. We note that the expression for the kick is the beginning of a multidimensional Taylor series expansion of the kick at a displaced point. Explicitly, denote the unmodified kick component functions by

$$k_j(x) = -\Delta t \left.\frac{\partial V_1}{\partial x_j}\right|_x \tag{10.5}$$

and let

$$x_j' = x_j + \frac{\Delta t}{12 m_j} k_j(x). \tag{10.6}$$

Then the modified kick is given simply by

$$p_j'' = p_j + k_j(x'). \tag{10.7}$$

The coordinates are unchanged in this step

$$x_j'' = x_j. \tag{10.8}$$

The intermediate values x_j' are discarded. Higher order terms are also generated, but they are of order ϵ^3, and so are ignorable at this stage. Strictly speaking, the error term lowers the order from fourth order to third order. Whether this matters depends of the magnitude of the $\epsilon^3 \Delta t^4$ error term compared to the $\epsilon^2 \Delta t^5$ error term. By combining three force evaluations we can keep the method fourth order. For the lazy implementer, this expression gives an approximation of the modified kick with very little extra programming effort. The cost is that two force evaluations are required this way. This version of the map is still symplectic since the change in p is only a function of x, and x is not modified. The two-force-evaluation scheme is of order $(\infty, 4, 3)$, the three-force-evaluation scheme is of order $(\infty, 4, 4)$. We do not have a direct generalization of this method to higher order.

These schemes look a lot like Runge-Kutta schemes in the sense that forces are evaluated at intermediate positions which are discarded in the final step. So in this sense we have an explicit symplectic generalization of the Runge-Kutta method that is third or fourth order in stepsize (depending on the form used), with error terms proportional to ϵ^2.

11 Application to the Restricted Three-Body Problem

In this section we illustrate the use of the higher order mapping derived in the previous three sections. This mapping is a fourth order integrator that has truncation error proportional to ϵ^2. We apply the mapping to the integration of the planar circular restricted three-body problem: an infinitesimal mass moving in the orbit plane of two massive bodies which themselves move in a circular relative orbit. In this illustration, the mass of one of the bodies is small: 0.001 times the total mass of the system. This is a crude model for the motion of asteroids perturbed by Jupiter. We split the Hamiltonian into two-body Kepler motion with respect to the "Sun" and perturbations from "Jupiter," as in the n-body problem (see Wisdom and Holman [1991]).

In an inertial frame, the circular restricted problem has two degrees of freedom, with explicit time dependence. In the frame rotating with the massive bodies the time dependence is removed, and the value of the Hamiltonian in the rotating frame is consequently conserved. The Hamiltonian in the rotating frame is proportional to what is traditionally called the "Jacobi" constant. It is natural to test integrators by the extent to which they preserve this integral.

To be specific, we choose to study an initially circular orbit with semimajor axis 0.63 times the distance between the massive bodies. The test particle is started on the line between "Jupiter" and the "Sun." This orbit is close to an orbital resonance, and consequently develops moderate eccentricity.

The relative error in the Jacobi constant is plotted against the number of steps per orbit in Figure 5. The integrations spanned 200 orbits. The curve at the top is the maximum relative error in the Jacobi constant using the uncorrected, unmodified second order mapping (M2 - for second-order map). The next curve down is the maximum relative error for the second order map used in conjunction with a corrector (CM2 - Corrected Map of $\underline{2}^{nd}$ order). The same corrector was used here as in the n-body example. The bottom curve is actually three curves superimposed. This is the relative error using the modified kick in conjunction with the corrector (CMM4 - Corrected Modified Map of $\underline{4}^{th}$ order). It is also the error obtained using the two-force-evaluation Runge-Kutta implementation of the modified kick as described in the last section. Finally, the bottom curve is also the error using the general purpose product kernel. The differences between the errors of these latter three methods are too small to be visible on this scale. Note that the Jacobi constant relative error of the second order map and the corrected second order map are indeed quadratic in step-size as expected. Also note that the error of the last three methods are all fourth order, even the two-kick method which could have have a third order component proportional to ϵ^3. Of course, among the higher order methods, the most efficient is the mapping with the explicitly modified kick, followed by the two-force-evaluation Runge-Kutta approximation, followed by the general purpose five kick Lie series product kernel. Note how ineffective it would be to try to eliminate error by reducing the stepsize alone. Whereas, using the corrector with some version of the modified kick is very effective at reducing the error.

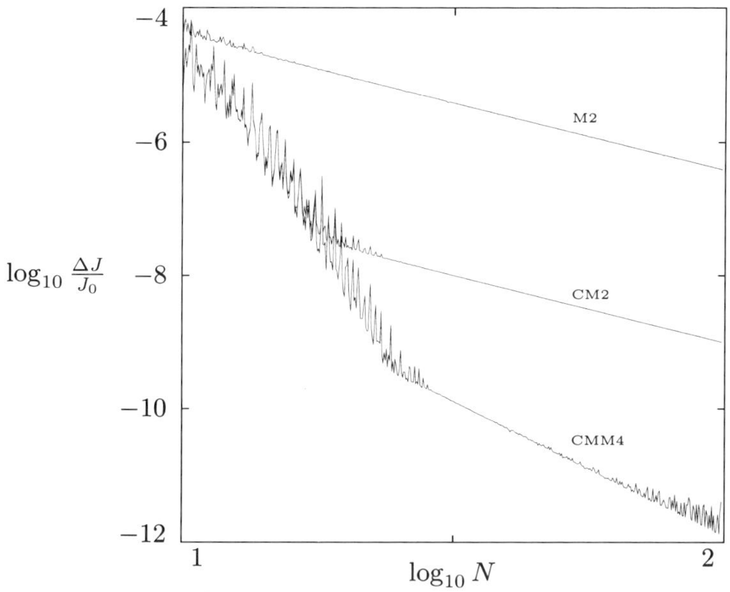

Figure 5 The relative error in the Jacobi constant versus number of steps per orbit, for several methods. The top curve is the error for the second order mapping (M2). The middle curve is the error for the second order mapping used in conjunction with a corrector (CM2). The bottom curve is the error for the fourth order corrected modified mapping (CMM4), the two-force-evaluation Runge-Kutta approximation to the modified kick with corrector, and the product kernel.

The most interesting result is not that the error behaved as expected, but rather how the observed error differs from that expected. The error of all the methods seem to be dominated by some "unexpected" error at large step sizes. This error could have several origins, and we are not sure which is the culprit. It could be a failure of the corrector to have high enough order. This does not seem to fit. The corrector we are using is correct to Δt^{17}, but the observed falloff is more like Δt^{13}. It could result from the fact that our derivation of the general purpose corrector formulas used here presumed that the mapping frequency is large compared to the frequencies present in the motion so that we could expand in the ratio of these frequencies. Thus we may be seeing what happens when that assumed condition is not satisfied. Another possibility is that we are seeing error introduced by stepsize resonances (Wisdom and Holman [1992]), which are most apparent at large stepsizes. We suspect that this is indeed a stepsize resonance problem.

We note that we have seen a similar high order falloff in a corresponding study of the energy error in the solar system as a function of stepsize, though the tests are much more limited. At the adopted stepsize of 7.2 days, the corrector seems to have less effect on the solar system integration error than we a priori expected (the corrector only improved the mean motion of Mercury by a factor of 10, and the energy by a factor of 100), but with a stepsize of 3.6 days the error dramatically drops and thereafter at smaller stepsizes the error falls off as a fourth order integrator should. The same phenomenon is seen here, but is more easily demonstrated since

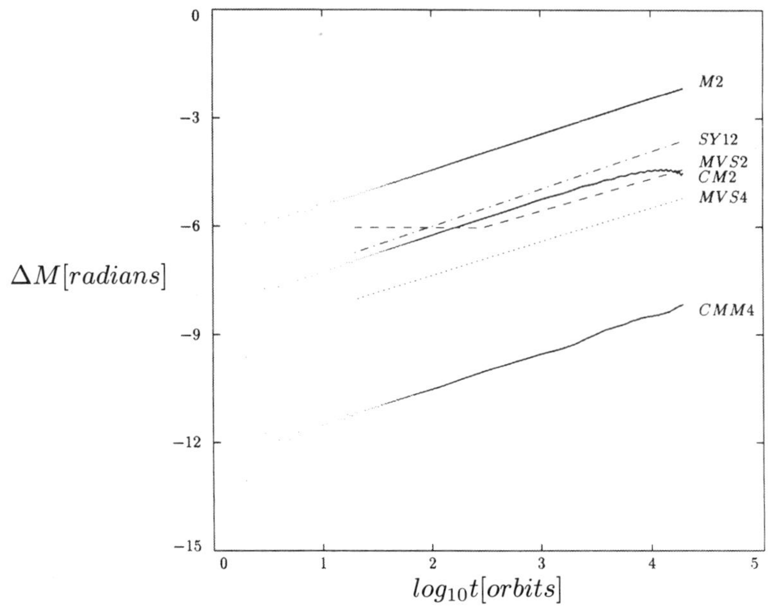

Figure 6 The error in the mean anomaly versus time of a fictitious asteroid in the three-dimensional circular restricted three-body problem for various methods.

the integrations take less time. In the case of the solar system we confirmed that higher order correctors did not solve the problem. Had we used only a factor of two smaller step in our 100 million year integration of the solar system, we would have had dramatically smaller errors!

Next, we consider the errors in longitude. Here we investigate the orbit studied by Saha and Tremaine [1992]. They were also interested in removing the effect of the spurious oscillations exhibited by the mapping trajectories. By "warming up" the trajectories they cleverly removed some of the ill effects of these oscillations without explicitly modelling the oscillations or removing them. By studying the same orbit we can compare the efficacy of our corrector method to the "warmup" method. The infinitesimal body orbits the "Sun" with a semimajor axis of about half the distance between the massive bodies. The orbit has moderate eccentricity $e = 0.25$, and moderate inclination $i = 0.2$ radians with respect to the plane of the massive bodies. They used roughly 100 force evaluations per orbit.

Figure 6 shows the error in mean anomaly of the test-particle for various methods as a function of time. In addition to the errors for the mapping methods, the errors for several other methods are also displayed. These are approximations to the data displayed in the paper by Saha and Tremaine. The method MVS2 refers to the second order Wisdom and Holman [1991] mapping with "warmup," the method MVS4 is the Forest and Ruth [1990] fourth order method with Wisdom and Holman [1991] components using "iterated start," and SY12 is a twelve step symmetric method (Quinlan and Tremaine [1990]). The performance of the corrector is comparable to that of the warmup procedure in eliminating spurious drift of the longitudes. Of course, the corrector does more than fix the longitude drift, it removes the spurious oscillations from the trajectory itself, so all aspects of the calculation are improved. Also shown is the error in longitude when the modified kick is used. Here the modified kick is programmed explicitly. The reduction in the error is dramatic. The modified kick reduces the error by roughly six orders of magnitude below the uncorrected second order map. The illustrations by Saha and Tremaine were all limited by truncation error rather than roundoff error; partial extended precision was used in this last calculation so that the error would continue to be dominated by truncation error rather than by roundoff error.

12 Suggestions for Future Research

The delta function formalism has suggested the use of correctors, and previous sections have directly explored this idea with a perturbative treatment of mapping Hamiltonians involving delta functions. This formalism leads to a particular set of integrators and a particular set of correctors. Of course, many other symplectic integrators are derived purely by matching Taylor series. Unfortunately, with integrators based solely on Taylor series matching there is no clear identification of which error terms are high-frequency and which would lead to secular growth of error. Nevertheless, we can also develop correctors for these integrators. In this case, since matching the Taylor series is the sole criterion for an integrator we could develop correctors that compensate for as many terms as possible. We may also develop composition integrators which are intended from the beginning to be used with a corrector. Presumably, the latter approach is the most economical path to high order, since the kernel must satisfy a much reduced set of constraints, only that the error terms be correctable, not that they be zero. Following this line of investigation, it would be interesting to determine whether the particular kernels suggested by the delta function formalism behave better or worse than kernels based purely on Taylor series order.

Another interesting avenue of investigation, in many ways related to the one presented here, is to look for composition methods that explicitly take into account the relative magnitudes of H_0 and H_1. Koseleff [1993] and McLachlan [1995b] have already pursued this line of investigation.

13 Summary and Exhortation

The spurious oscillations exhibited by symplectic integration schemes have a clear origin in the delta function formulation. We have developed canonical transformations that remove the spurious oscillations. For perturbations with simple Poisson series we give explicit closed form expressions for the generator of these canonical transformations. For more complicated problems, we have developed general expressions for the correctors that are coordinate independent and valid for any number of degrees of freedom.

We have illustrated the use of these correctors with applications to the n-body problem. Applied to the old data from the 100 million year integration of the solar system of Sussman and Wisdom [1992] the correctors have allowed us to reduce the apparent energy error in that integration by two orders of magnitude. Even more dramatic reduction of error is possible if the correctors are used with forethought.

The mapping method was originally based on first order averaging arguments. The mapping method is here extended to second order. A very efficient, very accurate fourth order mapping is obtained. Several different forms of the fourth order mapping are presented: a general purpose Lie series product formula, an explicit symplectic Runge-Kutta formula, and a single modified kick formulation. The latter two are special forms for problems such as the n-body problem. The modified kick mapping achieves fourth order with a single force evaluation. Application to the restricted three-body problem illustrates the dramatic decrease in the error in energy and longitude achieved by this higher order mapping.

The delta function formalism has been uniformly ignored in reviews of the symplectic integration literature, presumably because it is considered non-rigorous. Yet, the delta function formalism gives algorithms that are equivalent to the composition of Lie series approach. Furthermore, the delta function formalism gives clear insight into problems faced by symplectic integration schemes, and, more importantly, it provides solutions. Knowledge of their names gives power over demons. With the delta function formalism we can determine non-linear stability where other avenues lead only to linear results. With the delta function formalism we can quantitatively understand the origin of artifacts introduced by the algorithmic discretization. Practitioners of other methods simply shake their heads in despair. With the delta function formalism we understand the origin of the spurious oscillations induced by the discretization, and the delta function formalism has here provided a way to remove the oscillations and dramatically improve the accuracy of symplectic integrators. However it may appear to some, the delta function formalism is not necromancy. You might try whispering "delta functions" three times over your notes to see if the problems will disappear, but a better idea is to learn to use delta functions, and appreciate the wisdom they have to offer!

Acknowledgements

We thank Robert McLachlan for carefully reading the manuscript and making numerous insightful and scholarly suggestions. We also thank Presenjit Saha and Scott Tremaine for helpful discussions. This research was supported in part by grants from NASA and the National Science Foundation.

References

Belinfante, J. G. and Kolman, B. [1989], *A Survey of Lie Groups and Lie Algebras with Applications and Computational Methods.* SIAM, Philadelphia.

Butcher, J. C. [1969], In Lecture Notes in Mathematics, (Dold, A., Heidelberg, Z., and Eckmann, B., eds.), Springer-Verlag, New York.

Candy, J. and Rozmus, W. [1991], *A Symplectic Integration Algorithm for Separable Hamiltonian Functions*, J. Comp. Phys., **92**, 230-256.

Chirikov, B.V. [1979], *A Universal Instability of Many-Dimensional Oscillator Systems*, Phys. Rep., **52**, 263.

Forest, E. and Ruth, R. D. [1990], *Fourth-order Symplectic Integration*, Physica D, **43**, 105-117.

Hairer, E., Norsett, S. P., and Wanner, G. [1993], *Solving Ordinary Differential Equations I. Nonstiff Problems.* Springer-Verlag, Berlin.

Koseleff, P.-V. [1993], *Relations among Lie Formal Series and Construction of Symplectic Integrators*, In Applied Algebra, Algebraic Algorithms, and Error-correcting Codes (Cohen, G., Mora, T. and Moreno, O. eds.), AAECC-10, Springer-Verlag, New York.

Laskar, J. [1989], A *numerical experiment on the behaviour of the solar system*, Nature, **338**, 237-238.

McLachlan, R. I. [1995a], *On the numerical integration of ordinary differential equations by symmetric composition methods*, SIAM J. Sci. Comp., **16**, 151–168.

McLachlan, R. I. [1995b], *Composition methods in the presence of small parameters*, submitted to BIT.

Quinn, T. R., Tremaine, S. D. and Duncan, M. [1991], *A three million year integration of the Earth's orbit*, Astron. J., **101**, 2287-2305.

Quinlan, G. D. and Tremaine, S. D. [1990], *Symmetric multistep methods for the numerical integration of planetary orbits*, Astron. J., **100**, 1694-1700.

Saha, P. and Tremaine, S. D. [1992], *Symplectic integrators for solar system dynamics*, Astron. J., **104**, 1633-1640.

Sanz-Serna, J. M. [1992], *Symplectic integrators for Hamiltonian problems: an overview*, Acta Numer., **1**, 243-286.

Steinberg, S. [1988] *Lie Methods in Optics*, (Mondragon and Wolf, eds.). Springer-Verlag, New York.

Sussman, G. J. and Wisdom, J. [1988], *Numerical Evidence that the Motion of Pluto is Chaotic*, Science, 241:433.

Sussman, G. J. and Wisdom, J. [1992], *Chaotic Evolution of the Solar System*, Science, **257**, 56.

Suzuki, M. [1991], *General theory of fractal path integrals with applications to many-body theory and statistical physics*, J. Math. Phys., **32**, 400-407.

Tittemore, W. C. and Wisdom, J. [1988], *Tidal Evolution of the Uranian Satellites. Part I: Passage of Ariel and Umbriel through the 5:3 Mean-Motion Commensurability*, Icarus, **74**, 172.

Tittemore, W. C. and Wisdom, J. [1989]. *Tidal Evolution of the Uranian Satellites. Part II: Explanation of the Anomalously Large Inclination of Miranda*, Icarus, **78**, 63.

Tittemore, W. C. and Wisdom, J. [1990], *Tidal Evolution of the Uranian Satellites. Part III, Evolution through the Miranda-Umbriel 3:1, Miranda-Ariel 5:3, and Ariel-Umbriel 2:1 Mean Motion Commensurabilities*, Icarus, **85**, 394.

Wisdom, J. [1982], *The Origin of the Kirkwood Gaps: A Mapping for Asteroidal Motion Near the 3/1 Commensurability*, Astron. J., **87**, 577.

Wisdom, J. [1983], *Chaotic Behavior and the Origin of the 3/1 Kirkwood Gap*, Icarus, **56**, 51.

Wisdom, J. [1988], *Fire*, In Earth, Air, Fire, and Water (Roberts, P. and Schubert, G., eds.), Proceedings of the Fourth Annual University of California Summer School on Nonlinear Science. Institute of Geophysics and Planetary Physics, UCLA.

Wisdom, J. and Holman, M. [1991], *Symplectic Maps for the N-Body Problem*, Astron. J., **102**, 1528.

Wisdom, J. and Holman, M. [1992], *Symplectic Maps for the N-Body Problem: Stability Analysis*, Astron. J., **104**, 2022.

Yoshida, H. [1990]. *Construction of higher order symplectic integrators*, Phys. Lett. A, **150**, 262–268.

Yoshida, H. [1993], *Recent Progress in the Theory and Application of Symplectic Integrators*, Celest. Mech. Dyn. Ast., **56**, 27-43.